COMPUTER ANALYSIS OF STRUCTURES

MATRIX STRUCTURAL ANALYSIS
STRUCTURED PROGRAMMING

Siegfried M. Holzer

Professor of Civil Engineering,
Virginia Polytechnic Institute and State University,
Blacksburg, Virginia

ELSEVIER

New York • Amsterdam • Oxford

Elsevier Science Publishing Co., Inc.
52 Vanderbilt Avenue, New York, New York 10017

Sole distributors outside the United States and Canada:

Elsevier Science Publishers B.V.
P.O. Box 211, 1000 AE Amsterdam, The Netherlands

Library of Congress Cataloging in Publication Data

Holzer, Siegfried M.
    Computer analysis of structures.

    Includes index.
    1. Structures, Theory of—Matrix methods—Data processing.
    2. Structured programming. I. Title.
TA647.H65   1985   624.1'71'02854   85-6879
ISBN 0-444-00943-4

Current printing (last digit):
10   9   8   7   6   5   4   3   2   1

Manufactured in the United States of America

To Gail and Michael

CONTENTS

Preface   ix

CHAPTER 1 MATHEMATICAL MODELS OF ELEMENTS   1
    1.1 Introduction   1
    1.2 Structural Analysis   2
    1.3 Mathematical Models   2
    1.4 Generalized Displacements and Forces   3
    1.5 Continuum Models   11
    1.6 Discrete Models   22
    1.7 Discrete Models in Matrix Form   28
    1.8 Discrete Elements: Special Topics   38
    1.9 Finite Elements   48
    Problems   66
    References   71
    Additional References   73

CHAPTER 2 PREPARATION FOR MATRIX DISPLACEMENT METHOD   **74**
    2.1 Introduction   74
    2.2 Generalized Displacements   75
    2.3 Notation and Conventions   79
    2.4 Compatibility and Equilibrium   80
    2.5 Joint Forces   88
    2.6 Coordinate Transformations   90
    27. Compatibility and Equilibrium: Analytical Approach   99
    Problems   102
    References   107

**CHAPTER 3 MATRIX DISPLACEMENT METHOD: PLANE
STRUCTURES**   **108**

    3.1 Introduction   108
    3.2 Matrix Displacement Method   108
    3.3 Analysis of Continuous Beams   119
    3.4 Analysis of Frames   125
    3.5 Analysis of Trusses   130
    3.6 Element Actions   134
    3.7 Finite Element Formulation   143
    Problems   153
    References   158

**CHAPTER 4 MATRIX DISPLACEMENT METHOD: SPECIAL
TOPICS**   **160**

    4.1 Introduction   160
    4.2 Symmetric Structures   160
    4.3 Orthogonal Frames with Internal Constraints   176
    4.4 Internal Releases   184
    4.5 Assemblages of Distinct Elements   190
    4.6 Condensation and Substructuring   196
    4.7 Geometric Imperfections   214
    4.8 Temperature Changes   222
    4.9 Joint Reference Frames   232
    4.10 Influence Lines   238
    Problems   246
    References   252
    Additional Reference   252

**CHAPTER 5 MATRIX DISPLACEMENT METHOD: SPACE
STRUCTURES**   **253**

    5.1 Introduction   253
    5.2 Generalized Displacements   253
    5.3 Notation and Conventions   256
    5.4 Space Trusses   256
    5.5 Space Frames   262
    5.6 Grids   270
    Problems   276
    References   278
    Additional References   279

CHAPTER 6 SOLUTION OF SYSTEM EQUATIONS   **280**

    6.1 Introduction   280
    6.2 Band Matrices   280

6.3 Solution of Linear Equations 288
6.4 Band Solvers 302
6.5 Variable Band Solvers 307
6.6 Frontal Solutions 312
6.7 Solution Errors 316
Problems 323
References 325
Additional References 327

CHAPTER 7 PROGRAM DEVELOPMENT 328
7.1 Introduction 328
7.2 Structured Programming 328
7.3 Frame Program 337
Problems 360
References 363
Additional References 364

APPENDIX A ELEMENT ACTIONS AND RESPONSES 365
Problems 380
References 382

APPENDIX B SLOPE–DEFLECTION METHOD 383
Problems 389

APPENDIX C COORDINATE TRANSFORMATIONS 391
Problems 402
References 403

**APPENDIX D PRINCIPLES AND CONCEPTS OF ANALYTICAL
MECHANICS** 404
References 416

**APPENDIX E IMPOSITION OF CONSTRAINTS ON SYSTEM
MODEL** 417
Reference 421

INDEX 423

PREFACE

This book was written with two principal *objectives for the students* in mind: (1) to acquire a precise understanding of the *matrix displacement method* and its underlying concepts and principles; (2) to develop *well-structured programs* for the analysis of skeletal structures by the matrix displacement method. The displacement method was selected as the method of analysis because of its intrinsic modularity, good numerical properties, and popularity—matrix and finite element analysis programs are generally based on the displacement method. Structured programming is emphasized because it provides a systematic process for creating correct programs.

Students have demonstrated that they can use this programming knowledge to write special-purpose programs, such as computer-aided design programs and finite element programs.

The history of the development of matrix and finite element methods of analysis, motivated by the computer, is traced, for example, by Martin and Carey (1973). An overview of structured programming is provided in Bates (1976).

In the first three chapters, the matrix displacement method is presented in a form suitable for programming. The matrix displacement method is extended to special topics in Chapter 4 and to space structures in Chapter 5. Chapter 6 deals with the numerical solution of the system equations. Chapter 7 is concerned with structured programming. Five appendices include elementary methods of analysis, principles of analytical mechanics, and mathematical tools.

Discrete element models are formulated in Chapter 1 by three approaches: (1) by the solution of differential equations; (2) by force–deformation formulas,[1] and (3) by the finite element method. The students should adopt the approach that provides a natural link to their background. For example, students who have had a course in the mechanics of deformable bodies but who have not had a basic course in structural analysis should take the first approach. Students who have completed a traditional junior-level course in structural analysis might prefer the second approach, in which the extensional force–deformation relation and the slope–deflection equations are combined with conditions of equilibrium to construct the models for truss and beam elements. The finite element approach is recommended for students with a basic understanding of the matrix displacement method and the principle of virtual work.

Chapter 2 paves the way for the matrix displacement method appropriate for program development. The central task is the formulation of conditions of compatibility and equilibrium without a visual reference to the structure. For this purpose, the member code matrix is introduced (Section 2.4).

In Chapter 3 the matrix displacement method is formulated on the basis of the member code matrix, and it is illustrated for continuous beams, frames, and trusses. Joint loads and member loads are considered.

Chapter 4 illustrates how special features can be incorporated in the matrix displacement analysis. The topics covered can be divided into three groups: (1) the reduction of the degrees of freedom of an assembly of elements by utilizing symmetry, by introducing internal constraints, and by condensation; (2) the formulation of various element actions, such as geometric imperfections, temperature changes, and unit displacements imposed in the construction of influence lines by the Müller–Breslau principle; and (3) the formulation of assemblies with distinct elements, internal releases, and distinct joint reference frames.

In Chapter 5, the matrix displacement method is extended to space structures: space trusses, space frames, and grids.

Chapter 6 is concerned with the numerical solution of the system equations. It includes a literature review of solution techniques, a discussion of storage schemes for fixed and variable band matrices, the formulation of direct solution methods with reference to special structural analysis techniques for symmetric, positive definite band matrices, algorithms for fixed and variable band solvers and references to computer programs, the frontal solution technique and references to computer programs, and a study of solution errors and methods of error detection and error control.

Chapter 7 is concerned with structured programming. It includes discussions of the aims of structured programming, control structures, methods

---

[1] In the presentation, the first two approaches are not separated. The beginning of the second approach, which is contained in the first approach, is stated in Section 1.1.

of modularization, programming and coding languages, program correctness, and program efficiency. The principles of structured programming are applied in the design of a program for the matrix displacement analysis of plane frames. The program structure is represented by a tree chart, and the subprograms are described by structured flow charts (Nassi–Schneiderman diagrams) with lists of input and output arguments. FORTRAN 77 and the FORTRAN version of the WATFIV compiler are used to illustrate the coding of several subprograms.[2] A variety of programming problems is presented that includes the completion of the frame program (coding and testing) and extensions based on Chapters 4–6.

In Appendix A, fixed-end force formulas are presented for various element actions. In addition, displacement–deformation relations are derived that permit us to apply the moment–area method without a sketch of the deformed configuration. In Appendix B, the slope–deflection method is presented in a form that facilitates the transition to the matrix displacement method. Appendix C provides a comprehensive treatment of coordinate transformations. Appendix D is concerned with the principle of virtual work. In Appendix E, the imposition of joint constraints at the system level is addressed.

COURSE DESIGN. After the matrix displacement method in Chapters 1–3 has been studied, the remaining chapters can be studied in any order. It is recommended, however, to follow the formulation of the matrix displacement method in Chapter 3 with the program development as indicated in the course structure.

Upon the completion of the frame program in Chapter 7, the study of the material in Chapters 4–6 can be integrated with desirable program extensions.

The matrix displacement method and program development, with the completion of the frame program as a class project, are appropriate for a single semester of a senior-level course. In our quarter system, the textbook forms the basis for two senior-level courses. In the first course, we teach the matrix displacement method and, depending on the students' progress, one or two topics of Chapter 4. The second course is primarily concerned with program development, the frame program class project, and the matrix displacement analysis of space structures.

In designing and teaching a course, I find the two educational commandments of A. N. Whitehead (1967) reassuring:

Do not teach too many subjects.
What you teach, teach thoroughly.

---

[2] The completed frame program is available to the teacher.

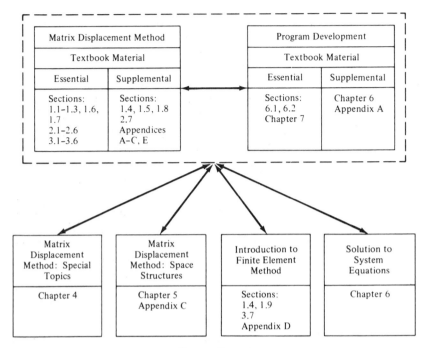

Course Structure

REFERENCES

Bates, D., ed. 1976. *Structured Programming, Infotech State of the Art Report.* Infotech International, England.

Martin, H. C., and G. F. Carey, 1973. *Introduction to Finite Element Analysis.* McGraw-Hill, New York.

Whitehead, A. N. 1967. *The Aims of Education.* Free Press, New York.

ACKNOWLEDGMENT

I would like to thank the following people for their contributions to this textbook: The authors of books and papers whose influence on my work is reflected in the textbook; David Pecknold and Thomas Baber for their careful reviews and valuable suggestions; Gene Somers for using the manuscript in his classes, for his contributions to the improvement of the textbook, and for his generous support; my former students for their stimulating interactions, among them Greg Katzenberger, who proofread the entire manuscript, Mohammad Hariri and Fawwaz Ghabra, who proofread portions of the manuscript, and Kim Basham, who used the manuscript in his classes; the staff at Elsevier, particularly Marjan Bace for suggesting the title of the textbook, Edmée Froment for the beautiful design of the book, and Louise Calabro Gruendel and Helene De Lorenzo for their thorough editing; Melissa Knocke for her expert and conscientious typing of the manuscript and Judy Brown for helping with the final revisions; and Richard Walker and Robert Krebs for their continual support.

S.M. Holzer

MATHEMATICAL MODELS OF ELEMENTS

1.1 INTRODUCTION

The prediction of the performance of a structure, which is the role of structural analysis, is generally based on the analysis of mathematical models. The accuracy of this prediction depends on how well the models approximate the behavior of the structure. Accordingly, it is important to know the limitations of the mathematical models used to represent the structure.

Insight into the limitations of a model can be gained from its construction. The first step in the construction of a model is to select generalized displacements, which define the configuration and determine the degrees of freedom of the model. If a model has infinitely many degrees of freedom, it is called a continuum model; otherwise it is called a discrete model. Next, the three basic components of a model, the conditions of compatibility, the conditions of equilibrium, and the constitutive law, are formulated. The synthesis of these components yields the mathematical model.

The primary purpose of this chapter is to formulate discrete element models for the analysis of skeletal structures by the matrix displacement method. The chapter is organized as follows: After some brief discussions of structural analysis and mathematical models (Sections 1.2 and 1.3), the concepts of generalized displacements and forces are defined and illustrated (Section 1.4). A modeling process is introduced and applied to obtain one-dimensional continuum models representing axial, flexural, and torsional deformations (Section 1.5). Discrete element models, which relate element-end displacements to element-end forces through stiffness matrices, are formulated by three approaches: (1) by the solution of continuum models (Sections 1.6 and 1.7); (2) by force–deformation formulas (Sections 1.6 and 1.7)—specifically, the axial deformation formula, Eq. (1.72), the slope–deflection equations, Eqs. (1.86), and the torsional deformation formula, Eq. (1.91); (3) by the finite element method (Section 1.9). It is recommended to select the approach that suits the students' background (see Preface). Special topics of discrete elements are presented in Section 1.8.

1.2 STRUCTURAL ANALYSIS

Structural engineering is concerned with the planning, designing, and building of structures. Structural analysis forms an integral part of the design process. Its function is to predict the behavior of a structure in its environment. This prediction is usually based on mathematical models. Physical models may be used if the reliability of a mathematical model is in doubt.

A model of a structure is defined as a mathematical representation of the behavior of the structure in its environment. It is expressed as an action–response relation. Actions are mathematical models of such environmental factors as loads, prescribed displacements, and temperature changes. The response is a measure of the change in state of the structure. It may be expressed, for instance, by displacements, strains, stresses, and forces.

In essence, structural analysis is concerned with the specification of actions, the construction of models of the structure, and the determination of the response to the imposed actions (Figure 1.1).

1.3 MATHEMATICAL MODELS

A mathematical model of a structure has three distinct components that represent signficant features of the structure: conditions of compatibility, conditions of equilibrium, and constitutive laws.

The conditions of compatibility reflect geometric properties of a structure, such as continuity of deformations of elements and assemblages of elements, and boundary constraints. In addition, restrictions on deformations are frequently imposed to reduce the three-dimensional structure to a two- or one-dimensional model. For example, the assumption concerning plane sections in the elementary beam theory (Crandall, Dahl, and Lardner, 1978; Freudenthal, 1966; Popov, 1968; Stippes, Wempner, Stern, and Beckett, 1961) transforms the beam into a one-dimensional element.

The conditions of equilibrium express the state of balance of a structure at rest. Newton's law or the principle of virtual work can be applied to

FIGURE 1.1 Analysis process

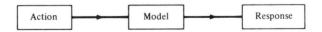

formulate the conditions of equilibrium. If the structure is in motion, the conditions of equilibrium are replaced by the laws of motion.

The constitutive laws model the behavior of materials. For example, Hooke's law is based on the idealization of a linearly elastic material.

It is important to keep in mind that mathematical models of structures represent idealizations. The results obtained from the analysis of the model can be valid only to the extent that the model approximates the behavior of the stucture.

1.4 GENERALIZED DISPLACEMENTS AND FORCES

Generalized Displacements

The formulation of a mathematical model of a structure centers on the selection of parameters that define the configuration of the model. The *configuration* is characterized by the simultaneous locations of all material points. The number of independent parameters required to define the configuration represents the *degrees of freedom* of the model. These parameters are called the *generalized displacements* (or generalized coordinates;[1] Langhaar, 1962) of the model.

In engineering analysis, the configuration of a model is generally described relative to its *initial state*, a reference configuration in which the model is not subjected to actions. Specifically, the configuration is defined by the displacements of each point from its initial position. This is illustrated in the following examples for one-dimensional elements without reference to actions that may correspond to these configurations.

Example 1. Consider the rigid bar in Figure 1.2, whose initial configuration coincides with the x axis of the rectangular frame of reference. Thus, the initial configuration is defined by the set of points $0 \leq x \leq L$. If the bar is confined to the x–y plane, it has three degrees of freedom. There is considerable freedom in the selection of generalized displacements. For example, the position of the bar in the x–y plane can be specified by the displacements—the deflections and rotation—of any initial point of the bar. Let us select the displacements at the a end of the bar, u_a, v_a, and θ_a, as generalized displacements and formulate the configuration in terms of them.

The deflections of the initial point P, located a distance x from the a end of the bar, in the directions of the x and y axes are denoted by $u(x)$ and $v(x)$, respectively. Since the bar is rigid, the distance from the a end of the bar to

---

[1] This definition is restricted to holonomic models in which the generalized displacements can be varied arbitrarily without violating kinematical constraints. Nonholonomic models are discussed by Lanczos (1970), Langhaar (1962), and Synge and Griffith (1959).

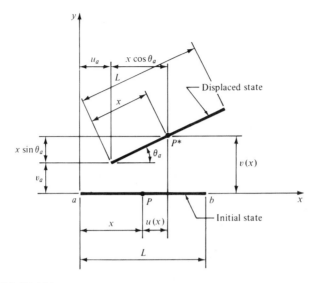

FIGURE 1.2 Rigid bar

the point P^*, the displaced position of point P, remains x. It follows from Figure 1.2 that

$$u(x) = u_a + x(\cos \theta_a - 1) \atop v(x) = v_a + x \sin \theta_a \left.\right\} \quad 0 \leq x \leq L \tag{1.1}$$

Equations (1.1) define the displaced position of every initial point of the bar in terms of the generalized displacements.

If the rotation of the bar is infinitesimal, that is, if

$$\theta_a^2 \cong 0 \tag{1.2}$$

relative to unity, we obtain (Thomas, 1956)

$$\cos \theta_a = 1 - \frac{\theta_a^2}{2!} + \frac{\theta_a^4}{4!} - \cdots \cong 1$$

$$\sin \theta_a = \theta_a - \frac{\theta_a^3}{3!} + \frac{\theta_a^5}{5!} - \cdots \cong \theta_a \tag{1.3}$$

and Eqs. (1.1) become

$$u(x) = u_a$$

$$v(x) = v_a + x\theta_a \tag{1.4}$$

Observe that for infinitesimal rotations, the configuration of the bar is a linear function of the generalized displacements. This is characteristic of *linear models* of structures.

Example 2. If the element in Figure 1.2 is free to experience a uniform axial deformation, it becomes the four-degree-of-freedom element shown in Figure 1.3a. Accordingly, the three rigid-body displacements, u_a, v_a, θ_a, and the deformation e represent a set of generalized displacements. However, if the element is part of an assemblage of elements, such as a truss, it is preferable to select the element-end deflections, u_a, v_a, u_b, v_b, as generalized displacements. The formulation of the element configuration in terms of the end deflections is illustrated.

It is convenient to resolve the configuration in Figure 1.3a into component

FIGURE 1.3 Axial deformation element

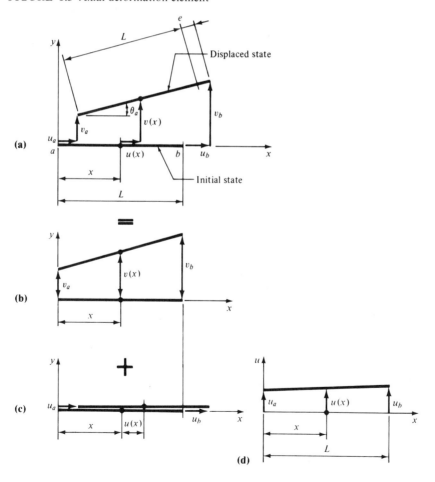

configurations, representing transverse and axial deflections separately (Figures 1.3b and c). Thus, the transverse deflections in Figure 1.3b are defined by the equation of the line

$$v(x) = v_a + \frac{x}{L}(v_b - v_a) \tag{1.5}$$

which can be expressed as

$$v(x) = \phi_a(x)\,v_a + \phi_b(x)\,v_b \tag{1.6}$$

where

$$\phi_a = 1 - \frac{x}{L}, \quad \phi_b = \frac{x}{L}, \qquad 0 \le x \le L \tag{1.7}$$

Since the axial deformation is uniform, the axial deflection varies linearly as shown in Figure 1.3d (see Problem 1.3). Hence, $u(x)$ can be expressed in the form of Eq. (1.6) as

$$u(x) = \phi_a(x)\,u_a + \phi_b(x)\,u_b \tag{1.8}$$

An alternative formulation of Eqs. (1.6) and (1.8) is based on function interpolation (Section 1.9). Specifically, since $u(x)$ and $v(x)$ must satisfy the boundary conditions

$$u(0) = u_a, \quad u(L) = u_b; \quad v(0) = v_a, \quad v(L) = v_b \tag{1.9}$$

we can express them as first-order polynomials,

$$u = b_0 + b_1 x, \quad v = c_0 + c_1 x \tag{1.10}$$

and determine the coefficients by imposing Eqs. (1.9). This approach is illustrated in the next example.

Example 3. Consider the element in Figure 1.4, whose configuration is defined by the polynomial function

$$v(x) = c_0 + c_1 x + c_2 x^2 + c_3 x^3 \tag{1.11}$$

By definition, the coefficients of Eq. (1.11) represent generalized displacements and the element has four degrees of freedom. The reasons for introducing Eq. (1.11) are that (1) it satisfies the homogeneous differential equation of a beam (Section 1.6), hence, Eq. (1.11) characterizes the configurations of beams subjected to boundary actions; and (2) polynomials form basic building blocks of many interpolation functions.

Analogous to Example 2, let us select the element-end displacements, $v_a, \theta_a, v_b, \theta_b$, as generalized displacements and express the configuration in terms of them. This can be accomplished by imposing the boundary conditions

$$v(0) = v_a, \quad \frac{dv(0)}{dx} = \theta_a$$
$$\tag{1.12}$$
$$v(L) = v_b, \quad \frac{dv(L)}{dx} = \theta_b$$

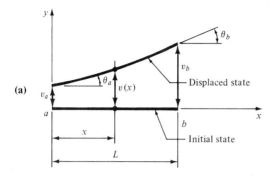

(a)

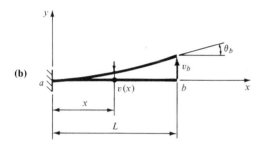

(b)

FIGURE 1.4 Flexural deformation element

on Eq. (1.11). The slope–rotation relations in Eqs. (1.12) are restricted to small rotations for which $\tan \theta \cong \theta$ [see Eqs. (1.3)]. Equations (1.11) and (1.12) yield the set of algebraic equations

$$c_0 = v_a$$
$$c_1 = \theta_a$$
$$c_0 + Lc_1 + L^2 c_2 + L^3 c_3 = v_b$$
$$c_1 + 2Lc_2 + 3L^2 c_3 = \theta_b$$

(1.13)

whose solution is

$$c_0 = v_a$$
$$c_1 = \theta_a$$
$$c_2 = \frac{1}{L^2}(-3v_a - 2L\theta_a + 3v_b - L\theta_b)$$

(1.14)

$$c_3 = \frac{1}{L^3}(2v_a + L\theta_a - 2v_b + L\theta_b)$$

Substituting Eqs. (1.14) into Eq. (1.11) and collecting terms of like displacements, we obtain

$$v = \phi_a v_a + \psi_a \theta_a + \phi_b v_b + \psi_b \theta_b$$

(1.15)

where

$$\phi_a = 1 - 3\xi^2 + 2\xi^3$$
$$\psi_a = (\xi - 2\xi^2 + \xi^3)L$$
$$\phi_b = 3\xi^2 - 2\xi^3$$
$$\psi_b = (-\xi^2 + \xi^3)L$$

(1.16)

and

$$\xi = \frac{x}{L}$$

(1.17)

Note that boundary conditions can be imposed directly on Eq. (1.15). For example, the conditions $\theta_a = v_a = 0$ reduce Eq. (1.15) to the equation

$$v = \phi_b v_b + \psi_b \theta_b$$

(1.18)

which describes the configuration of the cantilever in Figure 1.4b.

Example 4. Consider a beam whose configuration is defined by the power series (Thomas, 1956)

$$v(x) = \sum_{i=0}^{\infty} c_i \left(\frac{x}{L}\right)^i$$

(1.19)

It follows that the coefficients $c_0, c_1, c_2, \ldots$ are generalized displacements, since their values define the position of the beam, and the beam has infinitely many degrees of freedom. If the series is reduced to a finite number of terms, the configuration of the beam is approximated by a finite number of generalized displacements. This type of approximation forms the basis of many numerical methods of analysis, including the finite element method (Section 1.9).

Generalized Forces

Consider a system with n degrees of freedom whose configuration is defined by the generalized displacements $q_1, q_2, \ldots, q_n$. Let us impose on the generalized displacements the infinitesimal changes $\delta q_1, \delta q_2, \ldots, \delta q_n$ to produce a neighboring configuration defined by $q_1 + \delta q_1, q_2 + \delta q_2, \ldots, q_n + \delta q_n$. The infinitesimal work done by all forces acting on the system during the change in configuration can be expressed by the linear function

$$\delta W = R_1 \delta q_1 + R_2 \delta q_2 + \cdots + R_n \delta q_n = \sum_{i=1}^{n} R_i \delta q_i$$

(1.20)

The coefficients R_i are called generalized forces.[2]

---

[2] The δq_i's are called variations of the generalized displacements, and δW is the virtual work (Appendix D).

The units of generalized forces are determined by the product $R_i \, \delta q_i$, which has the unit of work. Accordingly, R_i has the unit of force if δq_i denotes a length, and R_i has the unit of moment of force if δq_i denotes an angle.

Example 5. For the rigid bar in Figure 1.2, let $q_1 = u_a$, $q_2 = v_a$, $q_3 = \theta_a$. Thus, the initial state is defined by the values of the generalized displacements $q_i = 0$; $i = 1, 2, 3$. If we impose the infinitesimal changes δq_i on the initial state, we obtain the neighboring configuration in Figure 1.5a, which by Eqs. (1.4) can be expressed as

$$\delta u(x) = \delta q_1$$
$$\delta v(x) = \delta q_2 + x \, \delta q_3 \tag{1.21}$$

Figures 1.5b and c show two distinct force distributions applied to the initial state of the bar. The formulation of the corresponding generalized forces is illustrated.

The infinitesimal work done by the forces on the bar in Figure 1.5b during the changes in configuration is

$$\delta W = f_1 \, \delta q_1 + f_2 \, \delta q_2 + f_3 \, \delta q_3 + f_4 \, \delta q_1 + f_5(\delta q_2 + L \, \delta q_3) + f_6 \, \delta q_3 \tag{1.22}$$

or

$$\delta W = (f_1 + f_4)\delta q_1 + (f_2 + f_5)\delta q_2 + (f_3 + f_6 + Lf_5) \, \delta q_3$$

$$= \sum_{i=1}^{3} R_i \, \delta q_i \tag{1.23}$$

Thus, the generalized forces of the bar in Figure 1.5b are

$$R_1 = f_1 + f_4$$
$$R_2 = f_2 + f_5 \tag{1.24}$$
$$R_3 = f_3 + f_6 + Lf_5$$

For the bar in Figure 1.5c

$$\delta W = f_1 \, \delta q_1 + f_2 \, \delta q_2 + f_3 \, \delta q_3 + \int_0^L p(x) \, \delta v(x) \, dx \tag{1.25}$$

where

$$p(x) = p_0 = \text{constant} \tag{1.26}$$

$$\int_0^L p(x) \, \delta v(x) \, dx = \int_0^L p_0(\delta q_2 + x \, \delta q_3) \, dx = p_0 L \delta q_2 + \tfrac{1}{2} p_0 L^2 \, \delta q_3 \tag{1.27}$$

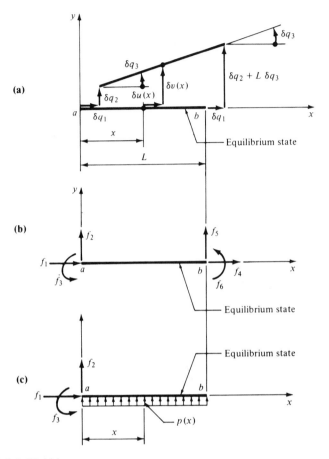

FIGURE 1.5 Rigid bar

Thus

$$\delta W = f_1 \delta q_1 + (f_2 + p_0 L)\delta q_2 + (f_3 + \tfrac{1}{2}p_0 L^2)\delta q_3 = \sum_{i=1}^{3} R_i \, \delta q_i \quad (1.28)$$

and the generalized forces of the bar in Figure 1.5c are

$$R_1 = f_1$$
$$R_2 = f_2 + p_0 L \qquad\qquad (1.29)$$
$$R_3 = f_3 + \tfrac{1}{2}p_0 L^2$$

Example 5 reveals an important principle of generalized forces: A configuration for which all generalized forces vanish is an equilibrium configuration. This principle is a generalization of the Newtonian law that a particle is in equilibrium if there is no unbalanced force acting on it (Langhaar, 1962). Specifically, the vanishing of the generalized forces in Eqs. (1.24) and (1.29) is equivalent to the conditions of equilibrium in the directions of the generalized displacements:

$$\sum F_x = 0, \quad \sum F_y = 0, \quad \sum M_a = 0 \qquad (1.30)$$

Characterization of Models

We have seen that an element may have a finite or an infinite number of degrees of freedom. This is a significant distinction, which is reflected in the form of the corresponding mathematical model. Specifically, finite-degree-of-freedom models, called *discrete models*, are represented by algebraic equations (provided the system is static), whereas infinite-degree-of-freedom models, called *continuum models*, are described by differential equations. The construction of continuum and discrete models of elements is considered in the following sections.

1.5 CONTINUUM MODELS

The field of structural mechanics contains a variety of continuum models, from simple string models to complex shell models. Although these models may appear unrelated, they have a common basis, which is displayed in the modeling process of Figure 1.6. This process defines the mathematical

FIGURE 1.6 Modeling process

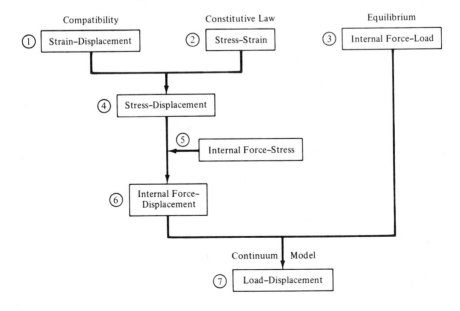

relations (steps 1–3) that form the components of the model (see Section 1.3), and it shows how these relations may be combined to produce the continuum model. Moreover, it identifies the geometric and material limitations of linear models. Specifically, the deformations must be sufficiently small (Stippes et al., 1961) to justify linear strain–displacement relations (step 1) and the formulation of equilibrium for the undeformed state (step 3); in addition, the stress must be sufficiently small (Stippes et al., 1961) to justify linear stress–strain relations (step 2).

The construction of continuum models according to Figure 1.6 is illustrated for members experiencing axial, flexural, and torsional deformations, which are the basic deformations of skeletal structures. The mathematical relations representing conditions of compatibility, conditions of equilibrium, and Hooke's law can be found in textbooks on mechanics of deformable bodies. (See, for example, Crandall et al., 1978; Freudenthal, 1966; Popov, 1968; Stippes et al., 1961.)

Axial Deformation Model

The bar in Figure 1.7a is subjected to a distributed load $p(x)$ and two concentrated forces N_a and N_b. The lines of action of all forces are collinear with the centroidal axis of the bar, which coincides with the x axis of the frame of reference. The bar is prismatic, homogeneous, and the material obeys Hooke's law.

The deformations of the bar are characterized by two features: (1) The bar remains straight, and (2) plane cross sections remain plane and normal to the centroidal axis. Thus, the configuration of the bar is determined by the configuration of the centroidal axis, which forms a one-dimensional model of the bar.

The relation between the load (the action) and the configuration (the response) of the bar constitutes the mathematical model. The construction of the model is based on the flow chart in Figure 1.6.

1. Strain–Displacement Relation

$$\varepsilon(x, y, z) = \varepsilon(x) = \frac{du(x)}{dx} \tag{1.31}$$

where ε is the extensional strain, which is positive in tension, and $u(x)$ is the axial deflection of any point on the centroidal axis located initially a distance x from the origin (see Figure 1.7a).

2. Stress–Strain Relation

According to Hooke's law,

$$\sigma = E\varepsilon \tag{1.32}$$

where σ is the normal stress and E is the modulus of elasticity.

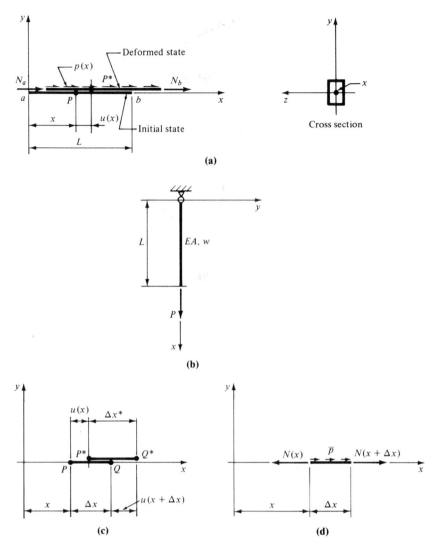

FIGURE 1.7 Axial deformation. **(a)** Element; **(b)** application; **(c)** line segment; **(d)** sign convention

3. Internal Force–Load Relation

$$\frac{dN}{dx} = -p \tag{1.33}$$

where N is the normal force on the cross section of the bar; it is positive in tension. Equation (1.33) represents the pointwise condition of equilibrium.

4. Stress–Displacement Relation

Equations (1.31) and (1.32) yield

$$\sigma = E\frac{du}{dx} \tag{1.34}$$

5. Internal Force–Stress Relation

$$N = \iint_A \sigma\, dA = \sigma A \tag{1.35}$$

because $\sigma(x, y, z) = \sigma(x)$; that is, the stress is constant over the cross-sectional area A.

6. Internal Force–Displacement Relation

Equations (1.34) and (1.35) yield

$$N = EA\frac{du}{dx} \tag{1.36}$$

7. Load–Displacement Relation

Combining Eqs. (1.33) and (1.36), we obtain the continuum model of a prismatic, homogeneous, elastic bar under axial loads:

$$EA\frac{d^2u}{dx^2} = -p \tag{1.37}$$

with labels: response, action, geometric property, material property.

Equation (1.37) indicates that the continuum model is a differential action–response relation involving geometric and material properties of the bar.

The application of the model is illustrated for the bar in Figure 1.7b. The bar is subjected to the concentrated load P and has the self-weight w (force per unit length of bar). Accordingly, $p(x) = w$, and the governing differential equation becomes

$$EA\frac{d^2u}{dx^2} = -w \tag{1.38}$$

which is subject to the boundary conditions [see Figure 1.7b and Eq. (1.36)]

$$u(0) = 0 \tag{1.39a}$$

$$N(L) = EA\frac{du(L)}{dx} = P \tag{1.39b}$$

If the self-weight of the bar is neglected, Eq. (1.38) reduces to the homogeneous differential equation

$$\frac{d^2u}{dx^2} = 0 \tag{1.40}$$

REVIEW. To encourage the student to review the concepts and principles underlying the components of the continuum model defined by Eqs. (1.31) and (1.33), their derivations are presented.

Consider the line segment PQ of the undeformed centroidal axis of the bar in Figure 1.7c. Its length is Δx. After deformation, the segment PQ becomes the segment P^*Q^* with length Δx^*. Strain is a measure of deformation. By definition (Strippes et al., 1961), the normal strain at point P in the direction of the x axis is

$$\varepsilon(x) = \lim_{\Delta x \to 0} \frac{\Delta x^* - \Delta x}{\Delta x} \tag{1.41}$$

Figure 1.7c indicates that $\Delta x^* = \Delta x + u(x + \Delta x) - u(x)$; this yields $\Delta x^* - \Delta x = u(x + \Delta x) - u(x)$. Thus

$$\varepsilon(x) = \lim_{\Delta x \to 0} \frac{u(x + \Delta x) - u(x)}{\Delta x} = \frac{du(x)}{dx} \tag{1.42}$$

which follows from differential calculus (Thomas, 1956).

Figure 1.7d shows an isolated section of the bar between the coordinates x and $x + \Delta x$. The condition of equilibrium of the section Δx is

$$N(x + \Delta x) - N(x) + \bar{p}\,\Delta x = 0 \tag{1.43}$$

where $\bar{p}$ is the average value of the distributed load over the section Δx. Rearranging Eq. (1.43) and taking the limit as $\Delta x \to 0$, we obtain

$$\lim_{\Delta x \to 0} \frac{N(x + \Delta x) - N(x)}{\Delta x} = \lim_{\Delta x \to 0} \bar{p} \tag{1.44}$$

which yields the pointwise condition of equilibrium

$$\frac{dN(x)}{dx} = -p(x) \tag{1.45}$$

Flexural Deformation Model

The beam in Figure 1.8a is subjected to the distributed transverse load $p(x)$ and the concentrated forces V_a, M_a, V_b, M_b; the term *force* is generalized to refer to forces and moments of forces. All forces act in the longitudinal plane of symmetry of the beam, which lies in the x–y plane of the frame of reference.

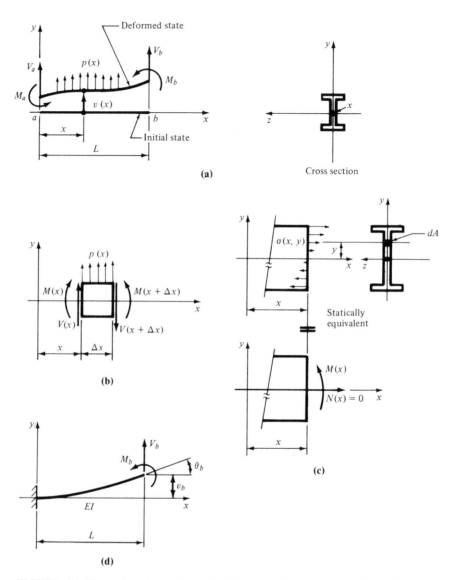

FIGURE 1.8 Flexural deformation. **(a)** Element; **(b)** sign convention; **(c)** stress resultants; **(d)** application

The initial configuration of the centroidal axis of the beam coincides with the x axis. The beam is prismatic, homogeneous, and composed of linearly elastic material.

The deformation of the beam is governed by the assumptions of the elementary beam theory: (1) The longitudinal plane of symmetry of the beam contains a neutral axis; in the absence of axial forces, the neutral axis passes through the centroidal axis of the beam. (2) Plane cross sections remain plane and normal to the neutral axis after deformation. It follows that the configuration of the beam is determined by the configuration of the centroidal axis.

The mathematical model of the beam is formulated on the basis of the flow chart in Figure 1.6.

1. Strain–Displacement Relation

$$\varepsilon(x, y, z) = \varepsilon(x, y) = -y \frac{d^2v(x)}{dx^2} \qquad (1.46)$$

where ε is the normal strain, which is positive in tension, and $v(x)$ defines the transverse deflection of the centroidal axis (see Figure 1.8a). Equation (1.46) is only valid for infinitesimal deformation (Stippes et al., 1961).

2. Stress–Strain Relation

It follows from Hooke's law that

$$\sigma = E\varepsilon \qquad (1.47)$$

where σ is the normal stress and E is the modulus of elasticity.

3. Internal Force–Load Relations

$$\frac{dV}{dx} = p, \quad \frac{dM}{dx} = V \qquad (1.48a)$$

or

$$\frac{d^2M}{dx^2} = p \qquad (1.48b)$$

where V is the shear force and M is the bending moment. The positive sense of the external and internal forces of Eqs. (1.48) is shown in Figure 1.8b.

4. Stress–Displacement Relation

Equations (1.46) and (1.47) yield

$$\sigma = -E \frac{d^2v}{dx^2} y \qquad (1.49)$$

5. Internal Force–Stress Relations

Figure 1.8c indicates that

$$N = \iint_A \sigma \, dA = 0 \qquad (1.50a)$$

and

$$M = - \iint_A \sigma \, dA \, y \qquad (1.50b)$$

6. Internal Force–Displacement Relations

Equations (1.49) and (1.50) yield

$$N = -E\frac{d^2v}{dx^2} \iint_A y\, dA = 0 \tag{1.51a}$$

and

$$M = E\frac{d^2v}{dx^2} \underbrace{\iint_A y^2\, dA}_{I} = EI\frac{d^2v}{dx^2} \tag{1.51b}$$

where I is the moment of inertia of the cross-sectional area with respect to the centroidal axis. It follows from Eq. (1.51a) that

$$\iint_A y\, dA = 0 \tag{1.52}$$

which verifies that the neutral axis passes through the centroidal axis.

7. Load–Displacement Relation

The substitution of Eq. (1.51b) into Eq. (1.48b) yields the continuum model of a prismatic, homogeneous, elastic beam:

$$EI\frac{d^4v}{dx^4} = p \tag{1.53}$$

with annotations: response, action, geometric property, material property.

Note that the configuration of the centroidal axis of the beam, which is defined by $v(x)$, determines any desired measure of response reflected by the model of the beam.[3] For example, from Eqs. (1.51b), (1.48a), and (1.49) we obtain

$$M(x) = EI\frac{d^2v(x)}{dx^2} \tag{1.54a}$$

$$V(x) = EI\frac{d^3v(x)}{dx^3} \tag{1.54b}$$

$$\sigma(x, y) = -\frac{M(x)\, y}{I} \tag{1.54c}$$

---

[3] The model does not represent shearing deformation, even though shear forces are required to establish equilibrium [see Eq. (1.48a)]. For a discussion of shearing deformation, see, for example, Langhaar (1962), Shames (1975), Stippes et al. (1961), and Timoshenko and Gere (1972).

The application of the continuum model is illustrated for the beam in Figure 1.8d. The model is defined by the homogeneous differential equation

$$\frac{d^4v}{dx^4} = 0 \tag{1.55a}$$

which is subject to the boundary conditions [see Figures 1.8b and d and Eqs. (1.54a,b)]

$$v(0) = \frac{dv(0)}{dx} = 0 \tag{1.55b}$$

$$M(L) = EI \frac{d^2v(L)}{dx^2} = M_b, \quad V(L) = EI \frac{d^3v(L)}{dx^3} = -V_b \tag{1.55c}$$

Torsional Deformation Model

The circular shaft in Figure 1.9a is subjected to the distributed torque $m(x)$ and the end torques T_a and T_b. The lines of action of all torques are collinear with the centroidal axis of the shaft. The sense of a torque is determined by the right-hand rule (Figure 1.9b). The material of the shaft is linearly elastic.

The deformations of the shaft are governed by the following assumptions (Stippes et al., 1961): (1) The centroidal axis remains straight, and (2) all radii of a cross section remain straight and rotate through the same angle about the centroidal axis.

The formulation of the mathematical model is based on the flow chart in Figure 1.6.

1. Strain-Displacement Relation

$$\gamma_{x\theta} = r \frac{d\omega(x)}{dx} \tag{1.56}$$

where $\gamma_{x\theta}$ is the shear strain at any point (r, θ) of the cross section and $\omega(x)$ is the angle of twist of the cross section (see Figure 1.9a).

2. Stress-Strain Relation

It follows from Hooke's law that

$$\tau_{x\theta} = G \gamma_{x\theta} \tag{1.57}$$

where $\tau_{x\theta}$ is the shear stress and G is the shear modulus of elasticity.

3. Internal Force-Load Relation

$$\frac{dT}{dx} = -m \tag{1.58}$$

Equation (1.58) represents a pointwise condition of equilibrium, which is based on the sign convention in Figure 1.9d.

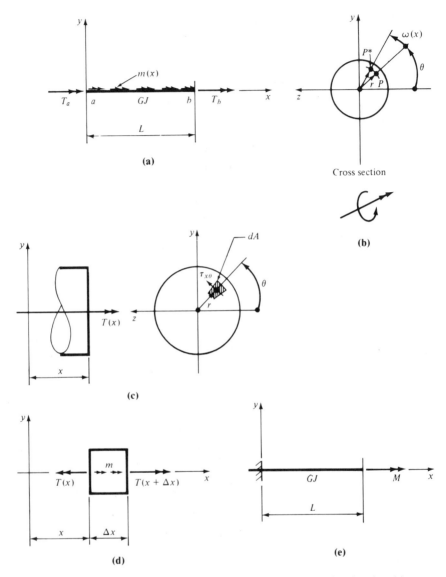

FIGURE 1.9 Torsional deformation. **(a)** Element; **(b)** right-hand rule; **(c)** stress resultant; **(d)** sign convention; **(e)** application

4. Stress–Displacement Relation

Equations (1.56) and (1.57) yield

$$\tau_{x\theta} = G \frac{d\omega}{dx} r \qquad (1.59)$$

5. Internal Force–Stress Relation

Figure 1.9c indicates that

$$T = \iint_A \tau_{x\theta} \, dA \, r \tag{1.60}$$

6. Internal Force–Displacement Relation

Equations (1.59) and (1.60) yield

$$T = G \frac{d\omega}{dx} \underbrace{\iint_A r^2 \, dA}_{J} = GJ \frac{d\omega}{dx} \tag{1.61}$$

where J is the polar moment of inertia of the cross section.

7. Load–Displacement Relation

Combining Eqs. (1.58) and (1.61), we obtain the continuum model of a circular, elastic shaft subjected to a distributed torque:

$$\underset{\substack{\big\downarrow\\ \text{material property}\\ \text{geometric property}}}{GJ} \frac{d^2\omega}{dx^2} = -m \tag{1.62}$$

response
action
geometric property
material property

To illustrate its application, Eq. (1.62) is specialized for the shaft in Figure 1.9e. Since $m = 0$, Eq. (1.62) reduces to the homogeneous differential equation

$$\frac{d^2\omega}{dx^2} = 0 \tag{1.63}$$

which is subject to the boundary conditions

$$\omega(0) = 0 \quad \text{and} \quad T(L) = GJ \frac{d\omega(L)}{dx} = M \tag{1.64}$$

Linear Continuum Models

The models in Eqs. (1.37), (1.53), and (1.62) are linear differential equations in the dependent variables, which determine the element configurations. The models are linear because their components (relations 1–3 in Figure 1.6) are linear. Specifically, the models are composed of linear strain–displacement relations, Hooke's law, and equations of equilibrium formulated for the undeformed state. Since the models are *linear*, the *principle of superposition* is valid; this means that the response to simultaneous actions can be expressed as the sum of the responses to each individual action.

1.6 DISCRETE MODELS

In the absence of interior loads, the continuum models of Section 1.5 [Eqs. (1.37), (1.53), and (1.62)] become homogeneous differential equations of order n ($n = 2, 4$, and 2) whose solutions contain n arbitrary constants. Since the solutions define element configurations, the constants represent n generalized displacements. Accordingly, the axial, flexural, and torsional continuum models of Section 1.5 can be reduced exactly to discrete models.

This reduction is carried out in the following way: First, the element configurations are expressed in terms of the element-end displacements by imposing geometric boundary conditions on the solutions of the homogeneous differential equations. Then, the element-end forces are related to the element-end displacements by means of the internal force–displacement relations of the continuum models (step 6 of Figure 1.6). The resulting sets of algebraic equations form the basic element models used in displacement method analyses of skeletal structures.

Axial Deformation Model

Consider the bar in Figure 1.10a that is subjected to the end forces N_a and N_b. We seek to establish relations between the end forces and the corresponding end displacements u_a and u_b.

Without interior loads, the differential equation of the bar in Eq. (1.37) becomes

$$\frac{d^2u}{dx^2} = 0 \tag{1.65}$$

According to Figure 1.10a, the geometric boundary conditions are

$$u(0) = u_a, \quad u(L) = u_b \tag{1.66}$$

The solution to Eq. (1.65),

$$u = c_0 + c_1 x \tag{1.67}$$

can be specialized by imposing the geometric boundary conditions. This gives

$$c_0 = u_a$$
$$c_1 = \frac{1}{L}(u_b - u_a) \tag{1.68}$$

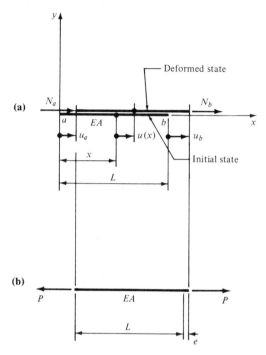

FIGURE 1.10 Axial deformation element

and, hence,

$$u = u_a + \frac{x}{L}(u_b - u_a) \tag{1.69}$$

The substitution of Eq. (1.69) into Eq. (1.36) results in the internal force–displacement relation

$$N(x) = \frac{EA}{L}(u_b - u_a) \tag{1.70}$$

The evaluation of Eq. (1.70) at $x = 0$ and L consistent with the sign convention in Figure 1.7d, that is $N_a = -N(0)$ and $N_b = N(L)$, yields the desired force–displacement relations

$$N_a = \frac{EA}{L}(u_a - u_b)$$
$$N_b = \frac{EA}{L}(-u_a + u_b) \tag{1.71}$$

Equations (1.71) can also be obtained from the familiar axial force–deformation relation (Popov, 1968)

$$P = \frac{EA}{L} e \tag{1.72}$$

by the condition of equilibrium

$$-N_a = N_b = P \tag{1.73}$$

and the deformation–deflection relation

$$e = u_b - u_a \tag{1.74}$$

which follow from Figures 1.10a and b.

Flexural Deformation Model

Figure 1.11 depicts a beam under the action of end forces but without interior loads. To formulate relations between the end forces and the corresponding end displacements, we start with the homogeneous differential equation

$$\frac{d^4 v}{dx^4} = 0 \tag{1.75}$$

which is obtained from Eq. (1.53) by setting $p = 0$. The geometric boundary conditions are

$$v(0) = v_a, \frac{dv(0)}{dx} = \theta_a$$

$$v(L) = v_b, \frac{dv(L)}{dx} = \theta_b \tag{1.76}$$

Equation (1.75) has the solution

$$v = c_0 + c_1 x + c_2 x^2 + c_3 x^3 \tag{1.77}$$

By imposing the geometric boundary conditions, we can express the configuration of the beam in the form

$$v = \phi_a v_a + \psi_a \theta_a + \phi_b v_b + \psi_b \theta_b \tag{1.78}$$

where

$$\begin{aligned}
\phi_a &= 1 - 3\xi^2 + 2\xi^3 \\
\psi_a &= (\xi - 2\xi^2 + \xi^3)L \\
\phi_b &= 3\xi^2 - 2\xi^3 \\
\psi_b &= (-\xi^2 + \xi^3)L
\end{aligned} \tag{1.79}$$

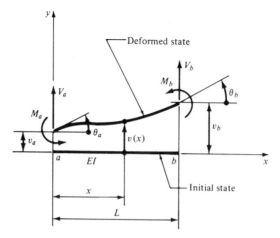

FIGURE 1.11 Flexural deformation element

and

$$\xi = \frac{x}{L}, \quad 0 \le \xi \le 1 \tag{1.80}$$

The transformation of Eq. (1.77) into Eq. (1.78) is carried out in detail in Example 3 of Section 1.4. The coordinate transformation in Eq. (1.80) was introduced to simplify the solution; it results in the following transformation of derivatives:

$$\frac{d^n v}{dx^n} = \left(\frac{1}{L}\right)^n \frac{d^n v}{d\xi^n}, \quad n = 1, 2, 3, \ldots \tag{1.81}$$

With the aid of Eqs. (1.54), (1.78), and (1.81), the bending moment and shear force can be expressed as

$$M(\xi) = \frac{EI}{L^2} (\phi_a'' v_a + \psi_a'' \theta_a + \phi_b'' v_b + \psi_b'' \theta_b)$$

$$V(\xi) = \frac{EI}{L^3} (\phi_a''' v_a + \psi_a''' \theta_a + \phi_b''' v_b + \psi_b''' \theta_b) \tag{1.82}$$

where

$$\begin{aligned}
\phi_a'' &= -6 + 12\xi, & \phi_a''' &= 12 \\
\psi_a'' &= (-4 + 6\xi)L, & \psi_a''' &= 6L \\
\phi_b'' &= 6 - 12\xi, & \phi_b''' &= -12 \\
\psi_b'' &= (-2 + 6\xi)L, & \psi_b''' &= 6L
\end{aligned} \tag{1.83}$$

On the basis of the beam sign convention of Figure 1.8b, the end forces of the beam in Figure 1.11 are defined as

$$M_a = -M(0), \quad V_a = V(0)$$
$$M_b = M(1), \quad V_b = -V(1) \tag{1.84}$$

The evaluation of Eqs. (1.82) and (1.83) according to Eqs. (1.84) results in the force–displacement relations

$$M_a = \frac{EI}{L^2} (6v_a + 4L\theta_a - 6v_b + 2L\theta_b) \tag{1.85a}$$

$$M_b = \frac{EI}{L^2} (6v_a + 2L\theta_a - 6v_b + 4L\theta_b) \tag{1.85b}$$

$$V_a = \frac{EI}{L^3} (12v_a + 6L\theta_a - 12v_b + 6L\theta_b) \tag{1.85c}$$

$$V_b = \frac{EI}{L^3} (-12v_a - 6L\theta_a + 12v_b - 6L\theta_b) \tag{1.85d}$$

Equations (1.85) can also be derived from the slope–deflection equations (Appendix B) for beams without interior loads (Beaufait, 1977; Gerstle, 1974; Ghali and Neville, 1972; Gutkowski, 1981; Hoff, 1956; Hsieh, 1982; Laursen, 1978; McCormac, 1975; Norris, Wilbur, and Utku, 1976; Timoshenko and Young, 1965; West, 1980; White, Gergely, and Sexsmith, 1976; Willems and Lucas, 1978)

$$M_a = \frac{2EI}{L} (2\theta_a + \theta_b - 3\psi_{ab})$$
$$M_b = \frac{2EI}{L} (2\theta_b + \theta_a - 3\psi_{ab}) \tag{1.86}$$

where

$$\psi_{ab} = \frac{1}{L} (v_b - v_a)$$

which are equivalent to Eqs. (1.85a,b), and the conditions of equilibrium

$$V_a = \frac{1}{L} (M_a + M_b)$$
$$V_b = -V_a \tag{1.87}$$

which combined with Eqs. (1.85a,b) yield Eqs. (1.85c,d).

For future use, it is convenient to express Eqs. (1.85) in the form

$$V_a = \alpha(12v_a + 6L\theta_a - 12v_b + 6L\theta_b)$$
$$M_a = \alpha(6Lv_a + 4L^2\theta_a - 6Lv_b + 2L^2\theta_b)$$
$$V_b = \alpha(-12v_a - 6L\theta_a + 12v_b - 6L\theta_b)$$
$$M_b = \alpha(6Lv_a + 2L^2\theta_a - 6Lv_b + 4L^2\theta_b)$$

$$(1.88)$$

where

$$\alpha = \frac{EI}{L^3} \tag{1.89}$$

Torsional Deformation Model

Since the differential equations of the axial and torsional deformation models have the same form [compare Eqs. (1.36) and (1.37) with Eqs. (1.61 and 1.62)], we need only substitute GJ for EA and ω for u in Eqs. (1.71) to obtain the torque–rotation relations (Figure 1.12)

$$T_a = \frac{GJ}{L}(\omega_a - \omega_b)$$
$$T_b = \frac{GJ}{L}(-\omega_a + \omega_b)$$

$$(1.90)$$

Equations (1.90) can also be obtained from the elementary torsion formula (Popov, 1968)

$$T = \frac{GJ}{L}\Delta\omega \tag{1.91}$$

FIGURE 1.12 Torsional deformation element

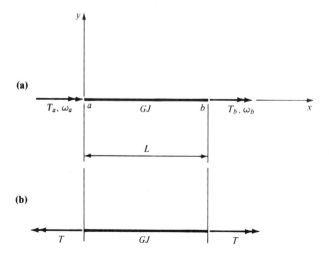

by the condition of equilibrium

$$- T_a = T_b = T \tag{1.92}$$

and the relative rotation expression

$$\Delta \omega = \omega_b - \omega_a \tag{1.93}$$

which follow from Figure 1.12.

Linear Discrete Models

The models in Eqs. (1.71), (1.88), and (1.90) are linear algebraic equations in the element-end displacements, which represent generalized displacements. As a consequence of linearity, the principle of superposition is valid (see Section 1.5).

1.7 DISCRETE MODELS IN MATRIX FORM

In preparation for matrix methods of analysis, the element models of Section 1.6 are expressed in matrix form. Matrix notation permits us to unify distinct structural elements in a mathematical form that reveals their common characteristics.

Specifically, the conventional x, y, z designations of the element coordinate axes (Figures 1.7–1.12) are replaced by the integers 1, 2, 3. Thus, the 1 axis, which is directed from the a end to the b end of the element, coincides with the centroidal axis of the element, and the 2 and 3 axes correspond to principal axes of the cross section (Figure 1.13). Since this coordinate system is intrinsic to the geometry of the element, it is called the *local coordinate system* or the *local frame of reference*.

FIGURE 1.13 Local coordinates

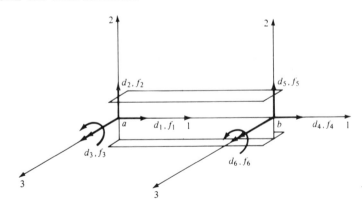

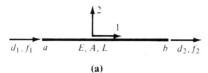

(a)

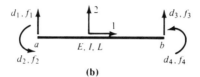

(b)

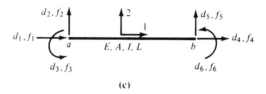

(c)

FIGURE 1.14 Elements. **(a)** Axial deformation; **(b)** flexural deformation; **(c)** axial and flexural deformations

The element-end displacements and forces, denoted by d_i and f_i, respectively, are numbered in the sequence of the coordinate axes from the a end to the b end of the element. The index $i = 1, 2, \ldots, m$, where m is the number of degrees of freedom of the element. This is illustrated in Figures 1.14a and b for elements with axial deformations and flexural deformations, respectively, and in Figures 1.13 and 1.14c for elements with axial and flexural deformations. Moreover, to regard the element-end displacements and forces as coordinates, we assign a local reference frame to each end of the element (Figure 1.13). This provides a simple sign convention: A displacement or force is positive in the direction of the coordinate axis. The positive sense of a rotation or moment about the 3 axis is determined by the right-hand rule; hence, the positive sense is counterclockwise in the 1–2 plane, the plane of the element.

According to these conventions, the elements in Figures 1.10 and 1.11 are represented as shown in Figures 1.14a and b and their models are expressed as

$$f_i = \sum_{j=1}^{m} k_{ij} d_j, \quad i = 1, 2, \ldots, m \tag{1.94a}$$

or

$$\mathbf{f} = \mathbf{kd} \tag{1.94b}$$

where k_{ij} is a stiffness coefficient [by Eq. (1.94a), k_{ij} has units of force per unit displacement], $\mathbf{k}$ is the stiffness matrix, $\mathbf{d}$ is the displacement vector, and $\mathbf{f}$ is the force vector.

Axial Deformation Model (Figure 1.14a)

The substitutions $\gamma = EA/L$ and

$$
\begin{aligned}
d_1 &= u_a, \quad d_2 = u_b \\
f_1 &= N_a, \quad f_2 = N_b
\end{aligned} \tag{1.95}
$$

transform Eqs. (1.71) into

$$
\begin{aligned}
f_1 &= \gamma(d_1 - d_2) \\
f_2 &= \gamma(-d_1 + d_2)
\end{aligned} \tag{1.96a}
$$

or

$$
\begin{bmatrix} f_1 \\ f_2 \end{bmatrix} = \gamma \begin{bmatrix} 1 & -1 \\ -1 & 1 \end{bmatrix} \begin{bmatrix} d_1 \\ d_2 \end{bmatrix}, \quad \gamma = \frac{EA}{L} \tag{1.96b}
$$

Flexural Deformation Model (Figure 1.14b)

The substitutions

$$
\begin{aligned}
d_1 &= v_a, \quad d_2 = \theta_a, \quad d_3 = v_b, \quad d_4 = \theta_b \\
f_1 &= V_a, \quad f_2 = M_a, \quad f_3 = V_b, \quad f_4 = M_b
\end{aligned} \tag{1.97}
$$

transform Eqs. (1.88) into

$$
\begin{aligned}
f_1 &= \alpha(12d_1 + 6Ld_2 - 12d_3 + 6Ld_4) \\
f_2 &= \alpha(6Ld_1 + 4L^2d_2 - 6Ld_3 + 2L^2d_4) \\
f_3 &= \alpha(-12d_1 - 6Ld_2 + 12d_3 - 6Ld_4) \\
f_4 &= \alpha(6Ld_1 + 2L^2d_2 - 6Ld_3 + 4L^2d_4)
\end{aligned} \tag{1.98a}
$$

or

$$
\begin{bmatrix} f_1 \\ f_2 \\ f_3 \\ f_4 \end{bmatrix} = \alpha \begin{bmatrix} 12 & 6L & -12 & 6L \\ 6L & 4L^2 & -6L & 2L^2 \\ -12 & -6L & 12 & -6L \\ 6L & 2L^2 & -6L & 4L^2 \end{bmatrix} \begin{bmatrix} d_1 \\ d_2 \\ d_3 \\ d_4 \end{bmatrix}, \quad \alpha = \frac{EI}{L^3} \tag{1.98b}
$$

Axial and Flexural Deformation Model (Figure 1.14c)

The axial and flexural deformation elements are sufficient to model plane skeletal structures. However, when the two deformations occur simultaneously—as in a frame—it is necessary to combine them into a single element. The incorporation of torsional deformations is deferred until we consider the analysis of space structures in Chapter 5.

To superimpose the models of the elements in Figures 1.14a and b to obtain the model of the element in Figure 1.14c, we introduce the following change in subscripts: The subscript 2 in Eqs. (1.96) is replaced by the subscript 4, and the subscripts 1, 2, 3, 4 in Eqs. (1.98) are replaced by the subscripts 2, 3, 5, 6, respectively. This yields the combined element model

$$
\begin{bmatrix} f_1 \\ f_2 \\ f_3 \\ f_4 \\ f_5 \\ f_6 \end{bmatrix} = \alpha \begin{bmatrix} \beta & 0 & 0 & -\beta & 0 & 0 \\ 0 & 12 & 6L & 0 & -12 & 6L \\ 0 & 6L & 4L^2 & 0 & -6L & 2L^2 \\ -\beta & 0 & 0 & \beta & 0 & 0 \\ 0 & -12 & -6L & 0 & 12 & -6L \\ 0 & 6L & 2L^2 & 0 & -6L & 4L^2 \end{bmatrix} \begin{bmatrix} d_1 \\ d_2 \\ d_3 \\ d_4 \\ d_5 \\ d_6 \end{bmatrix} \tag{1.99}
$$

where

$$
\alpha = \frac{EI}{L^3}, \quad \beta = \frac{AL^2}{I}, \quad \gamma = \alpha\beta = \frac{EA}{L} \tag{1.100}
$$

Special Features

The element models formulated in this section possess properties and admit matrix operations that are used in matrix structural analysis. They are briefly introduced here; some of these properties are further discussed in Section 1.8.

Stiffness Coefficients

The physical meaning of stiffness coefficients can be deduced from Eqs. (1.94), which are expressed as

$$
f_i = k_{i1}d_1 + k_{i2}d_2 + \cdots + k_{ij}d_j + \cdots k_{im}d_m \tag{1.101}
$$

where $i = 1, 2, \ldots, m$. Observe that each term on the right-hand side of Eq. (1.101) represents a contribution to f_i; specifically, $k_{ij}d_j$ is the contribution to f_i due to d_j. If we impose the configuration

$$
d_l = \delta_{lj}, \quad l = 1, 2, \ldots, m \tag{1.102}
$$

where δ_{lj}, which is known as the *Kronecker delta*, is defined as

$$\delta_{lj} = \begin{cases} 1 & \text{if } l = j \\ 0 & \text{if } l \neq j \end{cases} \tag{1.103}$$

that is, $d_j = 1$ and all other end-displacements are zero, Eq. (1.101) yields

$$f_i = k_{ij} \tag{1.104}$$

Accordingly,

$$k_{ij} = \begin{cases} \text{force } f_i \text{ corresponding to the equilibrium} \\ \text{configuration } d_j = 1 \text{ and the remaining} \\ \text{end-displacements are zero} \end{cases} \tag{1.105}$$

force unit displacement

Note that the first subscript of k_{ij}, the row number of **k**, identifies the force, and the second subscript of k_{ij}, the column number of **k**, identifies the unit displacement.

The definition in Eq. (1.105) permits us to compute stiffness coefficients individually or by columns and in the process gain insight into structural behavior. This is illustrated in the following examples.

Example 1. To compute the coefficients of the first column of **k** for the axial deformation element in Figure 1.14a, we impose the configuration

$$d_1 = 1, \quad d_2 = 0 \tag{1.106}$$

which is equilibrated by the coefficients k_{11}, k_{21} as shown in Figure 1.15a. Since the element is shortened by one unit of length, the forces are compressive, as shown in Figure 1.15b, and of value $\gamma = EA/L$. This follows from Eq. (1.72). The comparison of Figures 1.15a and b yields

$$\begin{bmatrix} k_{11} \\ k_{21} \end{bmatrix} = \begin{bmatrix} \dfrac{EA}{L} \\ -\dfrac{EA}{L} \end{bmatrix} = \gamma \begin{bmatrix} 1 \\ -1 \end{bmatrix} \tag{1.107}$$

which agrees with the first column of **k** in Eq. (1.96b).

Example 2. Consider the flexural deformation element in Figure 1.14b. It follows from Eq. (1.105) that the coefficients of the first column of **k** correspond to the configuration in Figure 1.15c. The values of the forces required to equilibrate this configuration are shown in Figure 1.15d. They can be computed with the aid of the slope–deflection equations, Eqs. (1.86). Specifically, since $v_a = 1$ and $\theta_a = \theta_b = v_b = 0$ (compare Figures 1.11 and 1.15d), $\psi_{ab} = -1/L$ and $M_a = M_b = 6EI/L^2$. According to the slope–deflection sign convention (Figure 1.11), the positive sign means that the moments act in a

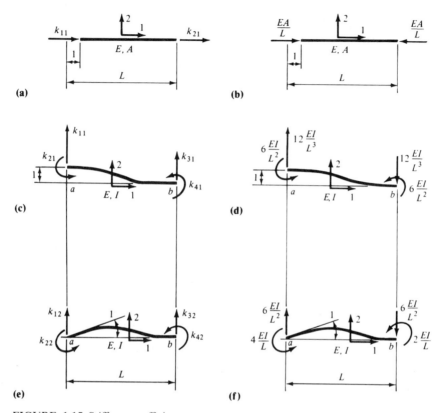

FIGURE 1.15 Stiffness coefficients

counterclockwise sense, as shown in Figure 1.15d. With the moments known, the shear forces in Figure 1.15d are determined by conditions of equilibrium. Equating the forces in Figures 1.15c and d, we obtain

$$
\begin{bmatrix} k_{11} \\ k_{21} \\ k_{31} \\ k_{41} \end{bmatrix} = \begin{bmatrix} \dfrac{12EI}{L^3} \\ \dfrac{6EI}{L^2} \\ -\dfrac{12EI}{L^3} \\ \dfrac{6EI}{L^2} \end{bmatrix} = \alpha \begin{bmatrix} 12 \\ 6L \\ -12 \\ 6L \end{bmatrix} \qquad (1.108a)
$$

which agrees with the first column of **k** in Eq. (1.98b).

The computation of the coefficients in the second column of **k** is based on Figures 1.15e and f. The moments in Figure 1.15f can again be computed with the aid of Eqs. (1.86) ($\theta_a = 1, v_a = v_b = \theta_b = 0$), and the shear forces follow from conditions of equilibrium. The comparison of Figures 1.15e and f yields

$$
\begin{bmatrix} k_{12} \\ k_{22} \\ k_{32} \\ k_{42} \end{bmatrix} = \begin{bmatrix} \dfrac{6EI}{L^2} \\ \dfrac{4EI}{L} \\ -\dfrac{6EI}{L^2} \\ \dfrac{2EI}{L} \end{bmatrix} = \alpha \begin{bmatrix} 6L \\ 4L^2 \\ -6L \\ 2L^2 \end{bmatrix}
\tag{1.108b}
$$

which agrees with the second column of **k** in Eq. (1.98b).

Partitioned Matrices

To partition a matrix means to divide it into submatrices consistent with the laws of matrix computation (Hohn, 1972). It is frequently desirable to partition the element models in Eqs. (1.96b), (1.98b), and (1.99) as indicated by the dashed lines to obtain

$$
\begin{bmatrix} \mathbf{f}_a \\ \mathbf{f}_b \end{bmatrix} = \begin{bmatrix} \mathbf{k}_{aa} & \mathbf{k}_{ab} \\ \mathbf{k}_{ba} & \mathbf{k}_{bb} \end{bmatrix} \begin{bmatrix} \mathbf{d}_a \\ \mathbf{d}_b \end{bmatrix}
\tag{1.109}
$$

where $\mathbf{d}_a$, $\mathbf{f}_a$ and $\mathbf{d}_b$, $\mathbf{f}_b$ are l-dimensional displacement and force vectors at the a and b end of the elements and $\mathbf{k}_{aa}, \mathbf{k}_{ab}, \mathbf{k}_{ba}, \mathbf{k}_{bb}$ are submatrices of order $(l \times l)$, where $l = 1, 2$, and 3. Thus, Eq. (1.109) is conformably partitioned for the matrix computation of the force vectors

$$
\mathbf{f}_a = \mathbf{k}_{aa}\mathbf{d}_a + \mathbf{k}_{ab}\mathbf{d}_b
\tag{1.110a}
$$

$$
\mathbf{f}_b = \mathbf{k}_{ba}\mathbf{d}_a + \mathbf{k}_{bb}\mathbf{d}_b
\tag{1.110b}
$$

As an application of Eqs. (1.110), consider the cantilever in Figure 1.16a with the boundary condition

$$
\mathbf{d}_a = \mathbf{0}
\tag{1.111}
$$

The substitution of Eq. (1.111) into Eq. (1.110b) yields the model of the cantilever, the relation between the applied forces and the corresponding displacements,

$$
\mathbf{f}_b = \mathbf{k}_{bb}\mathbf{d}_b
\tag{1.112}
$$

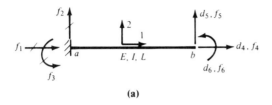

(a)

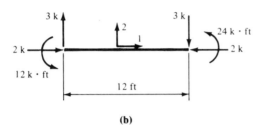

(b)

FIGURE 1.16 Frame element. **(a)** Cantilever; **(b)** free-body diagram

Similarly, the substitution of Eq. (1.111) into Eq. (1.110a) yields the reactive forces

$$\mathbf{f}_a = \mathbf{k}_{ab}\mathbf{d}_b \tag{1.113}$$

More efficiently, the reactive forces can be obtained from the conditions of equilibrium

$$f_1 = -f_4$$
$$f_2 = -f_5 \tag{1.114a}$$
$$f_3 = -Lf_5 - f_6$$

which become in matrix form

$$\begin{bmatrix} f_1 \\ f_2 \\ f_3 \end{bmatrix} = \begin{bmatrix} -1 & 0 & 0 \\ 0 & -1 & 0 \\ 0 & -L & -1 \end{bmatrix} \begin{bmatrix} f_4 \\ f_5 \\ f_6 \end{bmatrix} \tag{1.114b}$$

As an application of Eq. (1.114b), let $L = 12$ ft and

$$\mathbf{f}_b = \begin{bmatrix} -2 \\ -3 \\ 24 \end{bmatrix} \tag{1.115a}$$

where the force units are kip (kilopounds); then

$$\mathbf{f}_a = \begin{bmatrix} -1 & 0 & 0 \\ 0 & -1 & 0 \\ 0 & -12 & -1 \end{bmatrix} \begin{bmatrix} -2 \\ -3 \\ 24 \end{bmatrix} = \begin{bmatrix} 2 \\ 3 \\ 12 \end{bmatrix} \qquad (1.115b)$$

The free-body diagram of the element subjected to these forces is shown in Figure 1.16b.

Note that the force vectors $\mathbf{f}_a$ and $\mathbf{f}_b$ do not represent forces in a physical sense, since their components have distinct units. Specifically, the first two components are forces while the third component is a moment of force. However, the product

$$\mathbf{f}^T\mathbf{d} = \sum_{i=1}^{m} f_i d_i \qquad (1.116)$$

has units of work, which is in compliance with the definition of generalized forces in Section 1.4.

Symmetry

Observe that the stiffness matrices in Eqs. (1.96b), (1.98b), and (1.99) are symmetric (Hohn, 1972; Noble, 1969), that is,

$$k_{ij} = k_{ji} \qquad (1.117a)$$

or

$$\mathbf{k} = \mathbf{k}^T \qquad (1.117b)$$

Furthermore, if $\mathbf{k}$ is partitioned as in Eq. (1.109), Eq. (1.117b) yields

$$\begin{bmatrix} \mathbf{k}_{aa} & \mathbf{k}_{ab} \\ \mathbf{k}_{ba} & \mathbf{k}_{bb} \end{bmatrix} = \begin{bmatrix} \mathbf{k}_{aa}^T & \mathbf{k}_{ba}^T \\ \mathbf{k}_{ab}^T & \mathbf{k}_{bb}^T \end{bmatrix} \qquad (1.118)$$

Thus,

$$\mathbf{k}_{aa} = \mathbf{k}_{aa}^T, \quad \mathbf{k}_{ab} = \mathbf{k}_{ba}^T, \quad \mathbf{k}_{ba} = \mathbf{k}_{ab}^T, \quad \mathbf{k}_{bb} = \mathbf{k}_{bb}^T \qquad (1.119)$$

Equations (1.119) indicate that the submatrices $\mathbf{k}_{aa}$ and $\mathbf{k}_{bb}$ are symmetric. A proof of symmetry for linearly elastic models is presented in Section 1.8.

Singularity

A square matrix is singular, that is, it does not have an inverse, if it has linearly dependent rows. This is reflected by the rank of the matrix, which is equal to the number of linearly independent rows (Hohn, 1972; Noble, 1969). Accordingly, the rank of a square matrix is equal to its order minus the number of linearly dependent rows.

The stiffness matrices in Eqs. (1.96b), (1.98b), and (1.99) are singular and of rank one, two, and three, respectively. The apparent cause of the singularity is conditions of equilibrium that relate the element-end forces. However, this is a consequence of the existence of rigid-body displacements in the elements (see Section 1.8). For illustration consider the element in Figure 1.14c, whose forces are related by Eqs. (1.114a). These relations are reflected by the rows of the stiffness matrix in Eq. (1.99): Specifically, the first row is equal to the negative of the fourth row; the second row is equal to the negative of the fifth row; and the third row is equal to the sum of the negative of L times the fifth row and the negative of the sixth row.

Superposition

The stiffness matrix in Eq. (1.99) can be constructed directly from the component matrices in Eqs. (1.96b) and (1.98b) by a procedure that plays a key role in the assembly of stiffness matrices in Chapter 3. The procedure is as follows: First, the rows and columns of the component matrices, whose numbers correspond to the displacement numbers of the elements in Figures 1.14a and 1.14b, are matched with the displacement numbers of the combined element in Figure 1.14c as shown

$$
\mathbf{k} = \alpha \begin{bmatrix} \beta & -\beta \\ \text{sym.} & \beta \end{bmatrix} \begin{matrix} 1 \\ 4 \end{matrix}, \quad
\mathbf{k} = \alpha \begin{bmatrix} 12 & 6L & -12 & 6L \\ & 4L^2 & -6L & 2L^2 \\ & & 12 & -6L \\ \text{sym.} & & & 4L^2 \end{bmatrix} \begin{matrix} 2 \\ 3 \\ 5 \\ 6 \end{matrix} \qquad (1.120)
$$

Note that the symmetry property is utilized by specifying only the coefficients of the upper triangular portion of each matrix. Next, the coefficients in Eqs. (1.120) are transferred to the locations of the combined matrix defined by the row and column numbers in Eqs. (1.120). Finally, locations of the combined matrix that do not receive contributions are assigned the value zero. This yields

$$
\mathbf{k} = \alpha \begin{bmatrix}
\beta & 0 & 0 & -\beta & 0 & 0 \\
 & 12 & 6L & 0 & -12 & 6L \\
 & & 4L^2 & 0 & -6L & 2L^2 \\
 & & & \beta & 0 & 0 \\
\text{sym.} & & & & 12 & -6L \\
 & & & & & 4L^2
\end{bmatrix}
\begin{matrix} 1 \\ 2 \\ 3 \\ 4 \\ 5 \\ 6 \end{matrix} \qquad (1.121)
$$

which is the stiffness matrix defined in Eq. (1.99).

1.8 DISCRETE ELEMENTS: SPECIAL TOPICS

The properties of singularity and symmetry, identified in the previous section, form the basis of this section.

Singularity is a consequence of the presence of rigid-body displacements in the element model. The elimination of rigid-body displacements reduces the element stiffness matrix to a nonsingular matrix, which relates element deformation parameters to independent forces. Conversely, the nonsingular matrix can be combined with special (contragredient) transformations to generate the complete stiffness matrix of the element. In addition, the inverse of this nonsingular matrix yields the flexibility matrix of the element.

Symmetry, a fundamental property of structural models, is proved with the aid of Betti's law.

Independent Forces

We noted in Section 1.7 that not all forces of the element models are independent. In particular, the forces of the six-degree-of-freedom element in Figure 1.17a are related by three equations of equilibrium. Consequently, only three of the six forces are statically independent.

There is some freedom in the selection of independent forces. Specifically, we may choose any combination of three forces from the six forces in Figure 1.17a that does not include both axial forces or both shear forces; the axial forces and shear forces are dependent, as shown in Eqs. (1.114a). To facilitate the partitioning of the element force vector into dependent and independent forces, let us select the forces at the b end of the element to be independent. (The forces at the a end represent an alternative choice.) We use the symbol p_i to denote independent forces. Thus,

$$\begin{bmatrix} f_4 \\ f_5 \\ f_6 \end{bmatrix} = \begin{bmatrix} p_1 \\ p_2 \\ p_3 \end{bmatrix} \quad \text{or} \quad \mathbf{f}_b = \mathbf{p} \tag{1.122}$$

In addition, let

$$\mathbf{T}_{ba} = \begin{bmatrix} 1 & 0 & 0 \\ 0 & 1 & L \\ 0 & 0 & 1 \end{bmatrix} \tag{1.123}$$

Then Eqs. (1.114b) assume the form

$$\mathbf{f}_a = -\mathbf{T}_{ba}^T \mathbf{f}_b \tag{1.124}$$

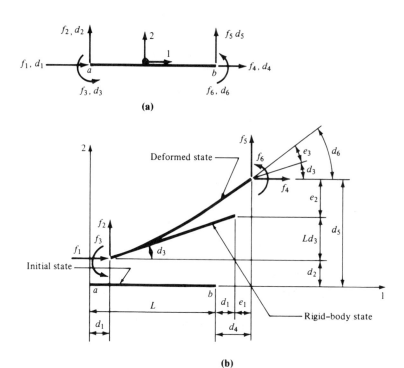

(a)

(b)

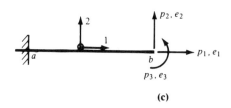

(c)

FIGURE 1.17 Independent forces and deformations. **(a)** Element forces; **(b)** element displacements and deformations; **(c)** independent forces and deformation parameters

The matrix $\mathbf{T}_{ba}$ has a geometric interpretation that will be given below. From Eqs. (1.122) and (1.124), we obtain the relation between the independent forces and the end forces of the element in Figure 1.17a

$$\begin{bmatrix} \mathbf{f}_a \\ \mathbf{f}_b \end{bmatrix} = \begin{bmatrix} -\mathbf{T}_{ba}^T \\ \mathbf{I} \end{bmatrix} \mathbf{p} \quad \text{or} \quad \mathbf{f} = \mathbf{C}^T \mathbf{p} \tag{1.125}$$

where

$$\mathbf{C} = \begin{bmatrix} -\mathbf{T}_{ba} & \mathbf{I} \end{bmatrix} \tag{1.126}$$

Deformations

A rigid bar in a plane has three degrees of freedom. Hence, the configuration of the six-degree-of-freedom element in Figure 1.17a can be defined by three rigid-body displacements and three deformations. Rigid-body displacements cause changes in the position of the element. Deformations, which are relative displacements of points constituting the configuration, cause changes in the shape of the element (Section 1.4).

We have the same freedom in the selection of deformation parameters as in the selection of independent forces. However, the choice of independent forces dictates the choice of deformation parameters because they correspond to each other in the same sense as generalized forces correspond to generalized displacements (Section 1.4).

The deformation parameters corresponding to the independent forces defined in Eqs. (1.122) are shown in Figure 1.17b. They are the relative displacements at the b end of the element denoted e_1, e_2, e_3. The displacements at the a end are rigid-body displacements, and the resulting rigid-body displacements at the b end are limited to infinitesimal rotations (Section 1.4). On the basis of Figure 1.17b, the deformation parameters can be expressed in terms of the element-end displacements as

$$
\begin{aligned}
e_1 &= d_4 - d_1 \\
e_2 &= d_5 - (d_2 + Ld_3) \\
e_3 &= d_6 - d_3
\end{aligned}
\tag{1.127}
$$

Equations (1.127) become, in matrix form,

$$
\begin{bmatrix} e_1 \\ e_2 \\ e_3 \end{bmatrix} =
\begin{bmatrix} -1 & 0 & 0 \\ 0 & -1 & -L \\ 0 & 0 & -1 \end{bmatrix}
\begin{bmatrix} d_1 \\ d_2 \\ d_3 \end{bmatrix} +
\begin{bmatrix} d_4 \\ d_5 \\ d_6 \end{bmatrix}
\tag{1.128}
$$

or by Eq. (1.123)

$$
\mathbf{e} = -\mathbf{T}_{ba}\mathbf{d}_a + \mathbf{d}_b
\tag{1.129}
$$

Equations (1.126) and (1.129) yield the element deformation–displacement relation

$$
\mathbf{e} = [-\mathbf{T}_{ba} \quad \mathbf{I}]\begin{bmatrix} \mathbf{d}_a \\ \mathbf{d}_b \end{bmatrix} = \mathbf{Cd}
\tag{1.130}
$$

Equation (1.129) can be expressed in the form

$$
\mathbf{d}_b = \underbrace{\mathbf{T}_{ba}\mathbf{d}_a}_{\text{rigid-body}} + \underbrace{\mathbf{e}}_{\text{deformation}}
\tag{1.131}
$$

rigid-body
displacement

which indicates that $\mathbf{T}_{ba}$ transforms the rigid-body displacements from the a end to the b end of the element. The deformation vector $\mathbf{e}$ is derived in Appendix A.

Contragredient Transformations

Equations (1.125) and (1.130) define transformations of the form

$$\mathbf{e} = \mathbf{Cd} \tag{1.132a}$$

$$\mathbf{f} = \mathbf{C}^T\mathbf{p} \tag{1.132b}$$

which are known as *contragredient transformations* (Hohn, 1972). Note that the force transformation can be obtained from the displacement transformation by cross relating the displacement vectors with the corresponding force vectors and replacing the displacement transformation matrix by its transpose. Similarly, the displacement transformation can be deduced from the force transformation. Contragredient transformations provide important links in the formulation of matrix methods of analysis.

Sufficient Condition

Contragredient transformations of Eqs. (1.132) exist if

$$\mathbf{d}^T\mathbf{f} = \mathbf{e}^T\mathbf{p} \tag{1.133}$$

the displacements d_i are linearly independent, and the forces p_i are linearly independent.

PROOF. Suppose that Eq. (1.132a) is given. The substitution of Eq. (1.132a) into Eq. (1.133) yields

$$\mathbf{d}^T\mathbf{f} = \mathbf{d}^T\mathbf{C}^T\mathbf{p} \tag{1.134a}$$

or

$$\mathbf{d}^T(\mathbf{f} - \mathbf{C}^T\mathbf{p}) = \mathbf{0} \tag{1.134b}$$

Equation (1.134b) can be expressed as

$$\sum_{i=1}^{m} d_i r_i = 0 \tag{1.135}$$

where $\mathbf{r} = \mathbf{f} - \mathbf{C}^T\mathbf{p}$. It follows that $r_i = 0$, $i = 1, 2, \ldots, m$, if the displacements d_i are linearly independent (Hohn, 1972). Thus, $\mathbf{r} = \mathbf{0}$, and Eq. (1.132b) is satisfied.

Conversely, if Eq. (1.132b) is given, the substitution of Eq. (1.132b) into the relation

$$\mathbf{f}^T\mathbf{d} = \mathbf{p}^T\mathbf{e} \tag{1.136}$$

which is equivalent to Eq. (1.133), since inner products are symmetric, yields

$$\mathbf{p}^T\mathbf{Cd} = \mathbf{p}^T\mathbf{e} \qquad (1.137a)$$

or

$$\mathbf{p}^T(\mathbf{Cd} - \mathbf{e}) = \mathbf{0} \qquad (1.137b)$$

Equation (1.132a) follows from Eq. (1.137b) if the forces p_i are linearly independent.   □

Work Equality

The work equality in Eq. (1.133) can be associated with two loading processes. For example, consider the element in Figure 1.17b. If the forces f_i are applied to the initial state and gradually increased to their final values in the deformed state, the external work done is

$$W_e^f = \sum_{i=1}^{6} \tfrac{1}{2} d_i f_i = \tfrac{1}{2}\mathbf{d}^T\mathbf{f} \qquad (1.138)$$

Alternatively, suppose that the element is first moved to the rigid-body state; in this phase no work is done, since all forces are zero. Next, the forces p_i are applied and gradually increased until the deformed state in Figure 1.17b is reached. During this loading phase, the a end is constrained against displacements, and the corresponding forces f_1, f_2, f_3 are reactive forces. Thus, the external work done is

$$W_e^p = \sum_{i=1}^{3} \tfrac{1}{2} e_i p_i = \tfrac{1}{2}\mathbf{e}^T\mathbf{p} \qquad (1.139)$$

Since the work done on a conservative system is path independent (Langhaar, 1962),

$$W_e^f = W_e^p \qquad (1.140)$$

which results in Eq. (1.133).

Independent Force–Deformation Relations

If we remove the rigid-body displacements from the element in Figure 1.17a, that is, if we impose the boundary condition in Eq. (1.111), we obtain from Eq. (1.131)

$$\mathbf{d}_b = \mathbf{e} \qquad (1.141)$$

and the element in Figure 1.17a reduces to the element in Figure 1.17c. The resulting independent force–deformation relation

$$\mathbf{p} = \mathbf{k}_{bb}\mathbf{e} \qquad (1.142)$$

is obtained from Eqs. (1.112), (1.122), and (1.141). The submatrix $\mathbf{k}_{bb}$ is non-singular because the forces p_1, p_2, p_3 are independent.

Stiffness Matrix: Submatrix Formulation

Equations (1.126), (1.132), and (1.142) permit us to define the element stiffness matrix in terms of the submatrix $\mathbf{k}_{bb}$ as follows:

$$\mathbf{f} = \mathbf{C}^T \mathbf{p} = \mathbf{C}^T \mathbf{k}_{bb} \mathbf{e} = \mathbf{C}^T \mathbf{k}_{bb} \mathbf{C} \mathbf{d} \tag{1.143a}$$

Thus

$$\mathbf{f} = \mathbf{k} \mathbf{d} \tag{1.143b}$$

where

$$\mathbf{k} = \mathbf{C}^T \mathbf{k}_{bb} \mathbf{C} = \begin{bmatrix} \mathbf{T}_{ba}^T \mathbf{k}_{bb} \mathbf{T}_{ba} & -\mathbf{T}_{ba}^T \mathbf{k}_{bb} \\ -\mathbf{k}_{bb} \mathbf{T}_{ba} & \mathbf{k}_{bb} \end{bmatrix} \tag{1.144}$$

Equating the stiffness matrices in Eqs. (1.109) and (1.144) and utilizing the symmetry property of Eqs. (1.119), we obtain

$$\mathbf{k}_{ab} = -\mathbf{T}_{ba}^T \mathbf{k}_{bb}$$
$$\mathbf{k}_{ba} = \mathbf{k}_{ab}^T \tag{1.145}$$
$$\mathbf{k}_{aa} = -\mathbf{T}_{ba}^T \mathbf{k}_{ba}$$

where $\mathbf{T}_{ba}$ is defined in Eq. (1.123). Equations (1.145) provide an efficient approach for the formulation of element stiffness matrices.

Example 1. Equations (1.145) are used to formulate the stiffness coefficients of the element with a hinge near the a end in Figure 1.18a. Since the hinge does not affect the axial stiffness of the element and axial and flexural deformations are not coupled in linear models, $\mathbf{k}_{bb}$ has the form [see Eq. (1.99)]

$$\mathbf{k}_{bb} = \begin{bmatrix} \dfrac{EA}{L} & 0 & 0 \\ 0 & k_{55} & k_{56} \\ 0 & k_{65} & k_{66} \end{bmatrix} \tag{1.146}$$

Moreover, due to symmetry, $k_{56} = k_{65}$. Thus, only three coefficients need to be computed. For this purpose it is convenient to use the approach based on the definition of stiffness coefficients in Eq. (1.105). Accordingly, the coefficients k_{55}, k_{65}, and k_{66} are the forces that equilibrate the configurations shown in Figures 1.18b and d, respectively. The values of these forces are specified in Figure 1.18c and e. They may be computed with the aid of the

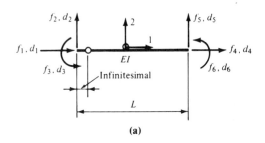

(a)

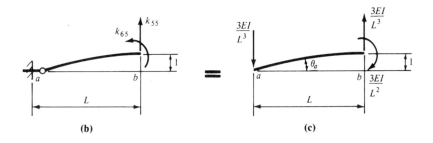

(b)   **(c)**

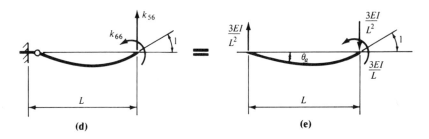

(d)   **(e)**

FIGURE 1.18 Formulation of stiffness matrix

slope–deflection equations, Eqs. (1.86). For example, consider the element in Figure 1.18c, which is subject to the conditions $M_a = 0$, $\theta_b = 0$, $\psi_{ab} = 1/L$. Thus,

$$M_a = \frac{2EI}{L}\left(2\theta_a - \frac{3}{L}\right) = 0 \qquad (1.147a)$$

which yields $\theta_a = 3/(2L)$ and, hence,

$$M_b = \frac{2EI}{L}\left(\theta_a - \frac{3}{L}\right) = -\frac{3EI}{L^2} \qquad (1.147b)$$

The shear forces follow from conditions of equilibrium. Similarly, M_b in Figure 1.18e is obtained from the conditions $M_a = 0$, $\psi_{ab} = 0$, $\theta_b = 1$. It follows from Figures 1.18b, c and 1.18d, e that

$$\mathbf{k}_{bb} = \alpha \begin{bmatrix} \beta & 0 & 0 \\ 0 & 3 & -3L \\ 0 & -3L & 3L^2 \end{bmatrix}, \quad \alpha = \frac{EI}{L^3}, \quad \beta = \frac{AL^2}{I} \tag{1.148}$$

Combining Eqs. (1.123), (1.145), and (1.148), we obtain the stiffness matrix of the element in Figure 1.18a:

$$\mathbf{k} = \alpha \left[\begin{array}{ccc|ccc} \beta & 0 & 0 & -\beta & 0 & 0 \\ 0 & 3 & 0 & 0 & -3 & 3L \\ 0 & 0 & 0 & 0 & 0 & 0 \\ \hline -\beta & 0 & 0 & \beta & 0 & 0 \\ 0 & -3 & 0 & 0 & 3 & -3L \\ 0 & 3L & 0 & 0 & -3L & 3L^2 \end{array} \right] \tag{1.149}$$

Flexibility Matrix

Because the stiffness matrix $\mathbf{k}_{bb}$ is nonsingular, Eq. (1.142) can be solved for the deformation parameters to yield the relation

$$\begin{bmatrix} e_1 \\ e_2 \\ e_3 \end{bmatrix} = \begin{bmatrix} h_{11} & h_{12} & h_{13} \\ h_{21} & h_{22} & h_{23} \\ h_{31} & h_{32} & h_{33} \end{bmatrix} \begin{bmatrix} p_1 \\ p_2 \\ p_3 \end{bmatrix} \tag{1.150a}$$

or

$$\mathbf{e} = \mathbf{hp} \tag{1.150b}$$

where

$$\mathbf{h} = \mathbf{k}_{bb}^{-1} \tag{1.151}$$

The matrix h is called the *flexibility matrix* because the coefficients h_{ij} are measures of flexibility. Specifically,

$$h_{ij} = \text{displacement at } i \text{ caused by a unit force at } j \tag{1.152}$$

This definition follows from Eq. (1.150a). For example, if $p_1 = 1$ and $p_2 = p_3 = 0$, Eq. (1.150a) yields $h_{11} = e_1, h_{21} = e_2, h_{31} = e_3$.

Example 2. We can compute the flexibility matrix of the element in Figure 1.19a by inverting the corresponding stiffness matrix

$$\mathbf{k}_{bb} = \alpha \begin{bmatrix} \beta & 0 & 0 \\ 0 & 12 & -6L \\ 0 & -6L & 4L^2 \end{bmatrix}, \quad \alpha = \frac{EI}{L^3}, \quad \beta = \frac{AL^2}{I} \tag{1.153}$$

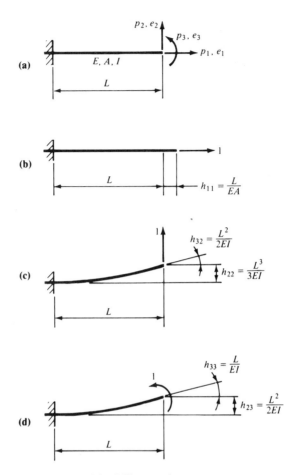

FIGURE 1.19 Formulation of flexibility matrix

which is taken from Eq. (1.99). Thus,

$$\mathbf{h} = \frac{1}{\alpha}\begin{bmatrix} \dfrac{1}{\beta} & 0 & 0 \\[2mm] 0 & \dfrac{1}{3} & \dfrac{1}{2L} \\[2mm] 0 & \dfrac{1}{2L} & \dfrac{1}{L^2} \end{bmatrix} = \begin{bmatrix} \dfrac{L}{EA} & 0 & 0 \\[2mm] 0 & \dfrac{L^3}{3EI} & \dfrac{L^2}{2EI} \\[2mm] 0 & \dfrac{L^2}{2EI} & \dfrac{L}{EI} \end{bmatrix} \qquad (1.154)$$

Alternatively, the flexibility matrix can be computed directly by means of the definition in Eq. (1.152). Accordingly, unit forces are applied one at a

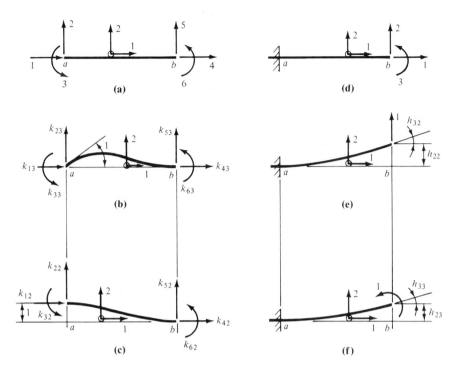

FIGURE 1.20 Symmetry of stiffness and flexibility coefficients

time, as shown in Figures 1.19b, c, and d. The resulting displacements[4] are flexibility coefficients of the first, second, and third column of **h**, respectively.

Note that symmetry, observed in the stiffness matrices, is also exhibited by the flexibility matrix, that is,

$$h_{ij} = h_{ji} \tag{1.155}$$

Symmetry

The symmetry of stiffness and flexibility coefficients is not peculiar to the models formulated in this chapter. It is a characteristic of linearly elastic models in general. This is proved with the aid of Betti's law (Laursen, 1978).

Suppose that the element in Figure 1.20a is linearly elastic but not re-stricted in any other way. For example, it need not be prismatic. To display

---

[4] They can be computed by elementary methods such as the moment–area method in Appendix A.

the type of element configurations involved in the application of Betti's law, the proof is first presented for a specific coefficient. Let us prove that

$$k_{23} = k_{32} \qquad (1.156)$$

It follows from the definition of stiffness coefficients in Eq. (1.105) that the coefficients in Eq. (1.156) are associated with the configurations in Figures 1.20b and c. According to Betti's law, the work done by the forces in Figure 1.20b acting through the corresponding displacements in Figure 1.20c is equal to the work done by the forces in Figure 1.20c acting through the corresponding displacements in Figure 1.20b. Thus,

$$k_{23} \cdot 1 = k_{32} \cdot 1 \qquad (1.157)$$

which proves Eq. (1.156).

Similarly, the symmetry of the flexibility coefficients

$$h_{23} = h_{32} \qquad (1.158)$$

follows from the application of Betti's law to Figures 1.20e and f.

To extend the proof of symmetry to the general condition

$$k_{ij} = k_{ji} \qquad (1.159)$$

we select two sets of displacements and their corresponding forces, which satisfy the relation $\mathbf{f} = \mathbf{kd}$:

$$
\begin{aligned}
d_e &= \delta_{ej}, \quad f_e = k_{ej} \\
d_e &= \delta_{ei}, \quad f_e = k_{ei}, \quad e = 1, 2, \ldots, 6
\end{aligned}
\qquad (1.160)
$$

where the symbols δ_{ej} and δ_{ei} are Kronecker deltas, defined in Eq. (1.103). Applying Betti's law, we obtain

$$\sum_{e=1}^{6} k_{ej}\delta_{ei} = \sum_{e=1}^{6} k_{ei}\delta_{ej} \qquad (1.161)$$

which reduces to Eq. (1.159).

1.9 FINITE ELEMENTS

The finite element method is a general technique for approximating a continuum by a discrete model. Its versatility is reflected by a variety of characterizations and applications (Bathe and Wilson, 1976; Cook, 1981; Desai, 1979; Gallagher, 1975; Huebner, 1975; Irons and Ahmad, 1980; Martin and Carey, 1973; Norrie and de Vries, 1978; Oden, 1972; Oliveira, 1968, 1975; Robinson, 1973; Tong and Rossettos, 1977; Zienkiewicz, 1977). For example, Huebner (1975) states: "*The finite element method* is a numerical analysis technique for obtaining approximate solutions to a wide variety of engineering problems." In regard to numerical solutions of nonlinear problems in continuum mechanics, Oden (1972) asserts,

> If numerical methods are employed in evaluating the results, the continuum is, in effect, approximated by a discrete model in the solution process. This observation suggests a logical alternative to the classical approach, namely, *represent the continuum by a discrete model at the onset*. Then further idealization in either the formulation or the solution may not be necessary. One such approach, based on the idea of piecewise approximating continuous fields, is referred to as *the finite element method*.

Oliveira (1975) states, "*The finite element method* is a discretization technique for the solution of differential equations, the characteristic feature of which is the superposition of coordinate fields piecewise defined on the domain."

The finite element method is introduced with reference to plane skeletal structures. In this context, we may regard the finite element method as a modeling process by which the structure is represented as an assembly of finite elements. The term *finite* refers to the *size* and to the *number* of degrees of freedom of the element. Due to its general discretization technique, the finite element method can be used to construct exact models, such as the discrete models of Section 1.7, or approximate models when exact models do not exist. Example 6 in Appendix D serves as an introduction to finite element formulation.

The finite element modeling process can be divided into three steps:

$$\text{Continuum} \rightarrow \begin{Bmatrix} \text{Discretization} \\ \text{Formulation of finite elements} \\ \text{Assembly of finite elements} \end{Bmatrix} \rightarrow \text{Discrete Model}$$

The description of these steps is confined to assemblies of one-dimensional finite elements.

Discretization

To discretize a continuum means to represent its configuration by a finite number of generalized displacements. This is accomplished as follows (Zienkiewicz, 1977):

a. The continuum is divided into a finite number of elements (subdomains).
b. The elements are assumed to be interconnected at boundary points, the end points of one-dimensional elements, called nodes.
c. The nodal displacements are selected as generalized displacements of the finite element.
d. Interpolation functions are selected to define the element configuration uniquely in terms of the nodal displacements.

The discretization of a continuum is illustrated for the cantilever beam in Figure 1.21a. The finite element representation might be chosen because the

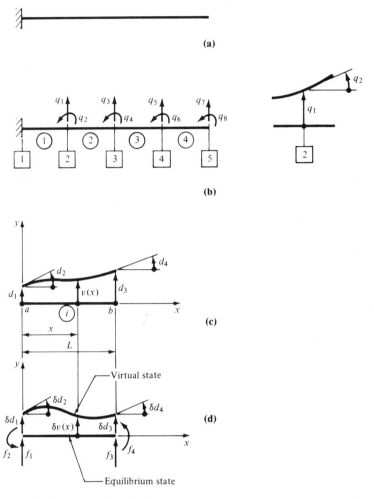

FIGURE 1.21 Discretization. **(a)** Cantilever beam; **(b)** assembly of beam elements; **(c)** beam element; **(d)** virtual displacements

geometric or material properties of the beam preclude the application of the elementary beam theory; for example, the beam may experience large displacements: (a) The beam is divided into four elements (Figure 1.21b). (b) The isolated element (Figure 1.21c) has two nodes, and the assembly of elements (Figure 1.21b) has five nodes. (c) The nodal displacements consist of a transverse deflection and a rotation; accordingly, the isolated element has four degrees of freedom and the assembly has eight degrees of freedom. (d) The configuration of the element (Figure 1.21c) can be expressed in the form

$$v(x) = \sum_{i=1}^{4} N_i(x)d_i = \mathbf{Nd} \qquad (1.162)$$

where N_i is an interpolation function (also called shape or displacement function) and $\mathbf{N}$ is the interpolation matrix. For example, Eqs. (1.16) represent suitable interpolation functions for this problem. Specifically, the substitutions

$$d_1 = v_a, \qquad d_2 = \theta_a, \qquad d_3 = v_b, \qquad d_4 = \theta_b$$
$$N_1 = \phi_a, \qquad N_2 = \psi_a, \qquad N_3 = \phi_b, \qquad N_4 = \psi_b \tag{1.163}$$

transform Eq. (1.15) into Eq. (1.162). Equation (1.162) defines piecewise approximation (element by element) of the complete solution, which is the essence of the finite element method.

Formulation of Finite Elements

Analogous to continuum models (Section 1.5), the construction of finite element models involves the formulation and synthesis of the three basic components of mathematical models (Sections 1.3 and 1.5). This is illustrated schematically for the beam finite element in Figure 1.21c:

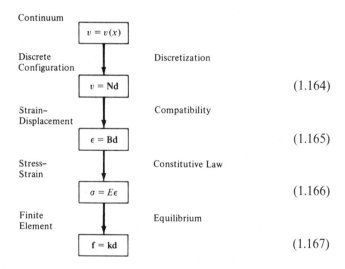

Continuum		
$v = v(x)$		
Discrete Configuration	Discretization	
$v = \mathbf{Nd}$		(1.164)
Strain–Displacement	Compatibility	
$\epsilon = \mathbf{Bd}$		(1.165)
Stress–Strain	Constitutive Law	
$\sigma = E\epsilon$		(1.166)
Finite Element	Equilibrium	
$\mathbf{f} = \mathbf{kd}$		(1.167)

In Eqs. (1.165) and (1.167), $\mathbf{B}$ is the compatibility matrix, which transforms the nodal displacements into the strain function, and $\mathbf{k}$ is the stiffness matrix, which is defined by the volume integral of the element

$$\mathbf{k} = \iiint_V \mathbf{B}^T \mathbf{E} \mathbf{B} \, dV \tag{1.168}$$

The formulation of $\mathbf{B}$ is illustrated in subsequent examples. First, Eq. (1.168) is derived by formulating equilibrium by the principle of virtual work (Appendix D).

PROOF. According to Eqs. (D.28) and (D.46), a sufficient condition of equilibrium is that

$$\delta W = \delta W_e - \delta U = 0 \tag{1.169}$$

for any virtual displacement. In Eq. (1.169), δW_e is the virtual work of the external forces acting on the isolated element, and δU is the first variation of the strain energy of the element. It follows from Figure 1.21d that

$$\delta W_e = \sum_{i=1}^{4} \delta d_i f_i = \delta \mathbf{d}^T \mathbf{f} \tag{1.170}$$

The virtual displacement is imposed on the initial state in Figure 1.21d because in the linear theory (Section 1.5) equilibrium is formulated for the undeformed state. By Eq. (D.41)

$$\delta U = \iiint_V \delta \varepsilon^T \sigma \, dV \tag{1.171}$$

From Eq. (1.165) we obtain the virtual strain (see Appendix D for variational operations)

$$\delta \varepsilon = \mathbf{B} \, \delta \mathbf{d} \tag{1.172a}$$

hence,

$$\delta \varepsilon^T = \delta \mathbf{d}^T \mathbf{B}^T \tag{1.172b}$$

The substitutions of Eqs. (1.165), (1.166), and (1.172b) into Eq. (1.171) yield

$$\delta U = \delta \mathbf{d}^T \left(\iiint_V \mathbf{B}^T E \mathbf{B} \, dV \right) \mathbf{d} \tag{1.173a}$$

or

$$\delta U = \delta \mathbf{d}^T \mathbf{k} \mathbf{d} \tag{1.173b}$$

where $\mathbf{k}$ is defined in Eq. (1.168). Combining Eqs. (1.169), (1.170), and (1.173b), we obtain

$$\delta W = \delta \mathbf{d}^T (\mathbf{f} - \mathbf{k} \mathbf{d}) = 0 \tag{1.174}$$

which yields Eq. (1.167) because the δd_i's are linearly independent [see argument following Eq. (1.135)]. $\qquad \square$

The stiffness matrix formulation by Eq. (1.168) is illustrated for several elements.

Example 1. Consider the axial deformation element in Figure 1.14a. The substitutions

$$d_1 = u_a, \quad d_2 = u_b$$
$$N_1 = \phi_a, \quad N_2 = \phi_b$$

$$(1.175)$$

into Eq. (1.8) yield the discrete element configuration

$$u = \sum_{i=1}^{2} N_i d_i = [N_1 \quad N_2] \begin{bmatrix} d_1 \\ d_2 \end{bmatrix} = \mathbf{Nd} \qquad (1.176)$$

where

$$N_1 = 1 - \xi, \quad N_2 = \xi, \quad \xi = \frac{x}{L} \qquad (1.177)$$

By Eqs. (1.131), (1.176), and (1.177), the strain–displacement relation can be expressed as

$$\varepsilon = \frac{du}{dx} = \frac{du}{d\xi}\frac{d\xi}{dx} = \frac{1}{L}\frac{du}{d\xi} = \frac{1}{L}\frac{d\mathbf{N}}{d\xi}\mathbf{d} \qquad (1.178)$$

or

$$\varepsilon = \mathbf{Bd} \qquad (1.179)$$

where

$$\mathbf{B} = \frac{1}{L}[-1 \quad 1] \qquad (1.180)$$

Rules for matrix differentiation and integration are presented in Appendix D. The substitution for **B** in Eq. (1.168) and integration yield

$$\mathbf{k} = \gamma \begin{bmatrix} 1 & -1 \\ -1 & 1 \end{bmatrix}, \quad \gamma = \frac{EA}{L} \qquad (1.181)$$

which is identical to the stiffness matrix in Eq. (1.96b) because Eq. (1.176) satisfies the homogeneous differential equation, Eq. (1.65).

Example 2. By Eqs. (1.15)–(1.17) and (1.163), the configuration of the beam element in Figure 1.14b can be expressed in the discrete form

$$v = \sum_{i=1}^{4} N_i d_i = \mathbf{Nd} \qquad (1.182)$$

where

$$N_1 = 2\xi^3 - 3\xi^2 + 1$$
$$N_2 = L(\xi^3 - 2\xi^2 + \xi)$$
$$N_3 = -2\xi^3 + 3\xi^2$$
$$N_4 = L(\xi^3 - \xi^2)$$

$$(1.183)$$

and

$$\xi = \frac{x}{L} \tag{1.184}$$

By Eqs. (1.182)–(1.184), the strain–displacement relation, Eq. (1.46), can be expressed as

$$\varepsilon = -y \frac{d^2v}{d\xi^2} \frac{d^2\xi}{dx^2} = -\frac{y}{L^2} \frac{d^2v}{d\xi^2} = -\frac{y}{L^2} \frac{d^2\mathbf{N}}{d\xi^2} \mathbf{d} \tag{1.185}$$

or

$$\varepsilon = \mathbf{Bd} \tag{1.186}$$

where

$$\mathbf{B} = -\frac{y}{L^2} \overline{\mathbf{B}}, \quad \overline{\mathbf{B}} = \frac{d^2\mathbf{N}}{d\xi^2} = [\bar{B}_1 \quad \bar{B}_2 \quad \bar{B}_3 \quad \bar{B}_4] \tag{1.187}$$

$$\begin{aligned}
\bar{B}_1 &= 6(2\xi - 1) \\
\bar{B}_2 &= 2L(3\xi - 2) \\
\bar{B}_3 &= 6(-2\xi + 1) \\
\bar{B}_4 &= 2L(3\xi - 1)
\end{aligned} \tag{1.188}$$

From Eqs. (1.168), (1.184), and (1.186), we obtain for

$$dV = dz\,dy\,dx = dA\,dx = dA\,L\,d\xi$$

$$\mathbf{k} = \frac{E}{L^3} \underbrace{\iint_A y^2\,dA}_{I} \int_0^1 \overline{\mathbf{B}}^T \overline{\mathbf{B}}\,d\xi \tag{1.189}$$

where I is the moment of inertia of the cross-sectional area with respect to the centroidal axis (Section 1.5). Thus,

$$\mathbf{k} = \alpha \int_0^1 \overline{\mathbf{B}}^T \overline{\mathbf{B}}\,d\xi, \quad \alpha = \frac{EI}{L^3} \tag{1.190}$$

Since

$$\overline{\mathbf{B}}^T \overline{\mathbf{B}} = [\bar{B}_i \bar{B}_j] \tag{1.191}$$

the coefficients of $\mathbf{k}$ can be computed by the relation

$$k_{ij} = \alpha \int_0^1 \bar{B}_i \bar{B}_j\,d\xi, \quad i, j = 1, 2, \dots, 4 \tag{1.192}$$

If we take advantage of symmetry, which is reflected by Eq. (1.190), and use Eqs. (1.145), we need only evaluate Eq. (1.192) for the coefficients k_{33}, k_{34}, and k_{44}. Accordingly,

$$k_{33} = \alpha \int_0^1 \bar{B}_3^2 \, d\xi = 12\alpha$$

$$k_{34} = \alpha \int_0^1 \bar{B}_3 \bar{B}_4 \, d\xi = -6L\alpha \qquad (1.193)$$

$$k_{44} = \alpha \int_0^1 \bar{B}_4^2 \, d\xi = 4L^2\alpha$$

Hence,

$$\mathbf{k}_{bb} = \alpha \begin{bmatrix} 12 & -6L \\ -6L & 4L^2 \end{bmatrix} \qquad (1.194)$$

and according to Eq. (1.123)

$$\mathbf{T}_{ba} = \begin{bmatrix} 1 & L \\ 0 & 1 \end{bmatrix} \qquad (1.195)$$

Equations (1.145), (1.194), and (1.195) yield the stiffness matrix of Eq. (1.98b). The exact stiffness matrix (for the linear model) is obtained because Eqs. (1.182), (1.183) satisfy the homogeneous differential equation, Eq. (1.75).

Example 3. Consider the tapered beam in Figure 1.22a. The cross section is rectangular, and the dimensions at a are half of those at b. The moment of inertia can be expressed as

$$I(x) = I_b \left(\frac{x}{2L}\right)^4, \quad L \leq x \leq 2L \qquad (1.196)$$

where I_b is the moment of inertia at b. By Eq. (1.184)

$$I(\xi) = \frac{I_b}{16} \xi^4, \quad 1 \leq \xi \leq 2 \qquad (1.197)$$

The element stiffness matrix representing flexural deformations is desired. This problem was posed and solved exactly by the differential equation approach of Section 1.6 by Hall and Woodhead (1965). We want to illustrate (1) the differential equation approach of Section 1.6 and (2) the finite element approach based on exact and approximate interpolation functions.

1. *Differential Equation Approach.* From Eqs. (1.48b) and (1.51b), we obtain the governing differential equation

$$\frac{d^2}{dx^2} \left(EI \frac{d^2 v}{dx^2} \right) = 0 \qquad (1.198)$$

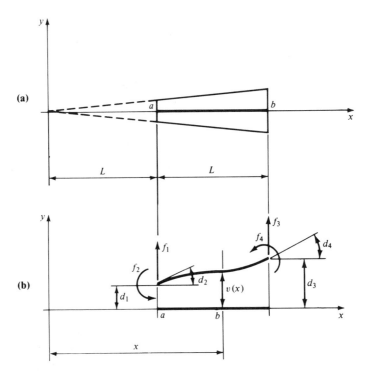

FIGURE 1.22 Tapered-beam element. **(a)** Tapered beam; **(b)** element

which is subject to the boundary conditions (Figure 1.22b)

$$v(L) = d_1, \quad \frac{dv(L)}{dx} = d_2, \quad v(2L) = d_3, \quad \frac{dv(2L)}{dx} = d_4 \qquad (1.199)$$

In terms of the nondimensional coordinate, Eq. (1.184), we can write Eqs. (1.198) and (1.199) as

$$\frac{d^2}{d\xi^2}\left(\frac{EI_b}{16L^2}\,\xi^4\,\frac{d^2v}{d\xi^2}\right) = 0 \qquad (1.200)$$

and

$$v_1(1) = d_1, \quad \frac{dv(1)}{d\xi} = Ld_2, \quad v(2) = d_3, \quad \frac{dv(2)}{d\xi} = Ld_4 \qquad (1.201)$$

The solution to Eqs. (1.200) and (1.201) can be expressed in the form

$$v = \sum_{i=1}^{4} N_i d_i = \mathbf{N}\mathbf{d} \qquad (1.202)$$

where

$$N_1 = -12\xi^{-2} + 28\xi^{-1} + 4\xi - 19$$
$$N_2 = (-4\xi^{-2} + 8\xi^{-1} + \xi - 5)L$$
$$N_3 = 12\xi^{-2} - 28\xi^{-1} - 4\xi + 20$$
$$N_4 = (-8\xi^{-2} + 20\xi^{-1} + 4\xi - 16)L$$

(1.203)

By Eqs. (1.48a), (1.51b), (1.184), (1.197), (1.202), and (1.203)

$$M(\xi) = \frac{EI}{L^2}\frac{d^2v}{d\xi^2} = \bar{\alpha}L\xi^4\frac{d^2\mathbf{N}}{d\xi^2}\mathbf{d}$$

$$V(\xi) = \frac{1}{L}\frac{dM}{d\xi} = \bar{\alpha}\frac{d}{d\xi}\left(\xi^4\frac{d^2\mathbf{N}}{d\xi^2}\right)\mathbf{d}$$

(1.204)

where

$$\bar{\alpha} = \frac{EI_b}{16L^3}$$

(1.205)

and

$$\left(\frac{d^2\mathbf{N}}{d\xi^2}\right)^T = \begin{bmatrix} -72\xi^{-4} + 56\xi^{-3*} \\ (-24\xi^{-4} + 16\xi^{-3})L \\ 72\xi^{-4} - 56\xi^{-3} \\ (-48\xi^{-4} + 40\xi^{-3})L \end{bmatrix}$$

(1.206)

The nodal forces are defined as (Figures 1.8b and 1.22b)

$$f_1 = V(1), \quad f_2 = -M(1), \quad f_3 = -V(2), \quad f_4 = M(2)$$

(1.207)

Equations (1.204)–(1.207) yield the element model

$$\mathbf{f} = \mathbf{kd}$$

(1.208)

where

$$\mathbf{k} = \bar{\alpha} \begin{bmatrix} 56 & 16L & -56 & 40L \\ & 8L^2 & -16L & 8L^2 \\ & & 56 & -40L \\ \text{sym.} & & & 32L^2 \end{bmatrix}$$

(1.209)

2. *Finite Element Approach.* The formulation of the stiffness coefficients differs from that in Example 2 only in the limits of integration and the variable moment of inertia. Accordingly, if we introduce these changes in Eq. (1.192), we obtain the formula

$$k_{ij} = \frac{E}{L^3}\int_1^2 I\bar{B}_i\bar{B}_j\,d\xi$$

(1.210)

which by Eqs. (1.197) and (1.205) becomes

$$k_{ij} = \bar{\alpha} \int_1^2 \xi^4 \bar{B}_i \bar{B}_j \, d\xi \tag{1.211}$$

where by Eqs. (1.187)

$$\bar{B} = \frac{d^2 N}{d\xi^2} \tag{1.212}$$

If we select Eqs. (1.203) as interpolation functions, we obtain the exact stiffness matrix defined in Eq. (1.209). For example, by Eqs. (1.206), (1.211), and (1.212)

$$k_{33} = \bar{\alpha} \int_1^2 \xi^4 \bar{B}_3^2 \, d\xi = \bar{\alpha} \int_1^2 \xi^4 (72\xi^{-4} - 56\xi^{-3})^2 \, d\xi = 56\bar{\alpha} \tag{1.213}$$

As in Example 2, the stiffness matrix can be computed by Eqs. (1.145).

If exact interpolation functions did not exist, we could select the prismatic beam functions in Eqs. (1.183) as interpolation functions. To specialize them for the beam in Figure 1.22b, we must replace ξ by $(\xi - 1)$. Thus,

$$\begin{aligned}
N_1 &= 2(\xi - 1)^3 - 3(\xi - 1)^2 + 1 \\
N_2 &= L[(\xi - 1)^3 - 2(\xi - 1)^2 + \xi - 1] \\
N_3 &= -2(\xi - 1)^3 + 3(\xi - 1)^2 \\
N_4 &= L[(\xi - 1)^3 - (\xi - 1)^2]
\end{aligned} \tag{1.214}$$

and

$$\bar{B}^T = \left(\frac{d^2 N}{d\xi^2}\right)^T = \begin{bmatrix} 6(2\xi - 3) \\ 2L(3\xi - 5) \\ 6(-2\xi + 3) \\ 2L(3\xi - 4) \end{bmatrix} \tag{1.215}$$

From Eqs. (1.211) and (1.215) we obtain

$$\mathbf{k}_{bb} = \bar{\alpha} \begin{bmatrix} 85.4 & -57.1L \\ -57.1L & 41.9L^2 \end{bmatrix} \tag{1.216}$$

which is combined with Eqs. (1.145) and (1.195) to yield the approximate stiffness matrix

$$\mathbf{k} = \bar{\alpha} \begin{bmatrix} 85.4 & 28.3L & -85.4 & 57.1L \\ & 13.1L^2 & -28.3L & 15.2L^2 \\ & & 85.4 & -57.1L \\ \text{sym.} & & & 41.9L^2 \end{bmatrix} \tag{1.217}$$

Assembly of Finite Elements

The assembly of elements represented by stiffness matrices is described in Chapters 2 and 3 in the context of the matrix displacement method. In addition, the finite element method is extended in Section 3.7 to structures with element loads and temperature changes.

Interpolation

Function interpolation forms the central task of the finite element method. To provide an introduction to this important field of mathematics (Atkinson, 1978; Forsythe, Malcolm, and Moler, 1977; Hildebrand, 1974), the classical Lagrange and Hermite interpolation polynomials are presented and illustrated.

Interpolation is concerned with approximating a function over a domain where the values of the function, and possibly its derivatives, are known at a finite number of points. For example, consider the function $f(\xi)$, whose values

$$f(\xi_i) = f_i \qquad (1.218)$$

are known at the distinct points $\xi_1 < \xi_2 < \cdots < \xi_n$, called the interpolation points. We are interested in polynomial interpolation of the form

$$p(\xi) = \sum_{i=1}^{n} l_i(\xi) f_i \qquad (1.219)$$

which in this case is subject to the constraints (see Figure 1.23a for $n = 3$)

$$p(\xi_j) = f_j, \quad j = 1, 2, \ldots, n \qquad (1.220)$$

Thus,

$$p(\xi_j) = l_1(\xi_j) f_1 + \cdots + l_j(\xi_j) f_j + \cdots + l_n(\xi_j) f_n = f_j \qquad (1.221)$$

To satisfy Eq. (1.221) the functions $l_i(\xi)$ must have the property (see Figures 1.23b for $n = 3$)

$$l_i(\xi_j) = \delta_{ij} = \begin{cases} 1 & \text{if } i = j \\ 0 & \text{if } i \neq j \end{cases}, \quad i, j = 1, 2, \ldots, n \qquad (1.222)$$

where δ_{ij} is the Kronecker delta.

Lagrange Interpolation

In Lagrange interpolation, the function $f(\xi)$ is approximated by the polynomial of degree $n - 1$

$$p(\xi) = \sum_{i=1}^{n} l_i(\xi) f_i \qquad (1.223)$$

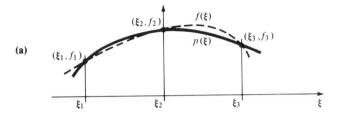

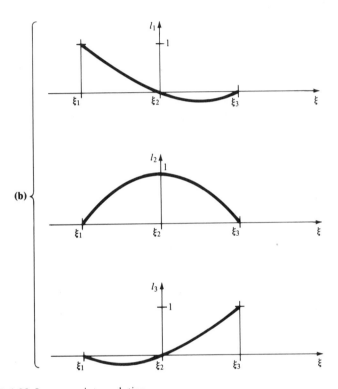

FIGURE 1.23 Lagrange interpolation

where $f_i = f(\xi_i)$, $\xi_1 < \xi_2 < \cdots < \xi_n$, and

$$l_i(\xi) = \frac{(\xi - \xi_1) \cdots (\xi - \xi_{i-1})(\xi - \xi_{i+1}) \cdots (\xi - \xi_n)}{(\xi_i - \xi_1) \cdots (\xi_i - \xi_{i-1})(\xi_i - \xi_{i+1}) \cdots (\xi_i - \xi_n)} \quad (1.224)$$

Note that $l_i(\xi_j)$ satisfies Eq. (1.222).

Example 4. Let $n = 3$, $\xi_1 = -1$, $\xi_2 = 0$, $\xi_3 = 1$. Then by Eq. (1.224)

$$l_1(\xi) = \frac{(\xi - \xi_2)(\xi - \xi_3)}{(\xi_1 - \xi_2)(\xi_1 - \xi_3)} = \frac{1}{2}\xi(\xi - 1)$$

$$l_2(\xi) = \frac{(\xi - \xi_1)(\xi - \xi_3)}{(\xi_2 - \xi_1)(\xi_2 - \xi_3)} = -(\xi + 1)(\xi - 1) \qquad (1.225)$$

$$l_3(\xi) = \frac{(\xi - \xi_1)(\xi - \xi_2)}{(\xi_3 - \xi_1)(\xi_3 - \xi_2)} = \frac{1}{2}(\xi + 1)\xi$$

and the quadratic Lagrange polynomial

$$p(\xi) = l_1(\xi)f_1 + l_2(\xi)f_2 + l_3(\xi)f_3 \qquad (1.226)$$

interpolates f at -1, 0, and 1. For example, if $f_1 = f_3 = 1$ and $f_2 = 0$, $p(\xi) = \xi^2$.

Example 5. Let us construct the Lagrange polynomial of degree 1 to interpolate the nodal displacements of the element in Figure 1.24a, where $n = 2$, $\xi_1 = 0$, $\xi_2 = 1$, $u(\xi_1) = d_1$, and $u(\xi_2) = d_2$. Thus,

$$l_1(\xi) = \frac{\xi - \xi_2}{\xi_1 - \xi_2} = 1 - \xi$$
$$\qquad (1.227)$$
$$l_2(\xi) = \frac{\xi - \xi_1}{\xi_2 - \xi_1} = \xi$$

and

$$p(\xi) = l_1(\xi)d_1 + l_2(\xi)d_2 \qquad (1.228)$$

The property $l_i(\xi_j) = \delta_{ij}$ is illustrated in Figures 1.24b. Note that the Lagrange polynomial describes the element configuration defined by Eq. (1.176).

Hermite Interpolation

Hermite interpolation, a generalization of Lagrange interpolation, permits us to interpolate f and consecutive derivatives of f at the interpolation points. For example, the Hermite interpolation polynomial of degree $2n - 1$

$$p(\xi) = \sum_{i=1}^{n} [\phi_i(\xi)\, f_i + \psi_i(\xi)\, f_i'] \qquad (1.229)$$

satisfies the constraints

$$p(\xi_j) = f(\xi_j) = f_j$$
$$\frac{dp(\xi_j)}{d\xi} = \frac{df(\xi_j)}{d\xi} = f_j' \qquad (1.230)$$

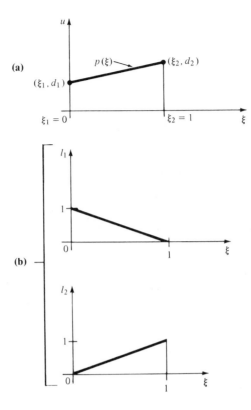

FIGURE 1.24 Lagrange interpolation

In Eq. (1.229)

$$\phi_i(\xi) = [1 - 2l_i'(\xi_i)(\xi - \xi_i)]l_i^2(\xi)$$
$$\psi_i(\xi) = (\xi - \xi_i)l_i^2(\xi)$$

(1.231)

where $l_i(\xi)$ is defined in Eq. (1.224), and

$$l_i'(\xi_i) = \frac{dl_i(\xi_i)}{d\xi}$$

(1.232)

The functions in Eqs. (1.231) have the property

$$\phi_i(\xi_j) = \delta_{ij}, \quad \phi_i'(\xi_j) = 0$$
$$\psi_i(\xi_j) = 0, \quad \psi_i'(\xi_j) = \delta_{ij}$$

(1.233)

Example 6. Consider the beam element in Figure 1.25a. The interpolation data are $n = 2$, $\xi_1 = 0$, $\xi_2 = 1$, $v(\xi_1) = v_1$, $dv(\xi_1)/d\xi = v'_1$, $v(\xi_2) = v_2$, $dv(\xi_2)/d\xi = v'_2$. Equation (1.229) yields the cubic Hermite interpolation polynomial

$$p(\xi) = \phi_1 v_1 + \psi_1 v'_1 + \phi_2 v_2 + \psi_2 v'_2 \tag{1.234}$$

where by Eqs. (1.227) and (1.231)

$$l_1 = 1 - \xi, \quad l'_1 = -1, \quad l_2 = \xi, \quad l'_2 = 1 \tag{1.235}$$

$$
\begin{aligned}
\phi_1 &= 2\xi^3 - 3\xi^2 + 1 \\
\psi_1 &= \xi^3 - 2\xi^2 + \xi \\
\phi_2 &= -2\xi^3 + 3\xi^2 \\
\psi_2 &= \xi^3 - \xi^2
\end{aligned}
\tag{1.236}
$$

The graphs of the functions in Eqs. (1.236) (Figures 1.25b) reflect the properties defined by Eqs. (1.233). Specifically, each function assumes a unit value in the direction of the corresponding interpolation parameter, a generalized displacement, and the value zero in the directions of all other interpolation parameters. For example, $\psi_1(\xi_1) = \psi_1(\xi_2) = \psi'_1(\xi_2) = 0$, $\psi'_1(\xi_1) = 1$.

Equation (1.234) can be specialized for the beam element in Figure 1.21c by the coordinate transformation in Eq. (1.184), which yields

$$\frac{dv}{d\xi} = L\frac{dv}{dx} \tag{1.237}$$

Thus,

$$v_1 = d_1, \quad v'_1 = Ld_2, \quad v_2 = d_3, \quad v'_2 = Ld_4 \tag{1.238}$$

In addition, let

$$N_1 = \phi_1, \quad N_2 = L\psi_1, \quad N_3 = \phi_2, \quad N_4 = L\psi_2 \tag{1.239}$$

then Eqs. (1.234) and (1.236) yield Eqs. (1.182) and (1.183), respectively.

Convergence

Consider the finite element representation of the beam in Figure 1.21b, which produces an approximation to the exact solution. We expect that successive finite element solutions corresponding to smaller and smaller elements will converge to the exact solution as the element size tends to zero.

64

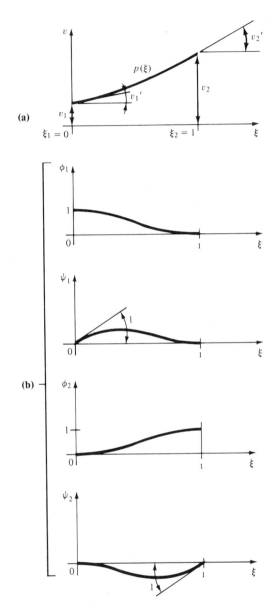

FIGURE 1.25 Hermite interpolation

However, to achieve convergence, the finite elements must satisfy certain requirements.

Various convergence criteria have been proposed and some have been proved rigorously. See, for example, discussions on convergence in Bathe and Wilson (1976), Cook (1981), Huebner (1975), Norrie and de Vries (1973), Schnobrich and Pecknold (1973), and Zienkiewicz (1977). A sufficient condition for monotonic energy convergence, which was proved by Oliveira (1968), is presented here. Monotonic convergence means that each successive reduction in element size (mesh refinement) produces a more accurate result. Convergence in energy means that in the limit as the element size tends to zero, the energy contained in the finite element assemblage approaches the energy level of the continuum.

Convergence in Energy

If the configuration of the finite element satisfies the conditions of *completeness* and *compatibility*, the finite element solution converges monotonically in energy. These conditions can be defined relative to

$$p = \text{highest-order derivative of strain energy function.}$$

Completeness is satisfied by a complete polynomial of degree p. A complete one-dimensional polynomial of degree n contains the $n + 1$ terms

$$1, \xi, \xi^2, \ldots, \xi^n$$

Compatibility, also called conformity, is satisfied if the displacements and derivatives up to order $(p - 1)$ are continuous at element boundaries (nodes for one-dimensional elements).

According to Eqs. (D.6) and (D.7), the strain energy can be expresssed as

$$U = \frac{1}{2} \iiint\limits_V E\varepsilon^2 \, dV \qquad (1.240)$$

Thus, p is determined by the strain–displacement function.

Example 7. For axial deformation elements, $p = 1$ [Eq. (1.31)]. Thus, completeness requires a complete polynomial of degree 1; this is satisfied by the Lagrange polynomial in Eq. (1.228), which is used in Example 1. Compatibility requires continuity in deflections at the nodes; this is satisfied for trusses during assembly (Chapters 2 and 3).

Example 8. For flexural deformation elements, $p = 2$ [Eq. (1.46)]. Thus, the complete cubic Hermite interpolation polynomial, Eq. (1.234), which is used in Examples 2 and 3, exceeds the completeness requirement. Compatibility requires continuity in deflections and rotations at the nodes; this is satisfied for frames during assembly (Chapters 2 and 3).

PROBLEMS

1.1   Define the configuration of the rigid bar, $u(x)$, $v(x)$, $-L/2 \leq x \leq L/2$, in terms of the generalized displacements q_1, q_2, q_3; q_3 is infinitesimal.

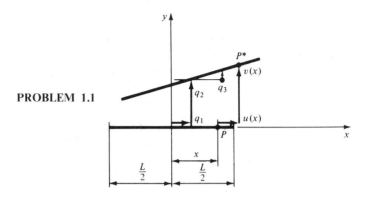

PROBLEM 1.1

1.2   Show that Eqs. (1.6) and (1.8) can be obtained from Eqs. (1.9) and (1.10). Equations (1.6) and (1.8) are Lagrange interpolation polynomials; they interpolate function values. Sketch ϕ_a and ϕ_b. Note that ϕ_a and ϕ_b assume a unit value in the directions of the corresponding generalized displacements and the value zero in the directions of the other generalized displacements.

1.3   The extensional strain of the element in Figure 1.3c is $\varepsilon = (u_b - u_a)/L$. Integrate the strain–displacement relation $\varepsilon = du/dx$ to obtain Eq. (1.8).

1.4   Equation 1.15 is an example of Hermite interpolation, where function values and function derivatives are interpolated. Sketch ϕ_a, ψ_a, ϕ_b, ψ_b. Note that these functions assume a unit value in the directions of the corresponding generalized displacements and the value zero in the directions of all other generalized displacements; for example, $\phi_a(0) = 1$, $\phi_a'(0) = \phi_a(1) = \phi_a'(1) = 0$.

1.5   Specialize Eq. (1.15) for the boundary conditions $v_a = \theta_b = 0$. Sketch the corresponding initial and deformed configurations, including the supports. (Note that the support at b is a shear release.)

1.6   Determine the generalized forces corresponding to the generalized displacements in Problem 1.1 for the force distributions shown:

$$p = 0, \ -L/2 \leq x \leq 0; \ p = -(2/L)p_0 x, \ 0 \leq x \leq L/2.$$

Set the generalized forces equal to zero to obtain the conditions of equilibrium.

1.7   The bar in Figure 1.7b is represented by Eqs. (1.38) and (1.39). Solve Eqs. (1.38), (1.39) and evaluate Eq. (1.36) at $x = 0$, $L/2$. Draw the free-body diagram of the bar segment $0 \leq x \leq L/2$; is equilibrium satisfied?

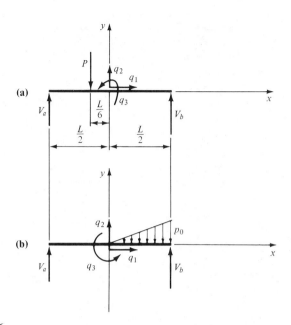

PROBLEM 1.6

1.8   Show that the solution to Eq. (1.40) subject to the boundary conditions $u(0) = u_a$, $u(L) = u_b$ is Eq. (1.8).

1.9   Derive Eq. (1.46); state the assumptions (kinematic constraints) used in the derivation.

1.10 Derive Eqs. (1.48) by formulating conditions of equilibrium for the beam segment in Figure 1.8b and taking the limit as Δx approaches zero, analogous to Eqs. (1.43)–(1.45).

1.11 Show that Eq. (1.18) satisfies Eqs. (1.55a) and (1.55b). Use Eqs. (1.16)–(1.18) and Eqs. (1.55c) to obtain the force–displacement relation of the beam in Figure 1.8d expressed in matrix form as

$$\begin{bmatrix} V_b \\ M_b \end{bmatrix} = \alpha \begin{bmatrix} 12 & -6L \\ -6L & 4L^2 \end{bmatrix} \begin{bmatrix} v_b \\ \theta_b \end{bmatrix}, \quad \alpha = \frac{EI}{L^3}$$

[Note that by Eq. (1.17), $d^n v/dx^n = (1/L)^n\, d^n v/d\xi^n$.]

1.12 Show that the solution to Eq. (1.63) subject to the geometric boundary conditions $\omega(0) = \omega_a$, $\omega(L) = \omega_b$ can be expressed in the form of Eq. (1.8).

1.13 Combine Eqs. (1.72)–(1.74) to obtain Eqs. (1.71).

1.14 Transform Eqs. (1.86) and (1.87) into Eqs. (1.88).

1.15 Superimpose the models of the elements in Figures 1.14a and b, which are defined by Eqs. (1.96b) and (1.98b), to obtain the model of the element shown. Use both procedures illustrated in Section 1.7: (a) the procedure used to construct Eq. (1.99), and (b) the procedure used to construct Eq. (1.121). Note that the axial and flexural deformations are not coupled, that is, flexural displacements do not cause axial forces and vice versa. This is characteristic of linear models.

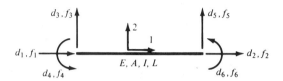

PROBLEM 1.15

1.16 Use the definition in Eq. (1.105) to compute the coefficients of the second column of **k** for the element in Figure 1.14a.

1.17 Use the definition in Eq. (1.105) to compute the coefficients of the third and fourth columns of **k** for the element in Figure 1.14b.

1.18 Consider the flexural deformation element in Figure 1.14b:
   a. Specialize Figure 1.17b for the element in Figure 1.14b,
   b. Use the figure in (a) to specialize Eq. (1.131) for the element in Figure 1.14b.
   c. Formulate the conditions of equilibrium for the element in Figure 1.14b and express them in the form of Eq. (1.124); let $\mathbf{f}_b = \mathbf{p}$ and express the equations of (b) and (c) in the form of Eqs. (1.132).

1.19 Follow the approach of Example 1 in Section 1.8 to formulate the stiffness matrix of the element shown. According to Problem 1.18,

$$\mathbf{T}_{ba} = \begin{bmatrix} 1 & L \\ 0 & 1 \end{bmatrix}$$

PROBLEM 1.19

1.20 Use the moment–area method (Appendix A) to verify the values of the flexibility coefficients in Figure 1.19.

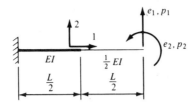

PROBLEM 1.21

1.21 Use the moment–area method to formulate the flexibility matrix, **h**, of the element shown.

1.22 Use the flexibility matrix of Problem 1.21 to derive the stiffness matrix of the element shown as follows:

a. Invert **h** to obtain the submatrix k_{bb}; for example, use the formula (Hohn, 1972)

$$\mathbf{h}^{-1} = \frac{\text{adj } \mathbf{h}}{|\mathbf{h}|}, \quad \mathbf{h} = \frac{1}{\alpha}\begin{bmatrix} \dfrac{3}{8} & \dfrac{5}{8L} \\[2mm] \dfrac{5}{8L} & \dfrac{3}{2L^2} \end{bmatrix}, \quad \alpha = \frac{EI}{L^3}$$

where adj **h** is the adjoint of **h** and $|\mathbf{h}|$ is the determinant of **h**.

b. Use Eqs. (1.145) to obtain the stiffness matrix; $\mathbf{T}_{ba}$ is defined in Problem 1.19.

c. Sketch the configurations corresponding to the coefficients of the first column and fourth column of **k**; is equilibrium satisfied?

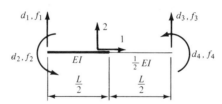

PROBLEM 1.22

1.23 Consider Figure 1.20a: Use Betti's law to prove that $k_{36} = k_{63}$; sketch appropriate configurations.

1.24 Consider the tapered beam of Example 3, Section 1.9, Figure 1.22a:

a. Show that the cross-sectional area $A(x) = \frac{1}{4}A_b(x/L)^2$, where A_b is the cross-sectional area at the b end.

b. Use Eqs. (1.33) and (1.36) to express the differential equation representing axial deformation as

$$\frac{d}{d\xi}\left(\bar{\gamma}\xi^2 \frac{du}{d\xi}\right) = 0, \quad \bar{\gamma} = \frac{1}{4}\frac{EA_b}{L}$$

c. Verify that the differential equation in (b) has the solution

$$u = \sum_{i=1}^{2} N_i(\xi)d_i, \qquad 1 \le \xi \le 2$$

$$N_1(\xi) = 2\xi^{-1} - 1, \quad N_2(\xi) = 2(-\xi^{-1} + 1)$$

$$d_1 = u(1), \qquad\qquad d_2 = u(2)$$

d. Use Eq. (1.31) and $u(\xi)$ defined in (c) to determine $\varepsilon(\xi)$; plot $\varepsilon(\xi)$ for $1 \le \xi \le 2$ and the extensional strain $\varepsilon = (1/L)(d_2 - d_1)$ of a prismatic element.

e. Use the differential equation approach (Section 1.6) to obtain the discrete model

$$\begin{bmatrix} f_1 \\ f_2 \end{bmatrix} = \frac{1}{2}\frac{EA_b}{L}\begin{bmatrix} 1 & -1 \\ -1 & 1 \end{bmatrix}\begin{bmatrix} d_1 \\ d_2 \end{bmatrix}$$

f. Use the finite element approach (Section 1.9) to obtain the discrete model in (e).

g. Show that the finite element approach based on the Lagrange interpolation functions (for $\xi_1 = 1$, $\xi_2 = 2$)

$$N_1 = 2 - \xi, \quad N_2 = -1 + \xi$$

yields the approximate stiffness matrix

$$\mathbf{k} = \frac{7}{12}\frac{EA_b}{L}\begin{bmatrix} 1 & -1 \\ -1 & 1 \end{bmatrix}$$

1.25 The configuration of the axial deformation element is approximated by the quadratic Lagrange polynomial

$$u(\xi) = \sum_{i=1}^{3} N_i(\xi)d_i, \quad \xi = 2\frac{x}{L} - 1$$

where $N_i(\xi) = l_i(\xi)$, $d_i = u(\xi_i)$, and the l_i's are defined in Eq. (1.225). Use the finite element approach (Section 1.9) to formulate the stiffness matrix of the three degree-of-freedom element

$$\mathbf{f} = \mathbf{kd}$$

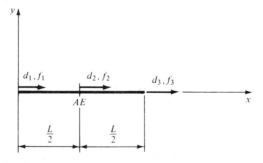

PROBLEM 1.25

1.26 The interpolation data are $\xi_1 = -1, \xi_2 = 1, u(\xi_1) = d_1, u(\xi_2) = d_2$.
   a. Construct the Lagrange interpolation polynomial.
   b. Use the interpolation functions in (a) to work Example 1 of Section 1.9.

1.27 The interpolation data are $\xi_1 = -1, \;\; \xi_2 = 1, \;\; v(\xi_1) = v_1, \;\; dv(\xi_1)/d\xi = v_1',$ $v(\xi_2) = v_2, dv(\xi_2)/d\xi = v_2'$.
   a. Construct the Hermite interpolation polynomial.
   b. Use the interpolation functions in (a) to work Example 2 of Section 1.9; $\xi = 2(x/L) - 1$.

REFERENCES

Atkinson, K. E. 1978. *An Introduction to Numerical Analysis*. Wiley, New York.

Bathe, K. J., and E. L. Wilson. 1976. *Numerical Methods in Finite Element Analysis*. Prentice-Hall, Englewood Cliffs, NJ.

Beaufait, F. W. 1977. *Basic Concepts of Structural Analysis*. Prentice-Hall, Englewood Cliffs, NJ.

Boresi, A. P., O. M. Sidebottom, F. B. Seely, and J. O. Smith. 1978. *Advanced Mechanics of Materials*. Wiley, New York.

Cook, R. D. 1981. *Concepts and Applications of Finite Element Analysis*, 2nd ed. Wiley, New York.

Crandall, S. H., N. C. Dahl, and T. J. Lardner. 1978. *An Introduction to the Mechanics of Solids*, 2nd ed. McGraw-Hill, New York.

Desai, C. S. 1979. *Elementary Finite Element Method*. Prentice-Hall, Englewood Cliffs, NJ.

Forsythe, G. E., M. A. Malcolm, and C. B. Moler. 1977. *Computer Methods for Mathematical Computations*. Prentice-Hall, Englewood Cliffs, NJ.

Freudenthal, A. M. 1966. *Introduction to the Mechanics of Solids*. Wiley, New York.

Gallagher, R. H. 1975. *Finite Element Analysis*. Prentice-Hall, Englewood Cliffs, NJ.

Gerstle, K. H. 1974. *Basic Structural Analysis*. Prentice-Hall, Englewood Cliffs, NJ.

Ghali, A., and A. M. Neville. 1972. *Structural Analysis*. Intext Educational Publishers, Scranton, PA.

Gutkowski, R. M. 1981. *Structures*. Van Nostrand Reinhold, New York.

Hall, A. S., and R. W. Woodhead. 1965. *Frame Analysis*, 2nd ed. Wiley, New York.

Hildebrand, F. B. 1974. *Introduction to Numerical Analysis*, 2nd ed. McGraw-Hill, New York.

Hoff, N. J. 1956. *The Analysis of Structures*. Wiley, New York.

Hohn, F. E. 1972. *Elementary Matrix Algebra*, 3d ed. Macmillan, New York.

Hsieh, Y. 1982. *Elementary Theory of Structures*. Prentice-Hall, Englewood Cliffs, NJ.

Huebner, K. H. 1975. *The Finite Element Method for Engineers*. Wiley, New York.

Irons, B., and S. Ahmad. 1980. *Techniques of Finite Elements*. Wiley, New York.

Lanczos, C. 1970. *The Variational Principles of Mechanics*, 4th ed. University of Toronto Press, Toronto.

Langhaar, H. L. 1962. *Energy Methods in Applied Mechanics*. Wiley, New York.

Laursen, H. I. 1978. *Structural Analysis*, 2nd ed. McGraw-Hill, New York.

Martin, H. C., and G. F. Carey. 1973. *Introduction to Finite Element Analysis*. McGraw-Hill, New York.

McCormac, J. C. 1975. *Structural Analysis*, 3d ed. Intext Educational Publishers, New York.

Noble, B. 1969. *Applied Linear Algebra*. Prentice-Hall, Englewood Cliffs, NJ.

Norrie, D. H., and G. de Vries. 1973. *The Finite Element Method*. Academic Press, New York.

Norrie, D. H., and G. de Vries. 1978. *An Introduction to Finite Element Analysis*. Academic Press, New York.

Norris, C. H., J. B. Wilbur, and S. Utku. 1976. *Elementary Structural Analysis*, 3d ed. McGraw-Hill, New York.

Oden, J. T. 1972. *Finite Elements of Nonlinear Continua*. McGraw-Hill, New York.

Oliveira, E. R. de A. 1968. "Theoretical Foundations of the Finite Element Method," *International Journal of Solids and Structures* 4, 929–952.

Oliveira, E. R. de A. 1975. *Foundations of the Mathematical Theory of Structures*. Springer-Verlag, Wien–New York.

Pilkey, W. D., and O. H. Pilkey. 1974. *Mechanics of Solids*. Quantum, New York.

Popov, E. P. 1968. *Introduction to Mechanics of Solids*. Prentice-Hall, Englewood Cliffs, NJ.

Robinson, J. 1973. *Integrated Theory of Finite Element Methods*. Wiley, New York.

Rubinstein, M. F. 1970. *Structural Systems—Statics, Dynamics and Stability*. Prentice-Hall, Englewood Cliffs, NJ.

Schnobrich, W. C., and D. A. W. Pecknold. 1973. *Introduction to the Finite Element Method*, Course Notes for CE 478, University of Illinois at Urbana-Champaign.

Shames, I. H. 1975. *Introduction to Solid Mechanics*. Prentice-Hall, Englewood Cliffs, NJ.

Stippes, M., G. Wempner, M. Stern, and R. Beckett. 1961. *An Introduction to the Mechanics of Deformable Bodies*. Merrill, Columbus, OH.

Synge, J. L., and B. A. Griffith. 1959. *Principles of Mechanics*, 3d ed. McGraw-Hill, New York.

Thomas, G. B., Jr. 1956. *Calculus and Analytic Geometry*, 2nd ed. Addison-Wesley, Reading, MA.

Timoshenko, S. P., and J. M. Gere. 1972. *Mechanics of Materials*. Van Nostrand, New York.

Timoshenko, S. P., and D. H. Young. 1965. *Theory of Structures*, 2nd ed. McGraw-Hill, New York.

Tong, P., and J. N. Rossettos. 1977. *Finite-Element Method*. MIT Press, Cambridge, MA.

West, H. H. 1980. *Analysis of Structures*. Wiley, New York.

White, R. N., P. Gergely, and R. G. Sexsmith. 1976. *Structural Engineering*, combined ed. Wiley, New York.

Willems, N., and W. M. Lucas, Jr., 1978. *Structural Analysis for Engineers*. McGraw-Hill, New York.

Zienkiewicz, O. C. 1977. *The Finite Element Method*, 3d ed. McGraw-Hill, New York.

Additional References

Bathe, K. J. 1982. *Finite Element Procedures in Engineering Analysis.* Prentice-Hall, Englewood Cliffs, NJ.

Becker, E. B., G. F. Carey, and J. T. Oden. 1981. *Finite Elements: An Introduction,* Vol. 1. Prentice-Hall, Englewood Cliffs, NJ.

Cheung, Y. K., and M. F. Yeo. 1979. *A Practical Introduction to Finite Element Analysis.* Pitman, London.

Desai, C. S., and J. F. Abel. 1972. *Introduction to the Finite Element Method.* Van Nostrand Reinhold, New York.

Fenner, R. T. 1975. *Finite Element Methods for Engineers.* Macmillan, London.

Hinton, E., and D. R. J. Owen, 1977. *Finite Element Programming.* Academic Press, London.

Hinton, E., and D. R. J. Owen. 1979. *An Introduction to Finite Element Computations.* Pineridge Press, Swansea, U.K.

Nath, B. 1974. *Fundamentals of Finite Elements for Engineers.* Athlone Press, London.

Prenter, P. M. 1975. *Splines and Variational Methods.* Wiley, New York.

Rao, S. S. 1982. *The Finite Element Method in Engineering.* Pergamon Press, Oxford.

Reddy, J. N. 1984. *An Introduction to the Finite Element Method.* McGraw-Hill, New York.

Segerlind, L. J. 1976. *Applied Finite Element Analysis.* Wiley, New York.

Weaver, W., Jr., and P. R. Johnston. 1984. *Finite Elements for Structural Analysis.* Prentice-Hall, Engelwood Cliffs, NJ.

Zienkiewicz, O. C., and K. Morgan. 1983. *Finite Elements and Approximation.* Wiley, New York.

PREPARATION FOR MATRIX DISPLACEMENT METHOD

2.1 INTRODUCTION

In matrix methods of analysis, the structure is represented as an assemblage of discrete elements interconnected at joints (nodes). A joint is a point to which at least one element is incident. The methods are called displacement, force, or mixed methods, depending on the choice of the unknowns in the analysis (Beaufait, Rowan, Hoadley, and Hackett, 1970; Kardestuncer, 1974; Livesley, 1975; Martin, 1966; McGuire and Gallagher, 1979; Przemieniecki, 1968; Robinson, 1966; Weaver and Gere, 1980).

In the matrix displacement method, the joint displacements are selected as unknowns. The elements are represented by stiffness matrices that relate the element-end displacements to element-end forces. The element models are assembled into a system model by imposing conditions of compatibility and equilibrium. The conditions of compatibility consist of continuity conditions between element and joint displacements, including joint constraints. Equilibrium is satisfied by balancing the applied joint forces with element-end forces. The system model relates the joint displacements to applied joint forces through the system stiffness matrix. Once the system model is solved for the joint displacements, any measure of response can be determined.

This chapter develops the foundation for the matrix displacement method. The degrees of freedom of plane skeletal structures are enumerated (Section 2.2). Conventions are introduced to represent plane skeletal structures in a form suitable for computer analysis (Section 2.3). Conditions of compatibility and equilibrium are formulated (Sections 2.4 and 2.5), and coordinate transformations required in the assembly of element models are presented (Section 2.6). As an alternative to the Newtonian approach, equilibrium of an assemblage of elements is formulated by the principle of virtual work (Section 2.7).

The slope–deflection method, a classical displacement method, is presented in Appendix B to focus attention on the basic steps common to all displacement methods. A review of the slope–deflection method in this framework provides a natural transition to the matrix displacement method.

2.2 GENERALIZED DISPLACEMENTS

By definition (Section 1.4), generalized displacements are independent parameters required to define the configuration of a model. The number of generalized displacements is equal to the number of degrees of freedom of the model. If the discrete element models of Chapter 1 are assembled into trusses, continuous beams, and frames, the resulting system models are also discrete; that is, they have a finite number of degrees of freedom. Specifically, the number of degrees of freedom is

$$n = l \cdot \text{NJ} - \text{NC} \tag{2.1}$$

with

$$l = \begin{cases} 2 & \text{for plane trusses} \\ 2 & \text{for plane continuous beams} \\ 3 & \text{for plane frames} \end{cases}$$

where l is the number of degrees of freedom of a free joint, NJ is the number of joints, and NC is the number of joint constraints. Equation (2.1) follows from conditions of compatibility, which are boundary conditions, expressed as joint constraints, and continuity conditions between element and joint displacements. Particularly, if an element has l degrees of freedom at each end, a free joint of an assemblage composed of these elements has l degrees of freedom. Thus, the number of degrees of freedom of the assemblage is equal to l times the number of joints minus the number of joint constraints.

Since the discrete element models do not represent element actions, assemblages of these elements can only represent joint actions, such as joint loads and prescribed joint displacements. Element actions will be incorporated in Section 3.6.

Equation (2.1) is illustrated next for plane trusses, plane frames, and continuous beams.

Plane Trusses

A truss is an assemblage of elements interconnected by frictionless pins. If we assume that the truss element can only experience uniform axial deformation, the element has four degrees of freedom and its configuration can be defined by the four end deflections. This is illustrated in Example 2 of Section 1.4.

Consider the truss in Figure 2.1a with joints 1 and 3 constrained and joint 2 free. The joint deflections q_1, q_2 are defined relative to *global coordinate axes*,

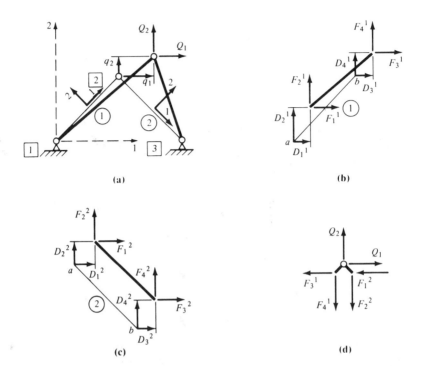

FIGURE 2.1 Truss

represented by dashed lines. Similarly, the element-end deflections D_j^i—$i = 1, 2; j = 1, \ldots, 4$ (Figures 2.1b and c)—are referred to the global axes. The symbol D_j^i has the following meaning: The uppercase letter signifies that the deflections are expressed in global coordinates, the superscript identifies the element, and the subscript denotes the direction of the deflection. On the basis of Figures 2.1a–c, the conditions of compatibility can be expressed as

$$D_1^1 = 0, \quad D_2^1 = 0, \quad D_3^1 = q_1, \quad D_4^1 = q_2$$
$$D_1^2 = q_1, \quad D_2^2 = q_2, \quad D_3^2 = 0, \quad D_4^2 = 0 \tag{2.2}$$

Equations (2.2) indicate that the joint deflections q_1, q_2 define all element-end deflections and, hence, the element configurations. Thus, the joint deflections represent generalized displacements, and the truss has two degrees of freedom. This agrees with Eq. (2.1), since $l = 2$, NJ $= 3$, and NC $= 4$.

Plane Frames

A frame is an assemblage of rigidly interconnected elements. If we select the element in Figure 1.14c, which represents axial and flexural deformations, the frame element has six degrees of freedom.

Consider the frame in Figure 2.2a with joint 1 constrained, joint 2 partially

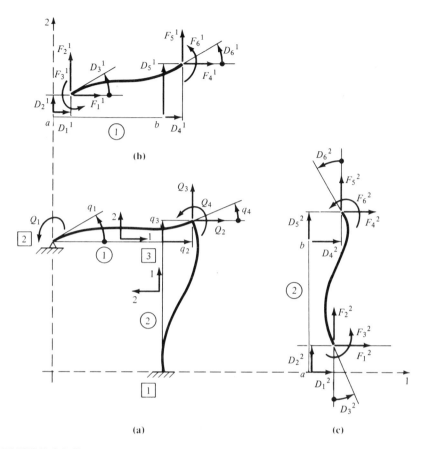

FIGURE 2.2 Frame

constrained, and joint 3 free. The joint displacements q_1–q_4 and the element-end displacements D_j^i—$i = 1, 2; j = 1, \ldots, 6$ (Figures 2.2b and c)—are referred to the global coordinate axes. The conditions of compatibility are

$$D_1^1 = 0, \quad D_2^1 = 0, \quad D_3^1 = q_1, \quad D_4^1 = q_2, \quad D_5^1 = q_3, \quad D_6^1 = q_4$$
$$D_1^2 = 0, \quad D_2^2 = 0, \quad D_3^2 = 0, \quad D_4^2 = q_2, \quad D_5^2 = q_3, \quad D_6^2 = q_4 \tag{2.3}$$

Accordingly, the joint displacements define, through the element-end displacements, the element configurations, and thus they represent generalized displacements for the frame. The four degrees of freedom of the frame in Figure 2.2a can be verified by Eq. (2.1) for $l = 3$, NJ = 3, and NC = 5.

Continuous Beams

Continuous beams are special frames in which the elements are collinear and the axial deformations of the element centroidal axes are zero or neglected

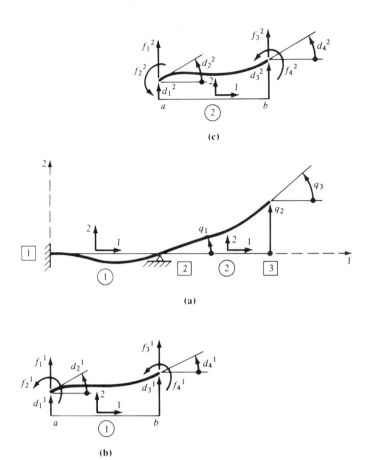

FIGURE 2.3 Continuous beams

(see Figure 2.3a). Thus, we may select as beam element the four degree-of-freedom element in Figure 1.14b.

For continuous beams, the local (Section 1.7) and global reference frames can be chosen to coincide as shown in Figure 2.3a. This permits us to express conditions of compatibility directly in terms of the local element displacements. In particular, it follows from Figures 2.3a–c that

$$d_1^1 = 0, \quad d_2^1 = 0, \quad d_3^1 = 0, \quad d_4^1 = q_1$$
$$d_1^2 = 0, \quad d_2^2 = q_1, \quad d_3^2 = q_2, \quad d_4^2 = q_3 \tag{2.4}$$

Thus, the continuous beams in Figure 2.3a have three degrees of freedom, which agrees with Eq. (2.1), since $l = 2$, NJ = 3, and NC = 3.

2.3 NOTATION AND CONVENTIONS

The symbols and conventions introduced thus far are summarized and augmented with new ones to represent plane skeletal structures in matrix form suitable for computer programming:

1. The structure is assumed to lie in the 1-2 global coordinate plane; otherwise the location and orientation of the global frame of reference is arbitrary. The global coordinate axes are represented by broken lines to differentiate them from the solid local coordinate axes (Figures 2.1-2.4).

2. The joints are numbered[1] in sequence from 1 to NJ, where NJ is the number of joints (Figures 2.1-2.3).

3. The joint displacements and the joint loads, denoted q_k and Q_k, respectively, are numbered in the sequence of the global coordinate axes from the lowest- to the highest-numbered joint (Figures 2.1-2.3); $k = 1, 2, \ldots, n$, where n is the number of joint displacements. In the absence of element actions, the joint displacements represent generalized displacements of the n degree-of-freedom system; n is defined by Eq. (2.1).

4. The elements are numbered in sequence from 1 to NE, where NE is the number of elements.

5. The local reference frame of each element is defined as follows: The local 1 axis is directed from the lower- to the higher-numbered joint to which the element is incident; the local 3 axis is oriented in the direction of the global 3 axis; and the local 2 axis is chosen to make the local frame of reference right-handed (Figures 2.1-2.3).

6. The local element displacements and forces, denoted by the lowercase letters d_j, f_j, are numbered in the sequence of the local coordinate axes from the a end to the b end of the element (Section 1.7). The global element displacements and forces, denoted by the uppercase letters D_j, F_j, are numbered in the sequence of the global coordinate axes from the a end to the b end of the element; $j = 1, 2, \ldots, m$; where m is the number of degrees of freedom of the element (Figure 2.4). When it is necessary to refer to a specific element in the assemblage, it is identified by a superscript [see, for example, Eqs. (2.2-2.4)].

7. The sign of a displacement or force at a point is determined by the direction of the corresponding coordinate axis. The right-hand rule determines the positive sense of the rotation or moment about the 3 axis. Accordingly, all displacements and forces shown in Figures 2.1-2.4 are positive.

---

[1] The assignment of joint numbers to produce band matrices is discussed in Chapter 6.

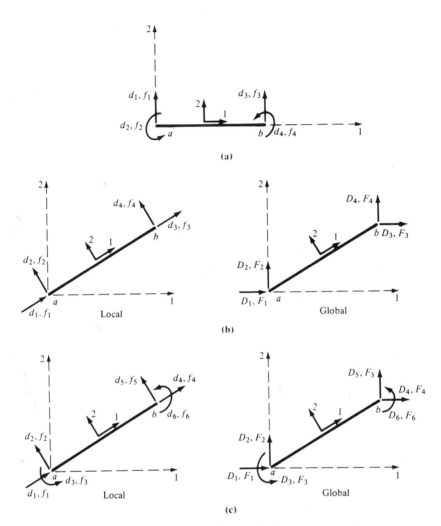

FIGURE 2.4 Element displacements and forces. **(a)** Beam element; **(b)** truss element; **(c)** frame element.

2.4 COMPATIBILITY AND EQUILIBRIUM

A mathematical model of a structure is composed of conditions of compatibility, conditions of equilibrium, and constitutive laws (Section 1.3). The constitutive laws, representing material behavior, are inherent in element models (Section 1.7). Hence, to assemble element models into a system model, we need only formulate conditions of compatibility and equilibrium.

In preparation for the automatic computer assembly of element models (Chapter 7), the conditions of compatibility and equlibrium are formulated by the code number technique that was introduced by Tezcan (1963).

Compatibility

The conditions of compatibility (Section 2.2) can be expressed in compact form by the member code matrix, $\mathbf{M}$, which is defined as

$$\mathbf{M}_{(m, \text{NE})} = [M_{li}] \tag{2.5}$$

where

$$M_{li} = \begin{cases} k & \text{if } D_l^i = q_k, \quad k = 1, 2, \ldots, n \\ 0 & \text{otherwise} \end{cases}$$

Thus, the ith column of $\mathbf{M}$ corresponds to the ith element, the lth row of $\mathbf{M}$ corresponds to the lth element displacement, and the entry in the location defined by the lth row and the ith column of $\mathbf{M}$ is k if the element displacement D_l^i corresponds to any joint displacement q_k; otherwise it is 0.

For example, the member code matrices for the truss in Figure 2.1, the frame in Figure 2.2, and the continuous beams in Figure 2.3 are, respectively,

$$\mathbf{M} = \begin{matrix} & 1 & 2 - \text{elements} \\ \begin{bmatrix} 0 & 1 \\ 0 & 2 \\ 1 & 0 \\ 2 & 0 \end{bmatrix} & \begin{matrix} 1 \\ 2 \\ 3 \\ 4 \end{matrix} \end{matrix} \tag{2.6}$$

displacements

$$\mathbf{M} = \begin{matrix} & 1 & 2 - \text{elements} \\ \begin{bmatrix} 0 & 0 \\ 0 & 0 \\ 1 & 0 \\ 2 & 2 \\ 3 & 3 \\ 4 & 4 \end{bmatrix} & \begin{matrix} 1 \\ 2 \\ 3 \\ 4 \\ 5 \\ 6 \end{matrix} \end{matrix} \tag{2.7}$$

displacements

$$\mathbf{M} = \begin{matrix} & 1 & 2 - \text{elements} \\ \begin{bmatrix} 0 & 0 \\ 0 & 1 \\ 0 & 2 \\ 1 & 3 \end{bmatrix} & \begin{matrix} 1 \\ 2 \\ 3 \\ 4 \end{matrix} \end{matrix} \tag{2.8}$$

displacements

Equations (2.6), (2.7), (2.8) are equivalent to Eqs. (2.2), (2.3), (2.4), respectively. Note that if the local and global reference frames coincide, as in Figure 2.3, $d_l^i = D_l^i$, and Eq. (2.5) can be expressed directly in local coordinates.

The significance of the member code matrix is that once it is defined, no visual reference to the structure is required in order to formulate conditions of compatibility or equilibrium for the assemblage. Specifically, the displacement vector of the ith element can be defined by the following transformation.

Displacement Transformation

$$\mathbf{q} \overset{M}{\to} \mathbf{D}^i$$

1. Match the rows of $\mathbf{D}^i$ with the entries in the ith column of $\mathbf{M}$.
2. Assign the value q_k to each component of $\mathbf{D}^i$ that corresponds to a positive integer k.
3. Assign the value 0 to each component of $\mathbf{D}^i$ that corresponds to a zero integer.

For example, on the basis of Eq. (2.6), this transformation yields

$$\mathbf{D}^1 = \begin{bmatrix} D_1^1 & 0 \\ D_2^1 & 0 \\ D_3^1 & 1 \\ D_4^1 & 2 \end{bmatrix} = \begin{bmatrix} 0 \\ 0 \\ q_1 \\ q_2 \end{bmatrix}, \quad \mathbf{D}^2 = \begin{bmatrix} D_1^2 & 1 \\ D_2^2 & 2 \\ D_3^2 & 0 \\ D_4^2 & 0 \end{bmatrix} = \begin{bmatrix} q_1 \\ q_2 \\ 0 \\ 0 \end{bmatrix} \tag{2.9}$$

which agrees with Eqs. (2.2).

Equilibrium

Any configuration of a system for which all generalized forces vanish is an equilibrium configuration (Langhaar, 1962). Since generalized forces correspond to generalized displacements in a work sense [Eq. (1.20)], the generalized forces of the assemblages defined in Section 2.3 are the net forces in the directions of the joint displacements. Particularly, the kth generalized force, R_k, is equal to the sum of the external force, Q_k, and the internal forces corresponding to the joint displacement, q_k.

For example, it follows from the free-body diagram in Figure 2.1d that the generalized forces of the truss in Figure 2.1a are

$$R_1 = Q_1 - F_3^1 - F_1^2, \quad R_2 = Q_2 - F_4^1 - F_2^2 \tag{2.10}$$

Setting $R_1 = R_2 = 0$, we obtain the condition of equilibrium expressed in matrix form as

$$\begin{bmatrix} Q_1 \\ Q_2 \end{bmatrix} = \begin{bmatrix} F_3^1 \\ F_4^1 \end{bmatrix} + \begin{bmatrix} F_1^2 \\ F_2^2 \end{bmatrix} \tag{2.11}$$

or

$$\mathbf{Q} = \mathbf{F}^{(1)} + \mathbf{F}^{(2)} \tag{2.12}$$

Observe that $\mathbf{F}^{(1)}$ and $\mathbf{F}^{(2)}$ are force vectors supplied by elements 1 and 2 to balance the applied joint force vector $\mathbf{Q}$.

If we generalize Eq. (2.12) for an n degree-of-freedom system composed of NE elements, we obtain the condition of equilibrium (a proof is presented in Section 2.7)

$$\mathbf{Q} = \sum_{i=1}^{NE} \mathbf{F}^{(i)} \tag{2.13}$$

where $\mathbf{Q}$ is the n-dimensional *applied joint force vector*, that is, the generalized external force vector; and $\mathbf{F}^{(i)}$ is the n-dimensional internal force vector supplied by element i, that is, the *generalized force vector of element i*. Since the member code matrix identifies the element displacements D_i^i that are equated with the joint displacements q_k, it also serves to identify the corresponding element forces F_i^i that balance the applied joint force Q_k [see Figure 2.1 and compare Eqs. (2.9) and (2.11)]. Accordingly, the generalized force vector of element i, $\mathbf{F}^{(i)}$, can be obtained from the global force vector of element i, $\mathbf{F}^i$, as follows.

Force Transformation

$$\mathbf{F}^i \overset{\mathbf{M}}{\to} \mathbf{F}^{(i)}$$

1. Match the rows of $\mathbf{F}^i$ with the entries in the ith column of $\mathbf{M}$.
2. Transfer every component of $\mathbf{F}^i$ that corresponds to a positive integer k to the kth row of $\mathbf{F}^{(i)}$.
3. Assign the value 0 to every row of $\mathbf{F}^{(i)}$ that does not receive a contribution from $\mathbf{F}^i$.

For example, if we apply this transformation via Eq. (2.6) to the global element force vectors of the truss in Figure 2.1, we obtain the generalized element force vectors defined in Eqs. (2.11) and (2.12):

$$\mathbf{F}^1 = \begin{bmatrix} F_1^1 \\ F_2^1 \\ F_3^1 \\ F_4^1 \end{bmatrix} \begin{matrix} 0 \\ 0 \\ 1 \\ 2 \end{matrix} \overset{\mathbf{M}}{\to} \mathbf{F}^{(1)} = \begin{bmatrix} F_3^1 \\ F_4^1 \end{bmatrix} \begin{matrix} 1 \\ 2 \end{matrix}, \quad \mathbf{F}^2 = \begin{bmatrix} F_1^2 \\ F_2^2 \\ F_3^2 \\ F_4^2 \end{bmatrix} \begin{matrix} 1 \\ 2 \\ 0 \\ 0 \end{matrix} \overset{\mathbf{M}}{\to} \mathbf{F}^{(2)} = \begin{bmatrix} F_1^2 \\ F_2^2 \end{bmatrix} \begin{matrix} 1 \\ 2 \end{matrix} \tag{2.14}$$

The following examples further illustrate that the member code matrix permits us to formulate conditions of compatibility and equilibrium without visual reference to the structure.

Example 1. Consider a truss with the member code matrix

$$\mathbf{M} = \begin{bmatrix} 0 & 1 & 0 \\ 0 & 2 & 0 \\ 1 & 3 & 3 \\ 2 & 0 & 0 \end{bmatrix} \tag{2.15}$$

It follows from Eqs. (2.5) and (2.15) that the truss is composed of three elements (**M** has three columns), and it has three degrees of freedom (the largest integer stored in **M**).

The displacement transformations yield

$$
\mathbf{D}^1 = \begin{bmatrix} D_1^1 & 0 \\ D_2^1 & 0 \\ D_3^1 & 1 \\ D_4^1 & 2 \end{bmatrix} = \begin{bmatrix} 0 \\ 0 \\ q_1 \\ q_2 \end{bmatrix}, \quad
\mathbf{D}^2 = \begin{bmatrix} D_1^2 & 1 \\ D_2^2 & 2 \\ D_3^2 & 3 \\ D_4^2 & 0 \end{bmatrix} = \begin{bmatrix} q_1 \\ q_2 \\ q_3 \\ 0 \end{bmatrix}
\tag{2.16}
$$

$$
\mathbf{D}^3 = \begin{bmatrix} D_1^3 & 0 \\ D_2^3 & 0 \\ D_3^3 & 3 \\ D_4^3 & 0 \end{bmatrix} = \begin{bmatrix} 0 \\ 0 \\ q_3 \\ 0 \end{bmatrix}
\tag{2.17}
$$

The force transformations yield

$$
\mathbf{F}^1 = \begin{bmatrix} F_1^1 & 0 \\ F_2^1 & 0 \\ F_3^1 & 1 \\ F_4^1 & 2 \end{bmatrix} \xrightarrow{\mathbf{M}} \mathbf{F}^{(1)} = \begin{bmatrix} F_3^1 & 1 \\ F_4^1 & 2 \\ 0 & 3 \end{bmatrix}, \quad
\mathbf{F}^2 = \begin{bmatrix} F_1^2 & 1 \\ F_2^2 & 2 \\ F_3^2 & 3 \\ F_4^2 & 0 \end{bmatrix} \xrightarrow{\mathbf{M}} \mathbf{F}^{(2)} = \begin{bmatrix} F_1^2 & 1 \\ F_2^2 & 2 \\ F_3^2 & 3 \end{bmatrix}
$$

$$
\mathbf{F}^3 = \begin{bmatrix} F_1^3 & 0 \\ F_2^3 & 0 \\ F_3^3 & 3 \\ F_4^3 & 0 \end{bmatrix} \xrightarrow{\mathbf{M}} \mathbf{F}^{(3)} = \begin{bmatrix} 0 & 1 \\ 0 & 2 \\ F_3^3 & 3 \end{bmatrix}
\tag{2.18}
$$

Combining Eqs. (2.13) and (2.18), we obtain the equations of equilibrium

$$
\mathbf{Q} = \sum_{i=1}^{3} \mathbf{F}^{(i)} \quad \text{or} \quad \begin{bmatrix} Q_1 \\ Q_2 \\ Q_3 \end{bmatrix} = \begin{bmatrix} F_3^1 + F_1^2 + 0 \\ F_4^1 + F_2^2 + 0 \\ 0 + F_3^2 + F_3^3 \end{bmatrix}
\tag{2.19}
$$

Thus

$$
Q_1 = F_3^1 + F_1^2, \quad Q_2 = F_4^1 + F_2^2, \quad Q_3 = F_3^2 + F_3^3
\tag{2.20}
$$

The truss represented by Eqs. (2.15)–(2.20) is shown in Figure 2.5a. Note that the 0 in the third row of $\mathbf{F}^{(1)}$ indicates that element 1 is not incident to q_3, that is, element 1 does not supply a force to balance Q_3. Similarly, the 0s in $\mathbf{F}^{(3)}$ indicate that element 3 is not incident to q_1 and q_2. Also note that the conditions of equilibrium in Eqs. (2.20) agree with the free-body diagrams of joints 2 and 3 in Figure 2.5b.

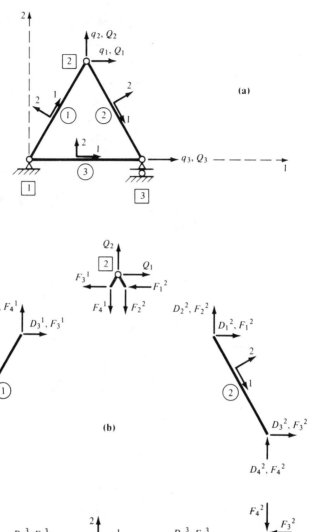

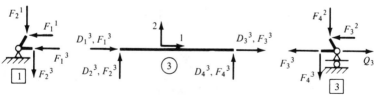

FIGURE 2.5 Compatibility and equilibrium. **(a)** Truss; **(b)** free-body diagrams.

Example 2. Consider a frame with the member code matrix

$$
\mathbf{M} = \begin{bmatrix} 0 & 1 \\ 0 & 2 \\ 0 & 3 \\ 1 & 4 \\ 2 & 0 \\ 3 & 5 \end{bmatrix} \tag{2.21}
$$

Thus, the frame is composed of two elements and has five degrees of freedom.
The displacement transformations yield

$$
\mathbf{D}^1 = \begin{bmatrix} D_1^1 & 0 \\ D_2^1 & 0 \\ D_3^1 & 0 \\ D_4^1 & 1 \\ D_5^1 & 2 \\ D_6^1 & 3 \end{bmatrix} = \begin{bmatrix} 0 \\ 0 \\ 0 \\ q_1 \\ q_2 \\ q_3 \end{bmatrix}, \quad \mathbf{D}^2 = \begin{bmatrix} D_1^2 & 1 \\ D_2^2 & 2 \\ D_3^2 & 3 \\ D_4^2 & 4 \\ D_5^2 & 0 \\ D_6^2 & 5 \end{bmatrix} \begin{bmatrix} q_1 \\ q_2 \\ q_3 \\ q_4 \\ 0 \\ q_5 \end{bmatrix} \tag{2.22}
$$

The force transformations yield

$$
\mathbf{F}^1 = \begin{bmatrix} F_1^1 & 0 \\ F_2^1 & 0 \\ F_3^1 & 0 \\ F_4^1 & 1 \\ F_5^1 & 2 \\ F_6^1 & 3 \end{bmatrix} \xrightarrow{\mathbf{M}} \mathbf{F}^{(1)} = \begin{bmatrix} F_4^1 & 1 \\ F_5^1 & 2 \\ F_6^1 & 3 \\ 0 & 4 \\ 0 & 5 \end{bmatrix}, \quad \mathbf{F}^2 = \begin{bmatrix} F_1^2 & 1 \\ F_2^2 & 2 \\ F_3^2 & 3 \\ F_4^2 & 4 \\ F_5^2 & 0 \\ F_6^2 & 5 \end{bmatrix} \xrightarrow{\mathbf{M}} \mathbf{F}^{(2)} = \begin{bmatrix} F_1^2 & 1 \\ F_2^2 & 2 \\ F_3^2 & 3 \\ F_4^2 & 4 \\ F_6^2 & 5 \end{bmatrix} \tag{2.23}
$$

The conditions of equilibrium are

$$
\mathbf{Q} = \sum_{i=1}^{2} \mathbf{F}^{(i)} \quad \text{or} \quad \begin{bmatrix} Q_1 \\ Q_2 \\ Q_3 \\ Q_4 \\ Q_5 \end{bmatrix} = \begin{bmatrix} F_4^1 + F_1^2 \\ F_5^1 + F_2^2 \\ F_6^1 + F_3^2 \\ 0 + F_4^2 \\ 0 + F_6^2 \end{bmatrix} \tag{2.24}
$$

The frame analyzed is shown in Figure 2.6a. Note that the 0s in the fourth and
fifth rows of $\mathbf{F}^{(1)}$ signify that element 1 is not incident to q_4 and q_5. In addition,
the conditions of equilibrium in Eqs. (2.24) are reflected by the free-body
diagrams of joints 2 and 3 in Figure 2.6b.

Example 3. Consider continuous beams with the member code matrix

$$
\mathbf{M} = \begin{bmatrix} 0 & 0 \\ 0 & 1 \\ 0 & 2 \\ 1 & 3 \end{bmatrix} \tag{2.25}
$$

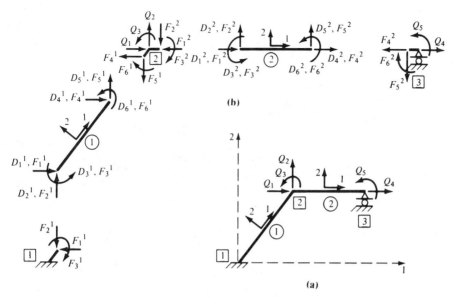

FIGURE 2.6 Compatibility and equilibrium. **(a)** Frame; **(b)** free-body diagrams.

Thus, the assemblage consists of two beams and has three degrees of freedom. The displacement transformations yield

$$\mathbf{D}^1 = \mathbf{d}^1 = \begin{bmatrix} d_1^1 & 0 \\ d_2^1 & 0 \\ d_3^1 & 0 \\ d_4^1 & 1 \end{bmatrix} = \begin{bmatrix} 0 \\ 0 \\ 0 \\ q_1 \end{bmatrix}, \quad \mathbf{D}^2 = \mathbf{d}^2 = \begin{bmatrix} d_1^2 & 0 \\ d_2^2 & 1 \\ d_3^2 & 2 \\ d_4^2 & 3 \end{bmatrix} = \begin{bmatrix} 0 \\ q_1 \\ q_2 \\ q_3 \end{bmatrix} \quad (2.26)$$

The force transformations yield

$$\mathbf{F}^1 = \mathbf{f}^1 = \begin{bmatrix} f_1^1 & 0 \\ f_2^1 & 0 \\ f_3^1 & 0 \\ f_4^1 & 1 \end{bmatrix} \overset{\mathbf{M}}{\to} \mathbf{F}^{(1)} = \begin{bmatrix} f_4^1 & 1 \\ 0 & 2 \\ 0 & 3 \end{bmatrix}, \quad \mathbf{F}^2 = \mathbf{f}^2 = \begin{bmatrix} f_1^2 & 0 \\ f_2^2 & 1 \\ f_3^2 & 2 \\ f_4^2 & 3 \end{bmatrix} \overset{\mathbf{M}}{\to} \mathbf{F}^{(2)} = \begin{bmatrix} f_2^2 & 1 \\ f_3^2 & 2 \\ f_4^2 & 3 \end{bmatrix}$$

$$(2.27)$$

The conditions of equilibrium are

$$\mathbf{Q} = \sum_{i=1}^{2} \mathbf{F}^{(i)} \quad \text{or} \quad \begin{bmatrix} Q_1 \\ Q_2 \\ Q_3 \end{bmatrix} = \begin{bmatrix} f_4^1 + f_2^2 \\ 0 + f_3^2 \\ 0 + f_4^2 \end{bmatrix} \quad (2.28)$$

The continuous beams analyzed are shown in Figure 2.7a, and the conditions of equilibrium in Eqs. (2.28) are reflected by the free-body diagrams of joints 2 and 3 in Figure 2.7b.

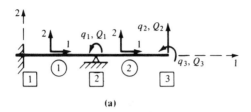

(a)

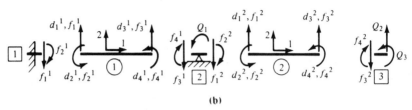

(b)

FIGURE 2.7 Compatibility and equilibrium. **(a)** Continuous beams; **(b)** free-body diagrams.

2.5 JOINT FORCES

The joint forces, the applied and reactive external forces at a joint, can be computed from element forces by conditions of joint equilibrium. In vector form, the conditions of joint equilibrium can be expressed as

$$\mathbf{P}_j = \sum_{i=1}^{NE} \mathbf{F}_j^i, \quad j = 1, 2, \ldots, NJ \tag{2.29}$$

where $\mathbf{P}_j$ is the force vector at joint j, whose components are ordered in the sequence of the global coordinate axes (Figure 2.8a),

$$\mathbf{F}_j^i = \begin{cases} \mathbf{F}_a^i & \text{if the } a \text{ end of element } i \text{ is incident to joint } j \\ \mathbf{F}_b^i & \text{if the } b \text{ end of element } i \text{ is incident to joint } j \\ \mathbf{0} & \text{otherwise} \end{cases}$$

and NE and NJ are the number of elements and the number of joints, respectively.

The derivation of Eq. (2.29) is illustrated for the frame in Figure 2.8a. For identification, the force vectors in the vector free-body diagrams of Figure 2.8b are represented by arrows whose orientations have no significance. Figure 2.8b leads to the conditions of joint equilibrium

$$\mathbf{P}_1 - \mathbf{F}_a^1 = 0$$

$$\mathbf{P}_2 - \mathbf{F}_b^1 - \mathbf{F}_a^2 = 0 \tag{2.30a}$$

$$\mathbf{P}_3 - \mathbf{F}_b^2 = 0$$

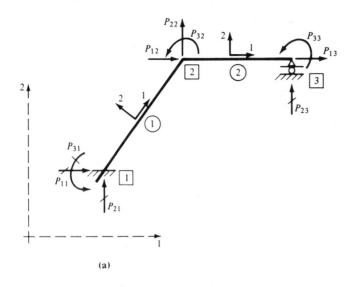

(a)

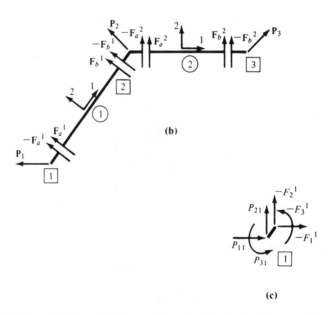

(b)

(c)

FIGURE 2.8 Joint equilibrium. (a) Frame; (b) vector free-body diagrams; (c) scalar free-body diagram.

or

$$P_1 = F_a^1$$
$$P_2 = F_b^1 + F_a^2 \qquad (2.30b)$$
$$P_3 = F_b^2$$

which are equivalent to Eq. (2.29).

Each vector equation in Eqs. (2.30a) represents the three conditions of equilibrium of a joint. For example, the equation $P_1 - F_a^1 = 0$ is equivalent to the conditions of equilibrium of joint 1 (Figure 2.8c):

$$P_{11} - F_1^1 = 0$$
$$P_{21} - F_2^1 = 0 \qquad (2.31)$$
$$P_{31} - F_3^1 = 0$$

2.6 COORDINATE TRANSFORMATIONS

Since element models are expressed in local coordinates (Section 1.7) and assembled in global coordinates (Section 2.4), coordinate transformations are required.

In Appendix C, transformations between local and global coordinates are derived and specialized for the local and global reference frames defined in Section 2.3. These results are used to formulate displacement and force transformations required in the matrix displacement analysis of frames and trusses.[2] Specifically, these transformations are of the form

$$\mathbf{d} = \mathbf{\Lambda D}, \quad \mathbf{F} = \mathbf{\Lambda}^T \mathbf{f} \qquad (2.32)$$

where $\mathbf{d}$ and $\mathbf{D}$ are the local and global element displacement vectors; $\mathbf{f}$ and $\mathbf{F}$ are the local and global element force vectors; and $\mathbf{\Lambda}$ is the transformation matrix from global to local coordinates. Observe that Eqs. (2.32) represent contragredient transformations (Section 1.8).

Frame Element

Figure 2.9 depicts the displacements and forces at the a end of a frame element referred to the local and global coordinate axes (Figure 2.4c). It follows from Eqs. (C.13), (C.19), and (C.20) that the displacement transformations from global to local coordinates are defined as (see Problem 2.11)

$$\begin{bmatrix} d_1 \\ d_2 \\ d_3 \end{bmatrix} = \begin{bmatrix} c_1 & c_2 & 0 \\ -c_2 & c_1 & 0 \\ 0 & 0 & 1 \end{bmatrix} \begin{bmatrix} D_1 \\ D_2 \\ D_3 \end{bmatrix} \quad \text{or} \quad \mathbf{d}_a = \lambda \mathbf{D}_a \qquad (2.33)$$

---

[2] For continuous beams, coordinate transformations can be avoided by selecting coincident local and global reference frames (Section 2.2).

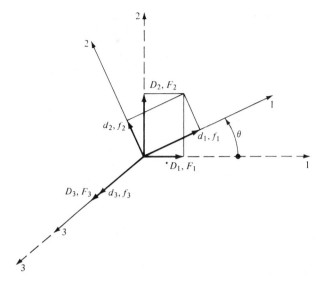

FIGURE 2.9 Local and global coordinates.

where the rotation matrix

$$\boldsymbol{\lambda} = \begin{bmatrix} c_1 & c_2 & 0 \\ -c_2 & c_1 & 0 \\ 0 & 0 & 1 \end{bmatrix}, \quad c_1 = \cos\theta, \quad c_2 = \sin\theta \tag{2.34}$$

The rotation angle θ is measured from the global 1 axis to the local 1 axis; its positive sense is counterclockwise. Similarly, the displacement transformations at the b end of a frame element (Figure 2.4c) can be expressed as

$$\begin{bmatrix} d_4 \\ d_5 \\ d_6 \end{bmatrix} = \begin{bmatrix} c_1 & c_2 & 0 \\ -c_2 & c_1 & 0 \\ 0 & 0 & 1 \end{bmatrix} \begin{bmatrix} D_4 \\ D_5 \\ D_6 \end{bmatrix} \quad \text{or} \quad \mathbf{d}_b = \boldsymbol{\lambda}\mathbf{D}_b \tag{2.35}$$

Figure 2.9 represents the displacements and forces at the b end if each subscript is increased by the number 3. Combining Eqs. (2.33) and (2.35), we obtain

$$\begin{bmatrix} \mathbf{d}_a \\ \mathbf{d}_b \end{bmatrix} = \begin{bmatrix} \boldsymbol{\lambda} & \mathbf{0} \\ \mathbf{0} & \boldsymbol{\lambda} \end{bmatrix} \begin{bmatrix} \mathbf{D}_a \\ \mathbf{D}_b \end{bmatrix} \quad \text{or} \quad \mathbf{d} = \boldsymbol{\Lambda}\mathbf{D} \tag{2.36}$$

Analogously, Eq. (C.16) yields the force transformations from local to global coordinates

$$\mathbf{F}_a = \boldsymbol{\lambda}^T\mathbf{f}_a \quad \text{and} \quad \mathbf{F}_b = \boldsymbol{\lambda}^T\mathbf{f}_b \tag{2.37}$$

Thus,

$$\begin{bmatrix} \mathbf{F}_a \\ \mathbf{F}_b \end{bmatrix} = \begin{bmatrix} \boldsymbol{\lambda}^T & \mathbf{0} \\ \mathbf{0} & \boldsymbol{\lambda}^T \end{bmatrix} \begin{bmatrix} \mathbf{f}_a \\ \mathbf{f}_b \end{bmatrix} \quad \text{or} \quad \mathbf{F} = \boldsymbol{\Lambda}^T\mathbf{f} \tag{2.38}$$

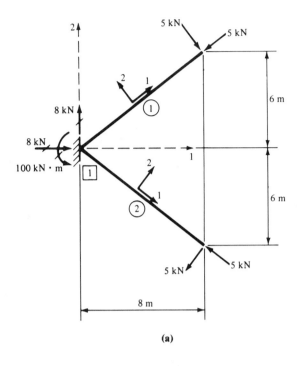

(a)

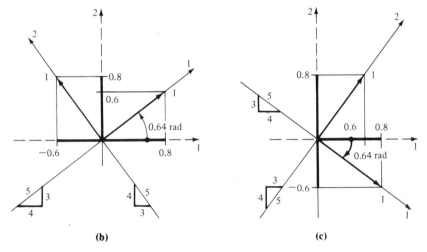

(b)

(c)

FIGURE 2.10 Computation of joint forces. **(a)** Frame; **(b)** verification of λ^1; **(c)** verification of λ^2.

Example 1. To illustrate the formulation of force transformations and equilibrium equations, the joint force vector $\mathbf{P}_1$ of the structure in Figure 2.10a is computed. The units are kilonewton (kN), meter (m), and radian(rad).

Figure 2.10a indicates that

$$\mathbf{f}_b^1 = \mathbf{f}_b^2 = \begin{bmatrix} -5 \\ -5 \\ 0 \end{bmatrix} \tag{2.39}$$

By Eqs. (2.29), (2.37), and Figure 2.10a[3]

$$\mathbf{P}_1 = \mathbf{F}_a^1 + \mathbf{F}_a^2 \tag{2.40}$$

$$\mathbf{F}_a^1 = \lambda^{1T}\mathbf{f}_a^1, \quad \mathbf{F}_a^2 = \lambda^{2T}\mathbf{f}_a^2 \tag{2.41}$$

The conditions of element equilibrium, Eq. (1.114b), and Eqs. (2.39) yield

$$\mathbf{f}_a^1 = \mathbf{f}_a^2 = \begin{bmatrix} -1 & 0 & 0 \\ 0 & -1 & 0 \\ 0 & -10 & -1 \end{bmatrix}\begin{bmatrix} -5 \\ -5 \\ 0 \end{bmatrix} = \begin{bmatrix} 5 \\ 5 \\ 50 \end{bmatrix} \tag{2.42}$$

Since the rotation angles for elements 1 and 2 are, respectively, $\theta^1 = 0.64$ rad and $\theta^2 = -0.64$ rad (Figures 2.10b and c), Eqs. (2.34) yield

$$\lambda^1 = \begin{bmatrix} 0.8 & 0.6 & 0 \\ -0.6 & 0.8 & 0 \\ 0 & 0 & 1 \end{bmatrix}, \quad \lambda^2 = \begin{bmatrix} 0.8 & -0.6 & 0 \\ 0.6 & 0.8 & 0 \\ 0 & 0 & 1 \end{bmatrix} \tag{2.43}$$

Figures 2.10b and c permit a simple check of Eqs. (2.43) because, by Eqs. (C.2a) and (C.14), the components of the jth row of λ are the projections of a unit length from the local j axis onto the global axes. For example, the projections of a unit length from the local 1 axis in Figure 2.10c onto the global axes are the entries in the first row of λ^2.

From Eqs. (2.40)–(2.43) we obtain

$$\mathbf{F}_a^1 = \begin{bmatrix} 1 \\ 7 \\ 50 \end{bmatrix}, \quad \mathbf{F}_a^2 = \begin{bmatrix} 7 \\ 1 \\ 50 \end{bmatrix}, \quad \mathbf{P}_1 = \begin{bmatrix} 8 \\ 8 \\ 100 \end{bmatrix} \tag{2.44}$$

Example 2. As a further illustration of force transformations, the joint forces of the frame in Figure 2.11a are computed. The element forces are given in the free-body diagrams of Figure 2.11b. The analysis units are kilopound (kip or k), foot (ft), and radian (rad).

---

[3] Equation (2.29) can be applied without reference to a visual image of the structure with the aid of a member incidence matrix (Chapter 7).

94

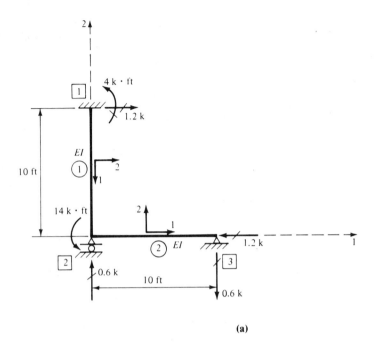

(a)

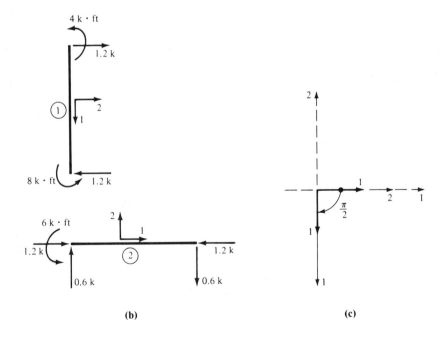

(b)   (c)

FIGURE 2.11 Computation of joint forces. **(a)** Frame; **(b)** element forces; **(c)** verification of λ^1.

According to Figure 2.11a and Eqs. (2.29) and (2.37),

$$\mathbf{P}_1 = \mathbf{F}_a^1 = \boldsymbol{\lambda}^{1T}\mathbf{f}_a^1$$
$$\mathbf{P}_2 = \mathbf{F}_b^1 + \mathbf{F}_a^2 = \boldsymbol{\lambda}^{1T}\mathbf{f}_b^1 + \boldsymbol{\lambda}^{2T}\mathbf{f}_a^2 \qquad (2.45)$$
$$\mathbf{P}_3 = \mathbf{F}_b^2 = \boldsymbol{\lambda}^{2T}\mathbf{f}_b^2$$

From Figure 2.11b we obtain

$$
\mathbf{f}^1 = \begin{bmatrix} \mathbf{f}_a^1 \\ \mathbf{f}_b^1 \end{bmatrix} = \begin{bmatrix} 0.0 \\ 1.2 \\ 4.0 \\ \hline 0.0 \\ -1.2 \\ 8.0 \end{bmatrix}, \quad
\mathbf{f}^2 = \begin{bmatrix} \mathbf{f}_a^2 \\ \mathbf{f}_b^2 \end{bmatrix} = \begin{bmatrix} 1.2 \\ 0.6 \\ 6.0 \\ \hline -1.2 \\ -0.6 \\ 0.0 \end{bmatrix} \qquad (2.46)
$$

The rotation angles are (Figure 2.11a) $\theta^1 = -\pi/2$, $\theta^2 = 0$. Thus, Eqs. (2.34) yield

$$
\boldsymbol{\lambda}^1 = \begin{bmatrix} 0 & -1 & 0 \\ 1 & 0 & 0 \\ 0 & 0 & 1 \end{bmatrix}, \quad \boldsymbol{\lambda}^2 = \mathbf{I} \qquad (2.47)
$$

Figure 2.11c provides a check for $\boldsymbol{\lambda}^1$ (see Example 1). Equations (2.45)–(2.47) yield

$$
\mathbf{F}_a^1 = \begin{bmatrix} 1.2 \\ 0.0 \\ 4.0 \end{bmatrix}, \quad \mathbf{F}_b^1 = \begin{bmatrix} -1.2 \\ 0.0 \\ 8.0 \end{bmatrix}, \quad \mathbf{F}_a^2 = \mathbf{f}_a^2, \quad \mathbf{F}_b^2 = \mathbf{f}_b^2
$$

$$ \qquad (2.48) $$

$$
\mathbf{P}_1 = \begin{bmatrix} 1.2 \\ 0.0 \\ 4.0 \end{bmatrix}, \quad \mathbf{P}_2 = \begin{bmatrix} 0.0 \\ 0.6 \\ 14.0 \end{bmatrix}, \quad \mathbf{P}_3 = \begin{bmatrix} -1.2 \\ -0.6 \\ 0.0 \end{bmatrix}
$$

The applied and reactive joint forces are shown in Figure 2.11a.

Truss Element

Although the truss element in Figure 2.4b has four degrees of freedom, the conditions of equilibrium require that $f_2 = f_4 = 0$. Moreover, the corresponding transverse deflections d_2 and d_4 are not required to compute the axial forces of the element [see Figure 1.14a and Eq. (1.96b)]. Thus, we can represent the local truss element as shown in Figure 2.12a. However, the global truss element must reflect the four degrees of freedom (Figure 2.12b) to formulate conditions of compatibility and equilibrium (see Example 1 of Section 2.4).

The global to local displacement transformation at the a end of the truss element (Figure 2.12c) can be obtained by solving Eq. (2.33) for d_1. Thus,

$$
d_1 = \begin{bmatrix} c_1 & c_2 \end{bmatrix} \begin{bmatrix} D_1 \\ D_2 \end{bmatrix} \quad \text{or} \quad d_a = \boldsymbol{\lambda}\mathbf{D}_a \qquad (2.49)
$$

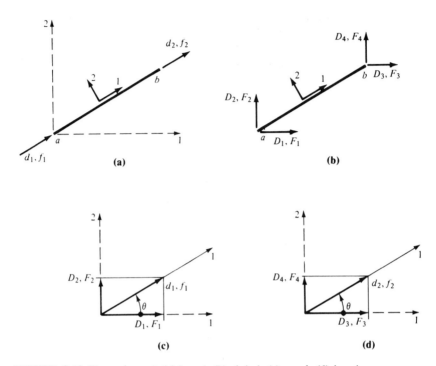

FIGURE 2.12 Truss element. **(a)** Local; **(b)** global; **(c)** a end; **(d)** b end.

Specializing Eqs. (2.49) for the b end (Figure 2.12d), we obtain

$$d_2 = [c_1 \quad c_2]\begin{bmatrix} D_3 \\ D_4 \end{bmatrix} \quad \text{or} \quad d_b = \lambda D_b \tag{2.50}$$

Equations (2.49) and (2.50) yield

$$\begin{bmatrix} d_a \\ d_b \end{bmatrix} = \begin{bmatrix} \lambda & 0 \\ 0 & \lambda \end{bmatrix}\begin{bmatrix} D_a \\ D_b \end{bmatrix} \quad \text{or} \quad \mathbf{d} = \mathbf{\Lambda} \mathbf{D} \tag{2.51}$$

where

$$\mathbf{\Lambda} = \begin{bmatrix} \lambda & 0 \\ 0 & \lambda \end{bmatrix} \tag{2.52a}$$

$$\lambda = [c_1 \quad c_2], \quad c_1 = \cos \theta, \quad c_2 = \sin \theta \tag{2.52b}$$

It follows from Appendix C that λ in Eqs. (2.52) is a unit vector on the local 1 axis defined relative to the global axes.

The force transformations can be obtained by specializing Eqs. (2.37). Thus, the transformation

$$
\begin{bmatrix} F_1 \\ F_2 \\ F_3 \end{bmatrix} = \begin{bmatrix} c_1 & -c_2 & 0 \\ c_2 & c_1 & 0 \\ 0 & 0 & 1 \end{bmatrix} \begin{bmatrix} f_1 \\ f_2 \\ f_3 \end{bmatrix}
\tag{2.53}
$$

subject to the condition $f_2 = 0$ yields

$$
\begin{bmatrix} F_1 \\ F_2 \end{bmatrix} = \begin{bmatrix} c_1 \\ c_2 \end{bmatrix} f_1 \quad \text{or} \quad \mathbf{F}_a = \boldsymbol{\lambda}^T f_a
\tag{2.54a}
$$

Note that Eq. (2.54a) follows directly from Figure 2.12c, which indicates that $F_1 = \cos \theta\, f_1 = c_1 f_1$ and $F_2 = \sin \theta\, f_1 = c_2 f_1$. Similarly, we obtain the force transformations at the b end (Figure 2.12d)

$$
\begin{bmatrix} F_3 \\ F_4 \end{bmatrix} = \begin{bmatrix} c_1 \\ c_2 \end{bmatrix} f_2 \quad \text{or} \quad \mathbf{F}_b = \boldsymbol{\lambda}^T f_b
\tag{2.54b}
$$

Thus,

$$
\begin{bmatrix} \mathbf{F}_a \\ \mathbf{F}_b \end{bmatrix} = \begin{bmatrix} \boldsymbol{\lambda}^T & \mathbf{0} \\ \mathbf{0} & \boldsymbol{\lambda}^T \end{bmatrix} \begin{bmatrix} f_a \\ f_b \end{bmatrix} \quad \text{or} \quad \mathbf{F} = \boldsymbol{\Lambda}^T \mathbf{f}
\tag{2.55}
$$

Example 3. Analogous to Examples 1 and 2 of this section, the element forces of the truss in Figure 2.13a are given in the free-body diagrams of Figure 2.13b, and the joint forces are computed. The analysis units are kilonewton (kN), meter (m), and radian (rad).

It follows from Figure 2.13a and Eqs. (2.29) and (2.54) that

$$
\begin{aligned}
\mathbf{P}_1 &= \mathbf{F}_a^1 = \boldsymbol{\lambda}^{1T} f_a^1 \\
\mathbf{P}_2 &= \mathbf{F}_a^2 = \boldsymbol{\lambda}^{2T} f_a^2 \\
\mathbf{P}_3 &= \mathbf{F}_b^1 + \mathbf{F}_b^2 = \boldsymbol{\lambda}^{1T} f_b^1 + \boldsymbol{\lambda}^{2T} f_b^2
\end{aligned}
\tag{2.56}
$$

The local element forces are (Figure 2.13b)

$$
\mathbf{f}^1 = \begin{bmatrix} f_a^1 \\ f_b^1 \end{bmatrix} = \begin{bmatrix} 5\sqrt{2} \\ \hline -5\sqrt{2} \end{bmatrix}, \quad \mathbf{f}^2 = \begin{bmatrix} f_a^2 \\ f_b^2 \end{bmatrix} = \begin{bmatrix} -5 \\ \hline 5 \end{bmatrix}
\tag{2.57}
$$

According to Figure 2.13a, $\theta^1 = \pi/4$ and $\theta^2 = 0$. Thus, Eqs. (2.52) yield

$$
\boldsymbol{\lambda}^1 = \begin{bmatrix} \dfrac{1}{\sqrt{2}} & \dfrac{1}{\sqrt{2}} \end{bmatrix}, \quad \boldsymbol{\lambda}^2 = \begin{bmatrix} 1 & 0 \end{bmatrix}
\tag{2.58}
$$

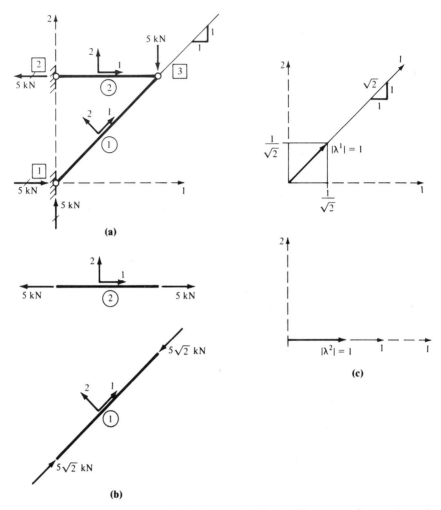

FIGURE 2.13 Computation of joint forces. **(a)** Truss; **(b)** element forces; **(c)** verifications of λ^1 and λ^2.

Alternatively, λ^1 and λ^2 can be obtained from Figure 2.13c, in which they are depicted as unit vectors on the local 1 axes. From Eqs. (2.56)–(2.58) we obtain

$$\mathbf{P}_1 = \begin{bmatrix} 5 \\ 5 \end{bmatrix}, \quad \mathbf{P}_2 = \begin{bmatrix} -5 \\ 0 \end{bmatrix}$$

$$\mathbf{P}_3 = \begin{bmatrix} -5 \\ -5 \end{bmatrix} + \begin{bmatrix} 5 \\ 0 \end{bmatrix} = \begin{bmatrix} 0 \\ -5 \end{bmatrix} \tag{2.59}$$

The joint forces are shown in Figure 2.13a.

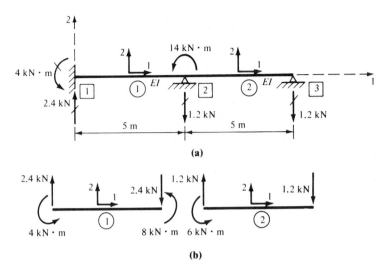

FIGURE 2.14 Computation of joint forces. (a) Continuous beams; (b) element forces.

Example 4. The element forces (Figure 2.14b) of the continuous beams in Figure 2.14a are given, and the joint forces are computed. The analysis units are kilonewton (kN) and meter (m).

Since the local and global reference frames coincide, no coordinate transformations are required. Thus, Figure 2.14a and Eq. (2.29) yield

$$P_1 = f_a^1, \quad P_2 = f_b^1 + f_a^2, \quad P_3 = f_b^2 \qquad (2.60)$$

where, according to Figure 2.14b,

$$\mathbf{f}^1 = \begin{bmatrix} \mathbf{f}_a^1 \\ \mathbf{f}_b^1 \end{bmatrix} = \begin{bmatrix} 2.4 \\ 4.0 \\ \hline -2.4 \\ 8.0 \end{bmatrix}, \quad \mathbf{f}^2 = \begin{bmatrix} \mathbf{f}_a^2 \\ \mathbf{f}_b^2 \end{bmatrix} = \begin{bmatrix} 1.2 \\ 6.0 \\ \hline -1.2 \\ 0.0 \end{bmatrix} \qquad (2.61)$$

Thus,

$$\mathbf{P}_1 = \begin{bmatrix} 2.4 \\ 4.0 \end{bmatrix}, \quad \mathbf{P}_2 = \begin{bmatrix} -1.2 \\ 14.0 \end{bmatrix}, \quad \mathbf{P}_3 = \begin{bmatrix} -1.2 \\ 0.0 \end{bmatrix} \qquad (2.62)$$

The joint forces are shown in Figure 2.14a.

2.7 COMPATIBILITY AND EQUILIBRIUM: ANALYTICAL APPROACH

The principle of virtual work (Appendix D) is used to prove Eq. (2.13). In the process, we obtain contragredient transformations that link the displacements and forces of an isolated element with the corresponding displacements and forces of an assemblage.

Compatibility

The conditions of compatibility (Sections 2.3 and 2.4) can be expressed by the transformation

$$\mathbf{D}^l = \mathbf{A}^l \mathbf{q}, \quad l = 1, 2, \ldots, \text{NE} \tag{2.63}$$

For example, Eqs. (2.9) yield

$$\mathbf{D}^1 = \begin{bmatrix} D_1^1 \\ D_2^1 \\ D_3^1 \\ D_4^1 \end{bmatrix} = \begin{bmatrix} 0 & 0 \\ 0 & 0 \\ 1 & 0 \\ 0 & 1 \end{bmatrix} \begin{bmatrix} q_1 \\ q_2 \end{bmatrix} = \mathbf{A}^1 \mathbf{q} \tag{2.64a}$$

$$\mathbf{D}^2 = \begin{bmatrix} D_1^2 \\ D_2^2 \\ D_3^2 \\ D_4^2 \end{bmatrix} = \begin{bmatrix} 1 & 0 \\ 0 & 1 \\ 0 & 0 \\ 0 & 0 \end{bmatrix} \begin{bmatrix} q^1 \\ q_2 \end{bmatrix} = \mathbf{A}^2 \mathbf{q} \tag{2.64b}$$

Equilibrium

It follows from Eqs. (D.24) and (D.46) that an assemblage is in a configuration of equilibrium if

$$\delta W = \delta W_\mathrm{e} + \delta W_\mathrm{i} = 0 \tag{2.65}$$

for any virtual displacement. The virtual work of the external forces is

$$\delta W_e = \sum_{k=1}^{n} \delta q_k Q_k = \delta \mathbf{q}^T \mathbf{Q} \tag{2.66}$$

The virtual work of the internal forces can be expressed as

$$\delta W_\mathrm{i} = \sum_{l=1}^{\text{NE}} \delta W_\mathrm{i}^l \tag{2.67}$$

where δW_i^l is the internal virtual work of element l. By Eqs. (D.3) and (D.4)

$$\delta W_\mathrm{i}^l = -\delta W_\mathrm{e}^l = -\sum_{j=1}^{m} \delta D_j^l F_j^l = -\delta \mathbf{D}^{lT} \mathbf{F}^l \tag{2.68}$$

From Eq. (2.63) we obtain

$$\delta \mathbf{D}^{lT} = \delta \mathbf{q}^T \mathbf{A}^{lT} \tag{2.69}$$

which is substituted into Eq. (2.68) to yield

$$\delta W_\mathrm{i}^l = -\delta \mathbf{q}^T \mathbf{F}^{(l)} \tag{2.70}$$

where

$$\mathbf{F}^{(l)} = \mathbf{A}^{lT} \mathbf{F}^l \tag{2.71}$$

Combining Eqs. (2.65), (2.66), (2.67), and (2.70), we obtain

$$\delta W = \delta \mathbf{q}^T \left(\mathbf{Q} - \sum_{l=1}^{NE} \mathbf{F}^{(l)} \right) = 0 \qquad (2.72)$$

Since the δq_k's are linearly independent [see argument following Eq. (1.135)], Eq. (2.72) yields the condition of equilibrium

$$\mathbf{Q} - \sum_{l=1}^{NE} \mathbf{F}^{(l)} = \mathbf{0} \qquad (2.73)$$

which proves Eq. (2.13).

Alternatively, we can use Eq. (D.3) to formulate the internal virtual work as follows:

$$\delta W_i = -\delta U = -\sum_{l=1}^{NE} \delta U^l \qquad (2.74)$$

Since $U^l = U^l(D_j^l)$,

$$\delta U^l = \sum_{j=1}^{m} \frac{\partial U^l}{\partial D_j^l} \delta D_j^l = \sum_{j=1}^{m} \delta D_j^l F_j^l = \delta \mathbf{D}^{lT} \mathbf{F}^l \qquad (2.75)$$

where

$$F_j^l = \frac{\partial U^l}{\partial D_j^l} \qquad (2.76)$$

Equations (2.69), (2.71), (2.74), and (2.75) yield

$$\delta W_i = -\sum_{l=1}^{NE} \delta \mathbf{q}^T \mathbf{F}^{(l)} \qquad (2.77)$$

Contragredience

From Eqs. (2.63) and (2.71) we obtain the contragredient transformations

$$\mathbf{D}^i = \mathbf{A}^i \mathbf{q}$$
$$\mathbf{F}^{(i)} = \mathbf{A}^{iT} \mathbf{F}^i, \quad i = 1, 2, \ldots, NE \qquad (2.78)$$

which relate the global element displacements with the generalized displacements of the assemblage and the global and generalized element forces.

PROBLEMS

2.1   Consistent with Sections 2.2 and 2.3, do the following:
   a. Compute n by Eq. (2.1).
   b. Complete the drawings by showing the joint displacements q_k, the applied joint forces Q_k, and the local reference frames.
   c. Isolate one element: Show the end displacements, and formulate the conditions of compatibility.

PROBLEM 2.1

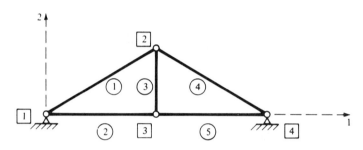

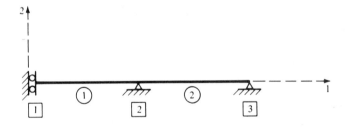

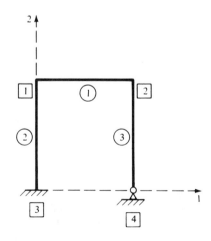

2.2 Consider Sections 2.3 and 2.4. The member code matrices for the truss, the frame, and the continuous beams are, respectively,

$$
\mathbf{M} = \begin{bmatrix} 0 & 0 & 0 \\ 0 & 0 & 1 \\ 0 & 0 & 0 \\ 1 & 2 & 2 \end{bmatrix}, \quad
\mathbf{M} = \begin{bmatrix} 0 & 2 \\ 0 & 0 \\ 1 & 3 \\ 2 & 4 \\ 0 & 0 \\ 3 & 0 \end{bmatrix}, \quad
\mathbf{M} = \begin{bmatrix} 0 & 0 \\ 1 & 2 \\ 0 & 3 \\ 2 & 0 \end{bmatrix}
$$

a. Use **M** to define the element displacement vectors in terms of the joint displacements.

PROBLEM 2.2

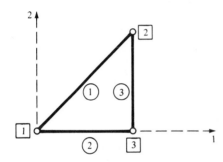

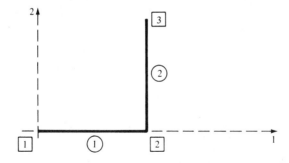

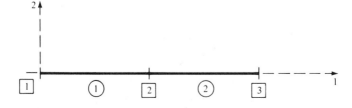

b. Formulate conditions of equilibrium by Eq. (2.13).

c. Complete the drawings by showing the local frames of reference, the boundary conditions, the joint displacements q_k, and the applied joint forces Q_k.

d. Verify the conditions of equilibrium of part (b) by drawing free-body diagrams of the appropriate joints.

2.3 Consider the frame in Figure 2.11a. Use the slope–deflection method to verify the element-end moments in Figure 2.11b. Draw the free-body diagrams of the joints to verify the joint forces in Figure 2.11a.

2.4 Consider the truss in Figure 2.13a. Use an elementary method (for example, the method of joints) to verify the element forces in Figure 2.13b. Draw the free-body diagrams of the joints to verify the joint forces in Figure 2.13a.

2.5 Consider the continuous beams in Figure 2.14a. Use the moment-distribution method to verify the element-end moments in Figure 2.14b. Draw the free-body diagrams of the joints to verify the joint forces in Figure 2.14a.

2.6 The local force vectors of the frame elements are

$$\mathbf{f}^{1T} = [-1.0 \quad -2.0 \quad -4.0 \ \vdots \ 1.0 \quad 2.0 \quad -8.0]$$
$$\mathbf{f}^{2T} = [\ 0.0 \quad -1.0 \quad -6.0 \ \vdots \ 0.0 \quad 1.0 \quad 0.0]$$
$$\mathbf{f}^{3T} = [\ 1.0 \quad -1.0 \quad -6.0 \ \vdots \ -1.0 \quad 1.0 \quad 0.0]$$

The units are kilonewton (kN), meter (m), and radian (rad).

a. Use Eq. (2.29) to compute the joint force vectors.

b. Draw the free-body diagrams of the elements and joints.

c. Use the slope–deflection method to verify the element-end moments.

PROBLEM 2.6

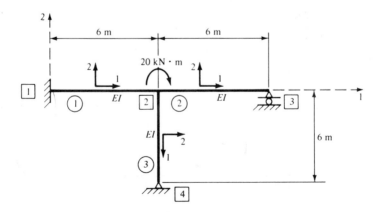

2.7 The frame and the load are symmetric; thus, the response is symmetric (Chapter 4). The local force vectors of elements 1 and 2 are

$$\mathbf{f}^{1T} = [18.0 \quad -3.0 \quad -8.0 \ \vdots \ -18.0 \quad 3.0 \quad -4.0]$$
$$\mathbf{f}^{2T} = [\ 3.0 \quad 18.0 \quad 8.0 \ \vdots \ -3.0 \quad 18.0 \quad -8.0]$$

The units are kilonewton (kN), meter (m), and radian (rad).
a. Use Eq. (2.29) to compute the joint force vectors $\mathbf{P}_1$ and $\mathbf{P}_2$.
b. Draw the free-body diagrams of elements 1, 2 and joints 1, 2.
c. Use the moment-distribution method to verify the end moments of elements 1 and 2 (take advantage of symmetry in the moment-distribution process).

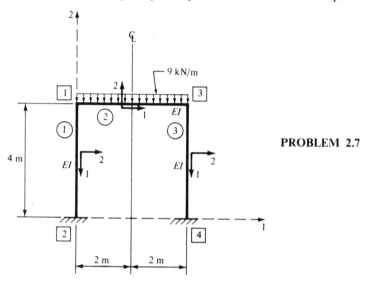

PROBLEM 2.7

2.8 The local element force vectors are

$$\mathbf{f}^{1T} = [-3.0 \quad -4.0 \ \vdots \ 3.0 \quad -8.0]$$
$$\mathbf{f}^{2T} = [\ 21.0 \quad 8.0 \ \vdots \ 27.0 \quad -20.0]$$

The units are kilonewton (kN), meter (m), and radian (rad).
a. Use Eq. (2.29) to compute the joint force vectors.
b. Draw the free-body diagrams of the elements and joints.
c. Verify the element-end moments by the moment-distribution method.

PROBLEM 2.8

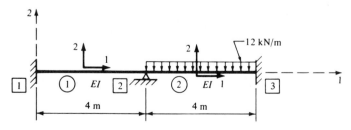

2.9   The local element force vectors are

$$\mathbf{f}^{1T} = [\ 10.4 \quad\quad 28.0 \ \vdots \ 5.6 \quad -4.0]$$

$$\mathbf{f}^{2T} = [-1.2 \quad\quad 4.0 \ \vdots \ 1.2 \quad -16.0]$$

The units are kilopound (k), foot (ft), and radian (rad).
a. Use Eq. (2.29) to compute the joint force vectors.
b. Draw the free-body diagrams of the elements and joints.
c. Verify the element-end moments by the slope–deflection method.

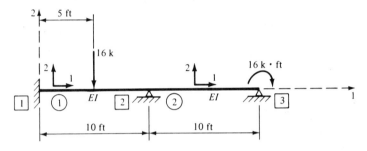

PROBLEM 2.9

2.10   The local element force vectors are

$$\mathbf{f}^1 = \begin{bmatrix} 10.0 \\ \text{-------} \\ -10.0 \end{bmatrix}, \quad \mathbf{f}^2 = \begin{bmatrix} -6.0 \\ \text{------} \\ 6.0 \end{bmatrix}, \quad \mathbf{f}^3 = \mathbf{0}$$

The units are kilonewton (kN) and radian (rad).
a. Use Eq. (2.29) to compute the joint force vectors.
b. Draw the free-body diagrams of the elements and the joints.
c. Verify element forces by an elementary method.

PROBLEM 2.10

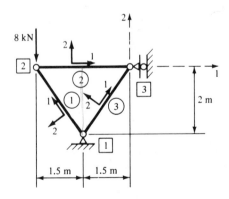

2.11 Derive the first two equations in Eq. (2.33) by expanding the relations

$$d_1 = l\cos(\gamma - \theta)$$
$$d_2 = l\sin(\gamma - \theta)$$

where $l = |\mathbf{d}| = |\mathbf{D}|$,

$$\mathbf{d} = \begin{bmatrix} d_1 \\ d_2 \end{bmatrix}, \quad \mathbf{D} = \begin{bmatrix} D_1 \\ D_2 \end{bmatrix}$$

PROBLEM 2.11

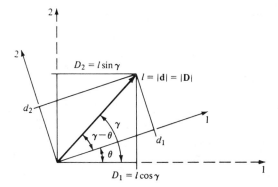

REFERENCES

Beaufait, F. W., W. H. Rowan, Jr., P. G. Hoadley, and R. M. Hackett. 1970. *Computer Methods of Structural Analysis.* Prentice-Hall, Englewood Cliffs, NJ.

Kardestuncer, H. 1974. *Elementary Matrix Analysis of Structures.* McGraw-Hill, New York.

Langhaar, H. L. 1962. *Energy Methods in Applied Mechanics.* Wiley, New York.

Livesley, R. K. 1975. *Matrix Methods of Structural Analysis*, 2nd ed. Pergamon Press, Oxford.

Martin, H. C. 1966. *Introduction to Matrix Methods of Structural Analysis.* McGraw-Hill, New York.

McGuire, W., and R. H. Gallagher. 1979. *Matrix Structural Analysis.* Wiley, New York.

Przemieniecki, J. S. 1968. *Theory of Matrix Structural Analysis.* McGraw-Hill, New York.

Robinson, J. 1966. *Structural Matrix Analysis for the Engineer.* Wiley, New York.

Tezcan, S. S. 1963. Discussion of "Simplified Formulation of Stiffness Matrices," by P. M. Wright, *Journal of the Structural Division, ASCE*, 89, No. ST6, 445–449.

Weaver, W., Jr., and J. M. Gere. 1980. *Matrix Analysis of Framed Structures*, 2nd ed. Van Nostrand, New York.

THREE

MATRIX DISPLACEMENT METHOD:
PLANE STRUCTURES

3.1 INTRODUCTION

The basic idea of the matrix displacement method is described in Section 2.1. The formulation of the method is based on the member code matrix (Schnobrich and Pecknold, 1973), Section 2.4, which is referred to by various names in the finite element literature, for example, the destination vectors (Irons and Ahmad, 1980), and the connectivity array (Bathe and Wilson, 1976).

The member code matrix controls the assembly of the element models into the system model. Moreover, it gives us the options of imposing the joint constraints (1) before and (2) after the elements are assembled. The first option is used primarily in this textbook. The second option is presented in Appendix E and the relative merits of both options are discussed in Appendix E.

The matrix displacement method is formulated in Section 3.2 and applied to structures with joint loads in Sections 3.3–3.5. Element actions are incorporated in Section 3.6 by a procedure similar to that used in the slope–deflection method. To further the transition to the finite element method, initiated in Section 1.9, finite elements with loads and temperature changes are formulated and assembled in Section 3.7.

3.2 MATRIX DISPLACEMENT METHOD

Introductory Problem

Since the slope–deflection method contains the essential features of the matrix displacement method to be presented, it is reviewed in Appendix B and used to introduce the matrix displacement method. Specifically, the analysis procedure in Figure B.1a is applied in conjunction with matrix

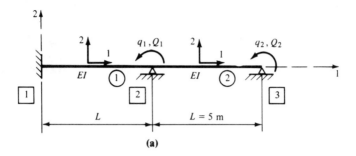

(a)

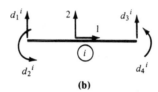

(b)

FIGURE 3.1 Introductory problem. **(a)** Continuous beams; **(b)** elements.

models and tools developed in Chapters 1 and 2 to the continuous beams in Figure 3.1a. The slope–deflection analysis of this problem is illustrated in Appendix B. The analysis units are kilonewton (kN), meter (m), and radian (rad). The elements have identical properties, and the joint loads are $Q_1 = 14\ \text{kN} \cdot \text{m}$, $Q_2 = 0$. Since the local and global reference frames coincide, no coordinate transformations are required.

1. Unknowns

The unknown joint displacements are

$$q_k, \quad k = 1, 2 \tag{3.1}$$

2. Element Models

According to Eqs. (1.94b) and (1.98b), the element models can be expressed as

$$\mathbf{f}^i = \mathbf{k}\mathbf{d}^i, \quad i = 1, 2 \tag{3.2}$$

where

$$\mathbf{k} = \alpha \begin{bmatrix} 12 & 6L & -12 & 6L \\ 6L & 4L^2 & -6L & 2L^2 \\ -12 & -6L & 12 & -6L \\ 6L & 2L^2 & -6L & 4L^2 \end{bmatrix}, \quad \alpha = \frac{EI}{L^3}$$

3. System Model

The element models are assembled into the system model by imposing conditions of compatibility and equilibrium (Section 2.4). The conditions of compatibility (Figures 3.1a and b) can be expressed by the member code matrix, Eq. (2.5),

$$
\mathbf{M} = \begin{bmatrix} 0 & 0 \\ 0 & 1 \\ 0 & 0 \\ 1 & 2 \end{bmatrix} \tag{3.3}
$$

which corresponds to the displacement relations

$$
\mathbf{D}^1 = \mathbf{d}^1 = \begin{bmatrix} d_1^1 & 0 \\ d_2^1 & 0 \\ d_3^1 & 0 \\ d_4^1 & 1 \end{bmatrix} = \begin{bmatrix} 0 \\ 0 \\ 0 \\ q_1 \end{bmatrix}, \quad \mathbf{D}^2 = \mathbf{d}^2 = \begin{bmatrix} d_1^2 & 0 \\ d_2^2 & 1 \\ d_3^2 & 0 \\ d_4^2 & 2 \end{bmatrix} = \begin{bmatrix} 0 \\ q_1 \\ 0 \\ q_2 \end{bmatrix} \tag{3.4}
$$

and yields the force transformations (Section 2.4)

$$
\mathbf{F}^1 = \mathbf{f}^1 = \begin{bmatrix} f_1^1 & 0 \\ f_2^1 & 0 \\ f_3^1 & 0 \\ f_4^1 & 1 \end{bmatrix} \xrightarrow{\mathbf{M}} \mathbf{F}^{(1)} = \begin{bmatrix} f_4^1 & 1 \\ 0 & 2 \end{bmatrix}
$$

$$
\mathbf{F}^2 = \mathbf{f}^2 = \begin{bmatrix} f_1^2 & 0 \\ f_2^2 & 1 \\ f_3^2 & 0 \\ f_4^2 & 2 \end{bmatrix} \xrightarrow{\mathbf{M}} \mathbf{F}^{(2)} = \begin{bmatrix} f_2^2 & 1 \\ f_4^2 & 2 \end{bmatrix} \tag{3.5}
$$

The generalized force vectors, $\mathbf{F}^{(i)}$, can be expressed in terms of the joint displacements, the generalized displacements of the assemblage, by substituting Eqs. (3.4) into Eqs. (3.2) and solving for the element forces appearing in Eqs. (3.5). Accordingly, we obtain

$$
\begin{bmatrix} f_1^1 \\ f_2^1 \\ f_3^1 \\ f_4^1 \end{bmatrix} = \alpha \begin{bmatrix} 0 & 0 & 0 & 1 \\ & & & \\ & & & \\ & & & 4L^2 \end{bmatrix} \begin{bmatrix} 0 & 0 \\ 0 & 0 \\ 0 & 0 \\ q_1 & 1 \end{bmatrix} \tag{3.6}
$$

which yields

$$
\mathbf{F}^{(1)} = \begin{bmatrix} f_4^1 \\ 0 \end{bmatrix} = \alpha \begin{bmatrix} 4L^2 & 0 \\ 0 & 0 \end{bmatrix} \begin{bmatrix} q_1 \\ q_2 \end{bmatrix} \tag{3.7}
$$

and

$$\begin{bmatrix} f_1^2 \\ f_2^2 \\ f_3^2 \\ f_4^2 \end{bmatrix} = \alpha \begin{matrix} & 0 & 1 & 0 & 2 \\ \end{matrix} \begin{bmatrix} & & & \\ & 4L^2 & & 2L^2 \\ & & & \\ & 2L^2 & & 4L^2 \end{bmatrix} \begin{bmatrix} 0 \\ q_1 \\ 0 \\ q_2 \end{bmatrix} \begin{matrix} 0 \\ 1 \\ 0 \\ 2 \end{matrix} \tag{3.8}$$

which yields

$$\mathbf{F}^{(2)} = \begin{bmatrix} f_2^2 \\ f_4^2 \end{bmatrix} = \alpha \begin{matrix} 1 & 2 \\ \end{matrix} \begin{bmatrix} 4L^2 & 2L^2 \\ 2L^2 & 4L^2 \end{bmatrix} \begin{bmatrix} q_1 \\ q_2 \end{bmatrix} \tag{3.9}$$

In Eqs. (3.6) and (3.8), only the coefficients that are transferred to Eqs. (3.7) and (3.9) are displayed. Equations (3.7) and (3.9) can be written in the form

$$\mathbf{F}^{(i)} = \mathbf{K}^{(i)}\mathbf{q}; \quad i = 1, 2 \tag{3.10}$$

where $\mathbf{K}^{(i)}$ is the generalized element stiffness matrix. The comparisons of Eq. (3.6) with Eq. (3.7) and of Eq. (3.8) with Eq. (3.9) suggest that the generalized element stiffness matrices can be obtained directly from the global element stiffness matrices (which in this case are equal to the local element stiffness matrices) by the following transformations:

$$\mathbf{K}^1 = \mathbf{k}^1 = \alpha \begin{matrix} 0 & 0 & 0 & 1 \\ \end{matrix} \begin{bmatrix} & & & 0 \\ & & & 0 \\ & & & 0 \\ & & & 4L^2 \end{bmatrix} \begin{matrix} 0 \\ 0 \\ 0 \\ 1 \end{matrix} \overset{\mathbf{M}}{\rightarrow} \mathbf{K}^{(1)} = \alpha \begin{matrix} 1 & 2 \\ \end{matrix} \begin{bmatrix} 4L^2 & 0 \\ 0 & 0 \end{bmatrix} \begin{matrix} 1 \\ 2 \end{matrix} \tag{3.11a}$$

$$\mathbf{K}^2 = \mathbf{k}^2 = \alpha \begin{matrix} 0 & 1 & 0 & 2 \\ \end{matrix} \begin{bmatrix} & & & 0 \\ & 4L^2 & 2L^2 & 1 \\ & & & 0 \\ & 2L^2 & 4L^2 & 2 \end{bmatrix} \overset{\mathbf{M}}{\rightarrow} \mathbf{K}^{(2)} = \alpha \begin{matrix} 1 & 2 \\ \end{matrix} \begin{bmatrix} 4L^2 & 2L^2 \\ 2L^2 & 4L^2 \end{bmatrix} \begin{matrix} 1 \\ 2 \end{matrix} \tag{3.11b}$$

This stiffness matrix transformation is analogous to the displacement and force transformations of Section 2.4; it will be defined explicitly below.

The substitution of Eq. (3.10) into the equation of equilibrium, Eq. (2.13), yields the system model

$$\sum_{i=1}^{2} \mathbf{K}^{(i)}\mathbf{q} = \mathbf{Q} \tag{3.12}$$

or

$$\mathbf{Kq} = \mathbf{Q} \tag{3.13}$$

where the system stiffness matrix

$$\mathbf{K} = \sum_{i=1}^{2} \mathbf{K}^{(i)} \tag{3.14}$$

Specifically, Eqs. (3.11)–(3.14) and the prescribed joint forces yield the system model in the form

$$\alpha \begin{bmatrix} 8L^2 & 2L^2 \\ 2L^2 & 4L^2 \end{bmatrix} \begin{bmatrix} q_1 \\ q_2 \end{bmatrix} = \begin{bmatrix} 14 \\ 0 \end{bmatrix} \tag{3.15}$$

which agrees with Eq. (B.7b) of the slope–deflection analysis.

4. Solution

The solution to Eq. (3.15) is

$$q_1 = \frac{2}{\alpha L^2}, \quad q_2 = -\frac{1}{\alpha L^2} \tag{3.16}$$

5. Element Forces

From Eqs. (3.2), (3.4), and (3.16), we obtain for $L = 5$ m

$$\mathbf{f}^1 = \begin{bmatrix} f_1^1 \\ f_2^1 \\ f_3^1 \\ f_4^1 \end{bmatrix} = \alpha \begin{bmatrix} 6L \\ 2L^2 \\ -6L \\ 4L^2 \end{bmatrix} \begin{bmatrix} 0 \\ 0 \\ 0 \\ \dfrac{2}{\alpha L^2} \end{bmatrix} = \begin{bmatrix} 2.4 \\ 4.0 \\ -2.4 \\ 8.0 \end{bmatrix} \tag{3.17a}$$

$$\mathbf{f}^2 = \begin{bmatrix} f_1^2 \\ f_2^2 \\ f_3^2 \\ f_4^2 \end{bmatrix} = \alpha \begin{bmatrix} 6L & 6L \\ 4L^2 & 2L^2 \\ -6L & -6L \\ 2L^2 & 4L^2 \end{bmatrix} \begin{bmatrix} 0 \\ \dfrac{2}{\alpha L^2} \\ 0 \\ -\dfrac{1}{\alpha L^2} \end{bmatrix} = \begin{bmatrix} 1.2 \\ 6.0 \\ -1.2 \\ 0.0 \end{bmatrix} \tag{3.17b}$$

In Eqs. (3.17), only the stiffness coefficients that correspond to nonzero displacements are displayed. The element forces are shown in the free-body diagrams of Figures 2.14b and B.2c.

6. Joint Forces

The computation of the joint forces is illustrated in Example 4 of Section 2.6.

General Approach

The matrix displacement method is presented in Figure 3.2. Aside from the notation, the matrix displacement method differs from the slope–deflection method (Figure B.1) in the form of presentation, which makes it suitable for the automatic computer analysis of any structure represented as an assemblage of discrete elements. Specifically, conditions of compatibility and equilibrium are formulated without reference to a visual image of the structure (Section 2.4), and an efficient procedure is employed for the assembly of element models into the system model. This assembly process, steps 2–3 in Figure 3.2, remains to be formulated.

If the local and global reference frames are distinct, the local element model

$$\mathbf{f}^i = \mathbf{k}^i \mathbf{d}^i, \quad i = 1, 2, \ldots, \text{NE} \tag{3.18}$$

FIGURE 3.2 Matrix displacement method without element actions (n is the number of system degrees of freedom, m the number of element degrees of freedom, and NE the number of elements).

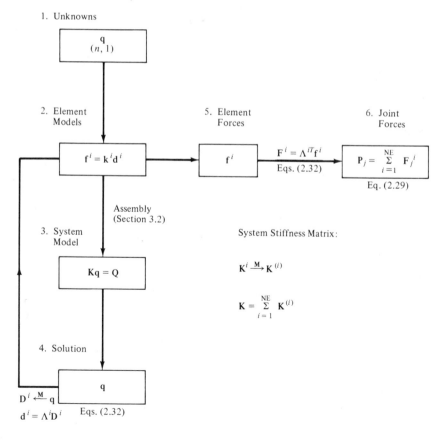

must be transformed into the global element model

$$\mathbf{F}^i = \mathbf{K}^i\mathbf{D}^i \tag{3.19}$$

where the global element stiffness matrix

$$\mathbf{K}^i = \mathbf{\Lambda}^{iT}\mathbf{k}^i\mathbf{\Lambda}^i \tag{3.20}$$

Equations (3.19) and (3.20) are obtained from Eq. (3.18) and the contragradient transformations, Eqs. (2.32), as follows:

$$\mathbf{F}^i = \mathbf{\Lambda}^{iT}\mathbf{f}^i = \mathbf{\Lambda}^{iT}\mathbf{k}^i\mathbf{d}^i = \mathbf{\Lambda}^{iT}\mathbf{k}^i\mathbf{\Lambda}^i\mathbf{D}^i = \mathbf{K}^i\mathbf{D}^i \tag{3.21}$$

The global element model is next transformed into the generalized element model

$$\mathbf{F}^{(i)} = \mathbf{K}^{(i)}\mathbf{q} \tag{3.22}$$

by conditions of compatibility [see Eqs. (3.3)–(3.10)]. Finally, the substitution of Eq. (3.22) into the equation of equilibrium, Eq. (2.13), yields the system model

$$\mathbf{Kq} = \mathbf{Q} \tag{3.23}$$

where the system stiffness matrix[1]

$$\mathbf{K} = \sum_{i=1}^{NE} \mathbf{K}^{(i)} \tag{3.24}$$

The generalized element stiffness matrix $\mathbf{K}^{(i)}$ can be constructed efficiently from the global element stiffness matrix $\mathbf{K}^i$ with the aid of the member code matrix as follows [see Eqs. (3.3) and (3.11)]:

Stiffness Matrix Transformation

$$\mathbf{K}^i \xrightarrow{\mathbf{M}} \mathbf{K}^{(i)}$$

1. Match the rows and columns of $\mathbf{K}^i$ with the entries in the ith column of $\mathbf{M}$.
2. Transfer every coefficient of $\mathbf{K}^i$ whose row and column correspond to positive integers j and k to the jth row and kth column of $\mathbf{K}^{(i)}$.
3. Assign the value 0 to every location of $\mathbf{K}^{(i)}$ that does not receive a contribution from $\mathbf{K}^i$.

Since the stiffness matrices are symmetric (see Special Features, which follows), only the upper or lower triangular portions need to be generated.

Special Features

Some important aspects of the generalized element model and the system model are discussed.

---

[1] The computation of the system stiffness matrix by the summation of the element stiffness matrices is called the direct stiffness method (Turner, Clough, Martin, and Topp, 1956).

Generalized Element Stiffness Matrix

The generalized element stiffness matrix can be derived formally from the global element model, Eq. (3.19), and the contragredient transformations, Eqs. (2.78), as follows:

$$\mathbf{F}^{(i)} = \mathbf{A}^{iT}\mathbf{F}^i = \mathbf{A}^{iT}\mathbf{K}^i\mathbf{D}^i = \mathbf{A}^{iT}\mathbf{K}^i\mathbf{A}^i\mathbf{q} \tag{3.25}$$

or

$$\mathbf{F}^{(i)} = \mathbf{K}^{(i)}\mathbf{q} \tag{3.26}$$

where

$$\mathbf{K}^{(i)}_{(n,\,n)} = \mathbf{A}^{iT}_{(n,\,m)}\ \mathbf{K}^i_{(m,\,m)}\ \mathbf{A}^i_{(m,\,n)} \tag{3.27}$$

Equation (3.27) clearly shows that the condition of compatibility, expressed in the form

$$\mathbf{D}^i_{(m,\,1)} = \mathbf{A}^i_{(m,\,n)}\ \mathbf{q}_{(n,\,1)} \tag{3.28}$$

links the (m, m) global element stiffness matrix with the (n, n) generalized element stiffness matrix.

Equation (3.27) could be used to construct the generalized element stiffness matrix. However, this approach is not efficient because of the required matrix multiplications. For illustration consider the introductory problem based on Figure 3.1. Expressing Eqs. (3.4) in the form of Eqs. (3.28), we obtain

$$\begin{bmatrix} d_1^1 \\ d_2^1 \\ d_3^1 \\ d_4^1 \end{bmatrix} = \begin{bmatrix} 0 & 0 \\ 0 & 0 \\ 0 & 0 \\ 1 & 0 \end{bmatrix} \begin{bmatrix} q_1 \\ q_2 \end{bmatrix} \quad \text{or} \quad \mathbf{d}^1 = \mathbf{A}^1\mathbf{q} \tag{3.29a}$$

and

$$\begin{bmatrix} d_1^2 \\ d_2^2 \\ d_3^2 \\ d_4^2 \end{bmatrix} = \begin{bmatrix} 0 & 0 \\ 1 & 0 \\ 0 & 0 \\ 0 & 1 \end{bmatrix} \begin{bmatrix} q_1 \\ q_2 \end{bmatrix} \quad \text{or} \quad \mathbf{d}^2 = \mathbf{A}^2\mathbf{q} \tag{3.29b}$$

The substitutions of the stiffness matrix from Eq. (3.2) and the transformation matrices from Eqs. (3.29) into Eq. (3.27) yield the generalized element stiffness matrices

$$\mathbf{K}^{(1)} = \alpha\begin{bmatrix} 4L^2 & 0 \\ 0 & 0 \end{bmatrix}, \quad \mathbf{K}^{(2)} = \alpha\begin{bmatrix} 4L^2 & 2L^2 \\ 2L^2 & 4L^2 \end{bmatrix} \tag{3.30}$$

Clearly, this approach requires considerably more effort than the equivalent transformations in Eqs. (3.11).

Symmetry

The symmetry of a matrix is preserved by the post- and premultiplication with a matrix and its transpose (Hohn, 1972); that is, if $\mathbf{C} = \mathbf{C}^T$,

$$\mathbf{B}^T\mathbf{C}\mathbf{B} = (\mathbf{B}^T\mathbf{C}\mathbf{B})^T = \mathbf{B}^T\mathbf{C}^T\mathbf{B} \qquad (3.31)$$

Thus, it follows from Eqs. (3.20), (3.27) and the symmetry of the local element stiffness matrices (Sections 1.7 and 1.8) that the global and generalized element stiffness matrices are symmetric.

The system stiffness matrix, Eq. (3.24), is symmetric because matrix addition preserves symmetry.

Stiffness Coefficients

Analogous to element stiffness coefficients (Section 1.7), the meaning of system stiffness coefficients can be deduced from the mathematical model of the assemblage, Eq. (3.23), which can be expressed as

$$Q_j = \sum_{k=1}^{n} K_{jk} q_k, \quad j = 1, 2, \ldots, n \qquad (3.32)$$

If we impose the configuration

$$q_l = \delta_{lk}, \quad l = 1, 2, \ldots, n \qquad (3.33)$$

where δ_{lk} is the Kronecker delta (Section 1.7), Eq. (3.32) yields

$$Q_j = K_{jk} \qquad (3.34)$$

Thus,

$$K_{jk} = \begin{cases} \text{force } Q_j \text{ corresponding to the equilibrium} \\ \text{configuration } q_k = 1 \text{ and the remaining} \\ \text{joint displacements are zero} \end{cases} \qquad (3.35)$$

$$\underset{\text{force}}{\diagup} \quad \underset{\text{unit displacement}}{\diagdown}$$

The definition in Eq. (3.35) can be used to compute system stiffness coefficients individually or by columns. This provides a way of checking the values of specific stiffness coefficients and of gaining insight into the behavior of assemblages.

In the application of this definition to plane skeletal structures, we encounter the element configurations shown in Figure 3.3. The corresponding

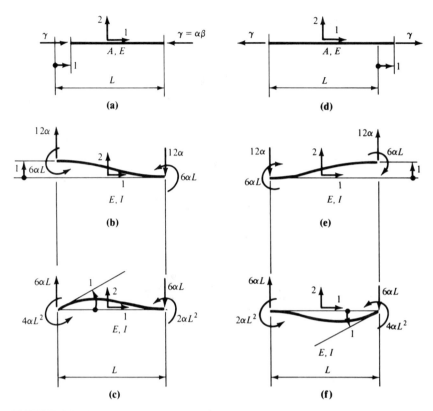

FIGURE 3.3 Element configurations.

element forces can be verified by substituting the values of the displacements into Eq. (1.99). For example, for Figure 3.3b

$$\mathbf{d} = \begin{bmatrix} 0 \\ 1 \\ 0 \\ 0 \\ 0 \\ 0 \end{bmatrix} \quad \text{and} \quad \mathbf{f} = \alpha \begin{bmatrix} 0 \\ 12 \\ 6L \\ 0 \\ -12 \\ 6L \end{bmatrix} \qquad (3.36)$$

Thus, the forces in Figures 3.3a, b, ..., f are the coefficients of columns 1, 2, ..., 6, respectively, of the stiffness matrix in Eq. (1.99).

For example, to compute the coefficients of the first column of the system stiffness matrix of the structure in Figure 3.1, we let $q_1 = 1$ and $q_2 = 0$ (Figure 3.4a). The resulting element forces that balance the applied joint forces are taken from Figure 3.3 and shown in Figure 3.4b. The conditions

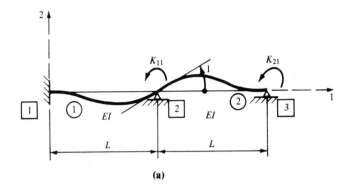

(a)

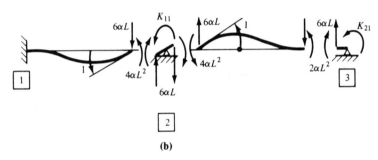

(b)

FIGURE 3.4 System stiffness coefficients. **(a)** Configuration: $q_1 = 1$, $q_2 = 0$; **(b)** joint equilibrium.

of joint equilibrium yield the coefficients $K_{11} = 8\alpha L^2$, $K_{21} = 2\alpha L^2$, which agree with the first column of Eq. (3.15).

Unique Solution

Equation (3.23) represents a set of simultaneous, nonhomogeneous, algebraic equations in the unknown joint displacements. The solution to Eq. (3.23) is unique if the system stiffness matrix is nonsingular, that is, if **K** has rank n. Physically, this means that the assemblage is sufficiently constrained to prevent (finite or infinitesimal) rigid-body displacements (Section 1.8). Accordingly, the initial configuration is a configuration of stable equilibrium.

On the basis of the theory of stability of equilibrium (Langhaar, 1962; Rubinstein, 1970), **K** can be characterized as a positive definite matrix.

The selection of efficient equation solvers is addressed in Chapter 6. In this chapter we are primarily concerned with the formulation and illustration of the matrix displacement method.

3.3 ANALYSIS OF CONTINUOUS BEAMS

In the matrix displacement analysis of continuous beams, we select coincident local and global reference frames (Figure 2.4a). Hence, no coordinate transformations are required.

The analysis is based on the flow chart in Figure 3.2. The element model is [Eq. (1.98b)]

$$\mathbf{f} = \mathbf{kd}$$

where

$$
\mathbf{k} = \alpha
\left[
\begin{array}{cc:cc}
12 & 6L & -12 & 6L \\
6L & 4L^2 & -6L & 2L^2 \\
\hdashline
-12 & -6L & 12 & -6L \\
6L & 2L^2 & -6L & 4L^2
\end{array}
\right],
\quad \alpha = \frac{EI}{L^3}
\tag{3.37}
$$

Example 1. The continuous beams in Figure 3.5a are subjected to the joint forces $Q_1 = 30\,\text{kN} \cdot \text{m}$, $Q_2 = 0$. The analysis units are kilonewton (kN), meter (m), and radian (rad). The analysis follows the flow chart in Figure 3.2.

1. *Unknowns*

$$q_k, \quad k = 1, 2 \tag{3.38}$$

2. *Element Models*

$$\mathbf{f}^i = \mathbf{kd}^i, \quad i = 1, 2 \tag{3.39}$$

$\mathbf{k}$ is defined in Eq. (3.37)

3. *System Model*

According to Figures 3.5a, b and Eq. (2.5)

$$
\mathbf{M}^T =
\begin{bmatrix}
0 & 0 & 0 & 1 \\
0 & 1 & 2 & 0
\end{bmatrix}
\tag{3.40}
$$

The stiffness matrix transformations (Section 3.2) yield

$$
\mathbf{K}^1 = \mathbf{k} = \alpha
\begin{matrix}
0 \quad 0 \quad 0 \quad 1 \\
\left[
\begin{array}{cccc}
 & & & 0 \\
 & & & 0 \\
 & & & 0 \\
\text{sym.} & & & 4L^2
\end{array}
\right]
\begin{array}{c} 0 \\ 0 \\ 0 \\ 1 \end{array}
\end{matrix}
\xrightarrow{\mathbf{M}}
\mathbf{K}^{(1)} = \alpha
\begin{matrix}
1 \quad 2 \\
\left[
\begin{array}{cc}
4L^2 & 0 \\
\text{sym.} & 0
\end{array}
\right]
\begin{array}{c} 1 \\ 2 \end{array}
\end{matrix}
\tag{3.41a}
$$

$$
\mathbf{K}^2 = \mathbf{k} = \alpha
\begin{matrix}
0 \quad 1 \quad 2 \quad 0 \\
\left[
\begin{array}{cccc}
 & & & 0 \\
 & 4L^2 & -6L & 1 \\
 & & 12 & 2 \\
\text{sym.} & & & 0
\end{array}
\right]
\end{matrix}
\xrightarrow{\mathbf{M}}
\mathbf{K}^{(2)} = \alpha
\begin{matrix}
1 \quad 2 \\
\left[
\begin{array}{cc}
4L^2 & -6L \\
\text{sym.} & 12
\end{array}
\right]
\begin{array}{c} 1 \\ 2 \end{array}
\end{matrix}
\tag{3.41b}
$$

120

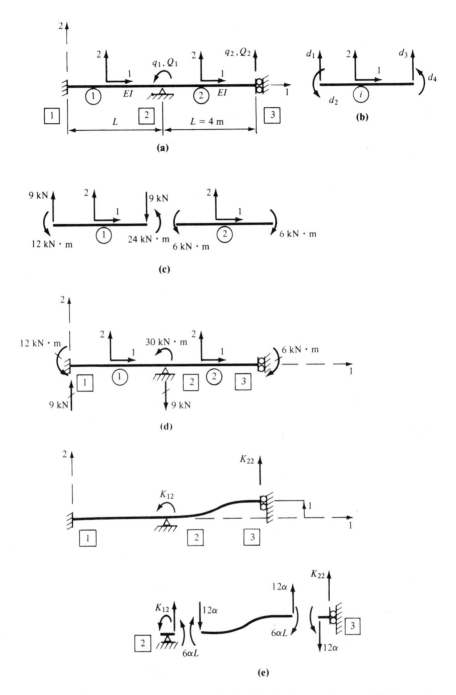

FIGURE 3.5 Matrix displacement analysis. (a) Continuous beams; (b) elements; (c) element forces; (d) joint forces; (e) verification of K_{12}, K_{22}.

Thus,

$$\mathbf{K} = \sum_{i=1}^{2} \mathbf{K}^{(i)} = \alpha \begin{bmatrix} 8L^2 & -6L \\ \text{sym.} & 12 \end{bmatrix} \tag{3.42}$$

and by Eq. (3.23)

$$\alpha \begin{bmatrix} 8L^2 & -6L \\ -6L & 12 \end{bmatrix} \begin{bmatrix} q_1 \\ q_2 \end{bmatrix} = \begin{bmatrix} 30 \\ 0 \end{bmatrix} \tag{3.43}$$

The system stiffness coefficients can be checked by using Eq. (3.35). For example, to compute the coefficients of the second column of $\mathbf{K}$, we impose the configuration $q_1 = 0, q_2 = 1$, which is shown in Figure 3.5e. The element forces are obtained from Figure 3.3. The conditions of joint equilibrium yield $K_{12} = -6\alpha L$, $K_{22} = 12\alpha$, which agree with Eq. (3.43).

4. *Solution*

$$q_1 = \frac{6}{\alpha L^2}, \quad q_2 = \frac{3}{\alpha L} \tag{3.44}$$

5. *Element Forces*
According to Eq. (3.40)

$$\mathbf{d}^1 = \begin{bmatrix} 0 & 0 \\ 0 & 0 \\ 0 & 0 \\ q_1 & 1 \end{bmatrix}, \quad \mathbf{d}^2 = \begin{bmatrix} 0 & 0 \\ q_1 & 1 \\ q_2 & 2 \\ 0 & 0 \end{bmatrix} \tag{3.45}$$

From Eqs. (3.37), (3.39), (3.44), and (3.45), we obtain for $L = 4$ m

$$\mathbf{f}^1 = \begin{bmatrix} f_1^1 \\ f_2^1 \\ f_3^1 \\ f_4^1 \end{bmatrix} = \alpha \begin{bmatrix} & 6L \\ & 2L^2 \\ & -6L \\ & 4L^2 \end{bmatrix} \begin{bmatrix} 0 \\ 0 \\ 0 \\ \frac{6}{\alpha L^2} \end{bmatrix} = \begin{bmatrix} 9.0 \\ 12.0 \\ \hline -9.0 \\ 24.0 \end{bmatrix} = \begin{bmatrix} \mathbf{f}_a^1 \\ \mathbf{f}_b^1 \end{bmatrix}$$

$$\tag{3.46a}$$

$$\mathbf{f}^2 = \begin{bmatrix} f_1^2 \\ f_2^2 \\ f_3^2 \\ f_4^2 \end{bmatrix} = \alpha \begin{bmatrix} 6L & -12 \\ 4L^2 & -6L \\ -6L & 12 \\ 2L^2 & -6L \end{bmatrix} \begin{bmatrix} \frac{6}{\alpha L^2} \\ \frac{3}{\alpha L} \end{bmatrix} = \begin{bmatrix} 0.0 \\ 6.0 \\ \hline 0.0 \\ -6.0 \end{bmatrix} = \begin{bmatrix} \mathbf{f}_a^2 \\ \mathbf{f}_b^2 \end{bmatrix}$$

$$\tag{3.46b}$$

The element forces are shown in Figure 3.5c.

6. *Joint Forces*

It follows from Figure 3.5a and Eq. (2.29) that

$$\mathbf{P}_1 = \mathbf{f}_a^2, \quad \mathbf{P}_2 = \mathbf{f}_b^1 + \mathbf{f}_a^2, \quad \mathbf{P}_3 = \mathbf{f}_b^2 \tag{3.47}$$

Thus, Eqs. (3.46) and (3.47) yield

$$\mathbf{P}_1 = \begin{bmatrix} 9.0 \\ 12.0 \end{bmatrix}, \quad \mathbf{P}_2 = \begin{bmatrix} -9.0 \\ 30.0 \end{bmatrix}, \quad \mathbf{P}_3 = \begin{bmatrix} 0.0 \\ -6.0 \end{bmatrix} \tag{3.48}$$

The joint forces are shown in Figure 3.5d.

Example 2. The continuous beams in Figure 3.6a are subjected to the joint forces $Q_1 = Q_2 = 0$, $Q_3 = 12\,\text{k} \cdot \text{ft}$. The analysis units are kilopound (k), foot (ft), and radian (rad). The analysis follows the flow chart in Figure 3.2.

1. *Unknowns*

$$q_k, \quad k = 1, 2, 3 \tag{3.49}$$

2. *Element Models*

$$\mathbf{f}^i = \mathbf{k}\mathbf{d}^i \tag{3.50}$$

$\mathbf{k}$ is defined in Eq. (3.37)

3. *System Model*

It follows from Figures 3.6a, b and Eq. (2.5) that

$$\mathbf{M}^T = \begin{bmatrix} 0 & 0 & 0 & 1 \\ 0 & 1 & 2 & 3 \end{bmatrix} \tag{3.51}$$

The stiffness matrix transformations (Section 3.2) yield

$$
\mathbf{K}^1 = \mathbf{k} = \alpha \begin{matrix} & 0 & 0 & 0 & 1 \\ & \begin{bmatrix} & & & 0 \\ & & & 0 \\ & & & 0 \\ \text{sym.} & & 4L^2 & 1 \end{bmatrix} \end{matrix} \xrightarrow{\text{M}} \mathbf{K}^{(1)} = \alpha \begin{matrix} & 1 & 2 & 3 \\ \begin{bmatrix} 4L^2 & 0 & 0 \\ & & 0 & 0 \\ \text{sym.} & & 0 \end{bmatrix} & \begin{matrix} 1 \\ 2 \\ 3 \end{matrix} \end{matrix}
$$
$$\tag{3.52a}$$

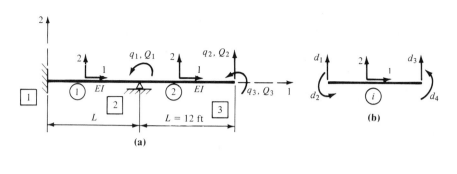

(a)

(b)

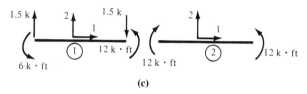

(c)

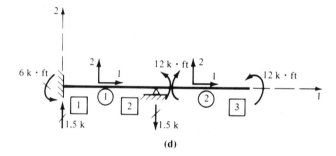

(d)

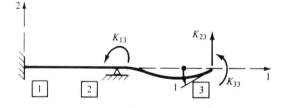

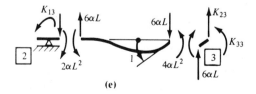

(e)

FIGURE 3.6 Matrix displacement analysis. **(a)** Continuous beams; **(b)** elements; **(c)** element forces; **(d)** joint forces; **(e)** verification of K_{13}, K_{23}, K_{33}.

$$
\mathbf{K}^2 = \mathbf{k} = \alpha
\begin{array}{cccc}
0 & 1 & 2 & 3 \\
\begin{bmatrix}
 & & & \\
 & 4L^2 & -6L & 2L^2 \\
 & & 12 & -6L \\
\text{sym.} & & & 4L^2
\end{bmatrix}
\begin{array}{l}
0 \\
1 \\
2 \\
3
\end{array}
\end{array}
\overset{M}{\to} \mathbf{K}^{(2)}
$$

$$
= \alpha
\begin{array}{ccc}
1 & 2 & 3 \\
\begin{bmatrix}
4L^2 & -6L & 2L^2 \\
 & 12 & -6L \\
\text{sym.} & & 4L^2
\end{bmatrix}
\begin{array}{l}
1 \\
2 \\
3
\end{array}
\end{array}
\qquad (3.52b)
$$

Thus,

$$
\mathbf{K} = \sum_{i=1}^{2} \mathbf{K}^{(i)} = \alpha
\begin{bmatrix}
8L^2 & -6L & 2L^2 \\
 & 12 & -6L \\
\text{sym.} & & 4L^2
\end{bmatrix}
\qquad (3.53)
$$

and

$$
\alpha
\begin{bmatrix}
8L^2 & -6L & 2L^2 \\
-6L & 12 & -6L \\
2L^2 & -6L & 4L^2
\end{bmatrix}
\begin{bmatrix}
q_1 \\
q_2 \\
q_3
\end{bmatrix}
=
\begin{bmatrix}
0 \\
0 \\
12
\end{bmatrix}
\qquad (3.54)
$$

The system stiffness coefficients can again be checked by Eq. (3.35). For example, from the conditions of joint equilibrium in Figure 3.6e we obtain the coefficients of the third column of $\mathbf{K}$: $K_{13} = 2\alpha L^2$, $K_{23} = -6\alpha L$, $K_{33} = 4\alpha L^2$.

4. *Solution*

$$
q_1 = \frac{3}{\alpha L^2}, \quad q_2 = \frac{9}{\alpha L}, \quad q_3 = \frac{15}{\alpha L^2}
\qquad (3.55)
$$

5. *Element Forces*
By Eq. (3.51)

$$
\mathbf{d}^1 =
\begin{bmatrix}
0 \\
0 \\
0 \\
q_1
\end{bmatrix}
\begin{array}{l}
0 \\
0 \\
0 \\
1
\end{array},
\quad
\mathbf{d}^2 =
\begin{bmatrix}
0 \\
q_1 \\
q_2 \\
q_3
\end{bmatrix}
\begin{array}{l}
0 \\
1 \\
2 \\
3
\end{array}
\qquad (3.56)
$$

From Eqs. (3.37), (3.50), (3.55), and (3.56), we obtain for $L = 12$ ft

$$
\mathbf{f}^1 =
\begin{bmatrix}
\mathbf{f}_a^1 \\
\mathbf{f}_b^1
\end{bmatrix}
=
\begin{bmatrix}
1.5 \\
6.0 \\
\hline
-1.5 \\
12.0
\end{bmatrix},
\quad
\mathbf{f}^2 =
\begin{bmatrix}
\mathbf{f}_a^2 \\
\mathbf{f}_b^2
\end{bmatrix}
=
\begin{bmatrix}
0.0 \\
-12.0 \\
\hline
0.0 \\
12.0
\end{bmatrix}
\qquad (3.57)
$$

The element forces are shown in Figure 3.6c

6. *Joint Forces*

Applying Eqs. (2.29) and (3.57) to Figure 3.6a, we obtain

$$\mathbf{P}_1 = \mathbf{f}_a^1 = \begin{bmatrix} 1.5 \\ 6.0 \end{bmatrix}, \quad \mathbf{P}_2 = \mathbf{f}_b^1 + \mathbf{f}_a^2 = \begin{bmatrix} -1.5 \\ 0.0 \end{bmatrix}, \quad \mathbf{P}_3 = \mathbf{f}_b^2 = \begin{bmatrix} 0.0 \\ 12.0 \end{bmatrix} \quad (3.58)$$

The joint forces are shown in Figure 3.6d. The results can easily be checked by elementary techniques. For example, the reactive moment at joint 1 (the carry-over moment in the moment-distribution method) must have the same sense and one-half the magnitude of the moment applied by element 2 to joint 2 (Figure 3.6d).

3.4 ANALYSIS OF FRAMES

To apply the matrix displacement method (Figure 3.2) to frames, we need to formulate the global element stiffness matrix. The local element model is [Eq. (1.99)]

$$\mathbf{f} = \mathbf{k}\mathbf{d}$$

$$\mathbf{k} = \begin{bmatrix} \mathbf{k}_{aa} & \mathbf{k}_{ab} \\ \mathbf{k}_{ba} & \mathbf{k}_{bb} \end{bmatrix} = \alpha \begin{bmatrix} \beta & 0 & 0 & -\beta & 0 & 0 \\ 0 & 12 & 6L & 0 & -12 & 6L \\ 0 & 6L & 4L^2 & 0 & -6L & 2L^2 \\ \hline -\beta & 0 & 0 & \beta & 0 & 0 \\ 0 & -12 & -6L & 0 & 12 & -6L \\ 0 & 6L & 2L^2 & 0 & -6L & 4L^2 \end{bmatrix} \quad (3.59)$$

where

$$\alpha = \frac{EI}{L^3}, \quad \beta = \frac{AL^2}{I}, \quad \gamma = \alpha\beta = \frac{EA}{L}$$

The substitutions of the partitioned Λ matrix [Eq. (2.36)] and the partitioned $\mathbf{k}$ matrix [Eq. (3.59)] into Eq. (3.20) yield the global element stiffness matrix in the form

$$\mathbf{K} = \begin{bmatrix} \mathbf{K}_{aa} & \mathbf{K}_{ab} \\ \mathbf{K}_{ba} & \mathbf{K}_{bb} \end{bmatrix} \quad (3.60)$$

where

$$\mathbf{K}_{aa} = \lambda^T \mathbf{k}_{aa} \lambda$$

$$\mathbf{K}_{ab} = \lambda^T \mathbf{k}_{ab} \lambda$$

$$\mathbf{K}_{ba} = \lambda^T \mathbf{k}_{ba} \lambda$$

$$\mathbf{K}_{bb} = \lambda^T \mathbf{k}_{bb} \lambda$$

$$(3.61)$$

Since $\mathbf{K}$ is symmetric (Section 3.2), it is sufficient to compute the coefficients of $\mathbf{K}_{ab}$ and the upper triangular portions of $\mathbf{K}_{aa}$ and $\mathbf{K}_{bb}$. For example, by Eqs. (2.34), (3.59), and (3.61)

$$\mathbf{K}_{aa} = \alpha \begin{bmatrix} c_1 & -c_2 & 0 \\ c_2 & c_1 & 0 \\ 0 & 0 & 1 \end{bmatrix} \begin{bmatrix} \beta & 0 & 0 \\ 0 & 12 & 6L \\ 0 & 6L & 4L^2 \end{bmatrix} \begin{bmatrix} c_1 & c_2 & 0 \\ -c_2 & c_1 & 0 \\ 0 & 0 & 1 \end{bmatrix}$$

$$\mathbf{K}_{aa} = \alpha \begin{bmatrix} \beta c_1^2 + 12c_2^2 & c_1 c_2(\beta - 12) & -6Lc_2 \\ & \beta c_2^2 + 12c_1^2 & 6Lc_1 \\ \text{sym.} & & 4L^2 \end{bmatrix} \qquad (3.62)$$

The resulting global stiffness matrix of the frame element can be expressed as

$$\mathbf{K} = \begin{bmatrix} g_1 & g_2 & g_4 & \vdots & -g_1 & -g_2 & g_4 \\ & g_3 & g_5 & \vdots & -g_2 & -g_3 & g_5 \\ & & g_6 & \vdots & -g_4 & -g_5 & g_7 \\ \cdots & \cdots & \cdots & \vdots & \cdots & \cdots & \cdots \\ & & & \vdots & g_1 & g_2 & -g_4 \\ & & & \vdots & & g_3 & -g_5 \\ \text{sym.} & & & \vdots & & & g_6 \end{bmatrix} \qquad (3.63)$$

where

$$g_1 = \alpha(\beta c_1^2 + 12c_2^2)$$
$$g_2 = \alpha c_1 c_2(\beta - 12)$$
$$g_3 = \alpha(\beta c_2^2 + 12c_1^2)$$
$$g_4 = -\alpha 6Lc_2 \qquad\qquad (3.64)$$
$$g_5 = \alpha 6Lc_1$$
$$g_6 = \alpha 4L^2$$
$$g_7 = \alpha 2L^2$$

$$\alpha = \frac{EI}{L^3}, \quad \beta = \frac{AL^2}{I}$$

Example 1. The frame in Figure 3.7a is subjected to the joint loads $Q_1 = Q_3 = 0$, $Q_2 = 14 \text{ k} \cdot \text{ft}$. The elements have identical properties: $A = 5 \text{ in}^2$, $I = 50 \text{ in}^4$, $L = 10 \text{ ft}$, $E = 30{,}000 \text{ ksi}$. The element rotation matrices are defined in Eqs. (2.47). The analysis is based on the flow chart in Figure 3.2. The analysis units are kilopound (k), foot (ft), and radian (rad).

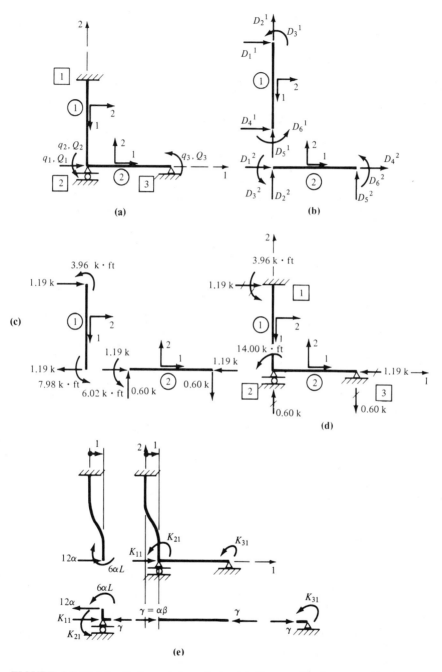

FIGURE 3.7 Matrix displacement analysis. **(a)** Frame; **(b)** elements; **(c)** element forces; **(d)** joint forces; **(e)** verification of K_{11}, K_{21}, K_{31}.

1. *Unknowns*

$$q_k, \quad k = 1, 2, 3 \tag{3.65}$$

2. *Element Models*

$$\mathbf{f}^i = \mathbf{k}\mathbf{d}^i, \quad i = 1, 2 \tag{3.66}$$

$\mathbf{k} =$ is defined in Eq. (3.59).

3. *System Model*

According to Figures 3.7a, b and Eq. (2.5)

$$\mathbf{M}^T = \begin{bmatrix} 0 & 0 & 0 & 1 & 0 & 2 \\ 1 & 0 & 2 & 0 & 0 & 3 \end{bmatrix} \tag{3.67}$$

Stiffness matrix transformations (Section 3.2): By Eqs. (2.34), (2.47), (3.63), (3.64), and (3.67)

$$\mathbf{K}^1 = \begin{array}{ccccccc} 0 & 0 & 0 & 1 & 0 & 2 \\ \begin{bmatrix} & & & & & \\ & & & & & \\ & & & g_1 & & -g_4 \\ & & & & & \\ \mathrm{sym.} & & & & & g_6 \end{bmatrix} \begin{array}{c} 0 \\ 0 \\ 0 \\ 1 \\ 0 \\ 2 \end{array} \end{array} \xrightarrow{\mathbf{M}} \mathbf{K}^{(1)} = \alpha \begin{array}{c} \\ \begin{array}{ccc} 1 & 2 & 3 \end{array} \\ \begin{bmatrix} 12 & -6L & 0 \\ & 4L^2 & 0 \\ \mathrm{sym.} & & 0 \end{bmatrix} \begin{array}{c} 1 \\ 2 \\ 3 \end{array} \end{array} \tag{3.68a}$$

because $c_1 = 0$, $c_2 = -1$, $g_1 = \alpha 12$, $g_4 = \alpha 6L$, $g_6 = \alpha 4L^2$. The local and global reference frames of element 2 coincide; hence,

$$\mathbf{K}^2 = \mathbf{k} = \alpha \begin{array}{cccccc} 1 & 0 & 2 & 0 & 0 & 3 \\ \begin{bmatrix} \beta & & 0 & & & 0 \\ & & & & & \\ & & 4L^2 & & & 2L^2 \\ & & & & & \\ & & & & & \\ \mathrm{sym.} & & & & & 4L^2 \end{bmatrix} \begin{array}{c} 1 \\ 0 \\ 2 \\ 0 \\ 0 \\ 3 \end{array} \end{array} \xrightarrow{\mathbf{M}} \mathbf{K}^{(2)} = \alpha \begin{array}{c} \begin{array}{ccc} 1 & 2 & 3 \end{array} \\ \begin{bmatrix} \beta & 0 & 0 \\ & 4L^2 & 2L^2 \\ \mathrm{sym.} & & 4L^2 \end{bmatrix} \begin{array}{c} 1 \\ 2 \\ 3 \end{array} \end{array} \tag{3.68b}$$

where $\mathbf{k}$ is defined in Eq. (3.59).

$$\mathbf{K} = \sum_{i=1}^{2} \mathbf{K}^{(i)} = \alpha \begin{bmatrix} \beta + 12 & -6L & 0 \\ & 8L^2 & 2L^2 \\ \mathrm{sym.} & & 4L^2 \end{bmatrix} \tag{3.69}$$

and

$$\alpha \begin{bmatrix} \beta + 12 & -6L & 0 \\ -6L & 8L^2 & 2L^2 \\ 0 & 2L^2 & 4L^2 \end{bmatrix} \begin{bmatrix} q_1 \\ q_2 \\ q_3 \end{bmatrix} = \begin{bmatrix} 0 \\ 14 \\ 0 \end{bmatrix} \tag{3.70}$$

Again, the system stiffness coefficients can be verified by Eq. (3.35). For example, the coefficients of the first column of $\mathbf{K}$ follow from Figure 3.7e, which depicts the configuration $q_1 = 1$, $q_2 = q_3 = 0$. The resulting element forces can be obtained from Figure 3.3. The conditions of joint equilibrium yield $K_{11} = \alpha(\beta + 12)$, $K_{21} = -\alpha 6L$, $K_{31} = 0$,

4. *Solution*

1490

$$q_1 = \frac{1}{\alpha} \frac{84}{(48 + 7\beta)L}, \quad q_2 = \frac{1}{\alpha} \frac{(12 + \beta)14}{(48 + 7\beta)L^2}, \quad q_3 = -\frac{1}{2}q_2 \quad (3.71)$$

$= ,0001$ $,0001$

5. *Element Forces*

By Eqs. (2.36), (2.47), and (3.67)

$$\mathbf{D}^1 = \begin{bmatrix} 0 & 0 \\ 0 & 0 \\ 0 & 0 \\ q_1 & 1 \\ 0 & 0 \\ q_2 & 2 \end{bmatrix}, \quad \mathbf{D}^2 = \begin{bmatrix} q_1 & 1 \\ 0 & 0 \\ q_2 & 2 \\ 0 & 0 \\ 0 & 0 \\ q_3 & 3 \end{bmatrix}; \quad \mathbf{d}^1 = \mathbf{\Lambda}^1 \mathbf{D}^1 = \begin{bmatrix} 0 \\ 0 \\ 0 \\ 0 \\ q_1 \\ q_2 \end{bmatrix}, \quad \mathbf{d}^2 = \mathbf{D}^2 \quad (3.72)$$

From Eqs. (3.59), (3.66), (3.71), and (3.72), we obtain the element forces

$$\mathbf{f}^1 = \alpha \begin{bmatrix} 0 \\ -12q_1 + 6Lq_2 \\ -6Lq_1 + 2L^2q_2 \\ \hline 0 \\ 12q_1 - 6Lq_2 \\ -6Lq_1 + 4L^2q_2 \end{bmatrix} = \begin{bmatrix} 0.00 \\ 1.19 \\ 3.96 \\ \hline 0.00 \\ -1.19 \\ 7.98 \end{bmatrix} = \begin{bmatrix} \mathbf{f}_a^1 \\ \mathbf{f}_b^1 \end{bmatrix} \quad (3.73a)$$

$$\mathbf{f}^2 = \alpha \begin{bmatrix} \beta q_1 \\ 3Lq_2 \\ 3L^2q_2 \\ \hline -\beta q_1 \\ -3Lq_2 \\ 0 \end{bmatrix} = \begin{bmatrix} 1.19 \\ 0.60 \\ 6.02 \\ \hline -1.19 \\ -0.60 \\ 0.00 \end{bmatrix} = \begin{bmatrix} \mathbf{f}_a^2 \\ \mathbf{f}_b^2 \end{bmatrix} \quad (3.73b)$$

which are shown in Figure 3.7c.

6. *Joint Forces*

By Eqs. (2.29), (2.37), (2.47), (3.73), and Figure 3.7a

$$\mathbf{P}_1 = \mathbf{F}_a^1 = \boldsymbol{\lambda}^{1T}\mathbf{f}_a^1 = \begin{bmatrix} 1.19 \\ 0.00 \\ 3.96 \end{bmatrix}, \quad \mathbf{P}_3 = \mathbf{F}_b^2 = \mathbf{f}_b^2 = \begin{bmatrix} -1.19 \\ -0.60 \\ 0.00 \end{bmatrix}$$

$$\mathbf{P}_2 = \mathbf{F}_b^1 + \mathbf{F}_a^2 = \boldsymbol{\lambda}^{1T}\mathbf{f}_b^1 + \mathbf{f}_a^2 = \begin{bmatrix} 0.00 \\ 0.60 \\ 14.00 \end{bmatrix} \tag{3.74}$$

The joint forces are shown in Figure 3.7d.

The comparison of Figures 2.11a, b and Figures 3.7c, d indicates slight differences in the reactive element and joint forces even though the frames have identical properties and loads. The differences are caused by the assumption in the slope–deflection method (which was used in the analysis of Figure 2.11a) that the element centroidal axes are inextensible, that is, their extensional stiffnesses are assumed to be infinite. If we impose this condition on Eqs. (3.71) by setting $\beta = \infty$, we obtain the joint displacements of the slope–deflection analysis

$$q_1 = 0, \quad q_2 = \frac{2}{\alpha L^2}, \quad q_3 = -\frac{1}{2}q_2 \tag{3.75}$$

which result in the element and joint forces of Figure 2.11.

In Section 4.3 the assumption of inextensible centroidal axes is incorporated in the matrix displacement analysis of orthogonal frames.

3.5 ANALYSIS OF TRUSSES

The matrix displacement analysis of trusses (Figure 3.2) is based on the local and global elements shown in Figures 2.12a and b. Accordingly, the local element model is [Eq. (1.96b)]

$$\mathbf{f} = \mathbf{kd}$$

$$\mathbf{k} = \begin{bmatrix} k_{aa} & k_{ab} \\ k_{ba} & k_{bb} \end{bmatrix} = \gamma \begin{bmatrix} 1 & \vdots & -1 \\ \cdots & \cdots & \cdots \\ -1 & \vdots & 1 \end{bmatrix} \tag{3.76}$$

where

$$\gamma = \frac{EA}{L}$$

In partitioned form, the global stiffness matrix of the truss element is defined by Eqs. (3.60) and (3.61). This follows from Eq. (3.20) and the partitioned $\boldsymbol{\Lambda}$

and **k** matrices in Eqs. (2.52) and (3.76). The substitutions for the submatrices in Eqs. (3.61) from Eqs. (2.52) and (3.76) yield the global stiffness submatrices

$$\mathbf{K}_{aa} = \mathbf{K}_{bb} = \gamma \begin{bmatrix} c_1^2 & c_1 c_2 \\ c_1 c_2 & c_2^2 \end{bmatrix} = -\mathbf{K}_{ab} = -\mathbf{K}_{ba} \tag{3.77}$$

Thus, the global element stiffness matrix is

$$\mathbf{K} = \gamma \begin{bmatrix} c_1^2 & c_1 c_2 & \vdots & -c_1^2 & -c_1 c_2 \\ & c_2^2 & \vdots & -c_1 c_2 & -c_2^2 \\ \cdots & \cdots & \cdots & \cdots & \cdots \\ & & \vdots & c_1^2 & c_1 c_2 \\ \text{sym.} & & \vdots & & c_2^2 \end{bmatrix}, \quad \gamma = \frac{EA}{L} \tag{3.78}$$

Example 1. The truss in Figure 3.8a is subjected to the joint loads $Q_1 = 0, Q_2 = -5 \text{ kN}$. The elements have identical extensional stiffness. The element rotation matrices are defined in Eqs. (2.58). The analysis follows the flow chart in Figure 3.2. The analysis units are kilonewton (kN), meter (m), and radian (rad).

1. *Unknowns*

$$q_k, \quad k = 1, 2 \tag{3.79}$$

2. *Element Models*

$$\mathbf{f}^i = \mathbf{k}\mathbf{d}^i, \quad i = 1, 2 \tag{3.80}$$

k is defined in Eq. 3.76

3. *System Model.*
It follows from Figures 3.8a, b and Eq. (2.5) that

$$\mathbf{M}^T = \begin{bmatrix} 0 & 0 & 1 & 2 \\ 0 & 0 & 1 & 2 \end{bmatrix} \tag{3.81}$$

Stiffness matrix transformations (Section 3.2): Eqs. (2.52), (2.58), (3.78), and (3.81) yield

$$\mathbf{K}^1 = \frac{\gamma}{2} \begin{bmatrix} & & & \vdots & 0 \\ & & & \vdots & 0 \\ & & 1 & & 1 \\ \text{sym.} & & & & 1 \end{bmatrix} \begin{matrix} 0 \\ 0 \\ 1 \\ 2 \end{matrix} \xrightarrow{\text{M}} \mathbf{K}^{(1)} = \frac{\gamma}{2} \begin{bmatrix} 1 & 2 \\ 1 & 1 \\ \text{sym.} & 1 \end{bmatrix} \begin{matrix} 1 \\ 2 \end{matrix} \tag{3.82a}$$

because for element 1 $c_1 = c_2 = 1/\sqrt{2}$, and

$$\mathbf{K}^2 = \gamma \begin{bmatrix} & & & \vdots & 0 \\ & & & \vdots & 0 \\ & & 1 & & 0 \\ \text{sym.} & & & & 0 \end{bmatrix} \begin{matrix} 0 \\ 0 \\ 1 \\ 2 \end{matrix} \xrightarrow{\text{M}} \mathbf{K}^{(2)} = \frac{\gamma}{2} \begin{bmatrix} 1 & 2 \\ 2 & 0 \\ \text{sym.} & 0 \end{bmatrix} \begin{matrix} 1 \\ 2 \end{matrix} \tag{3.82b}$$

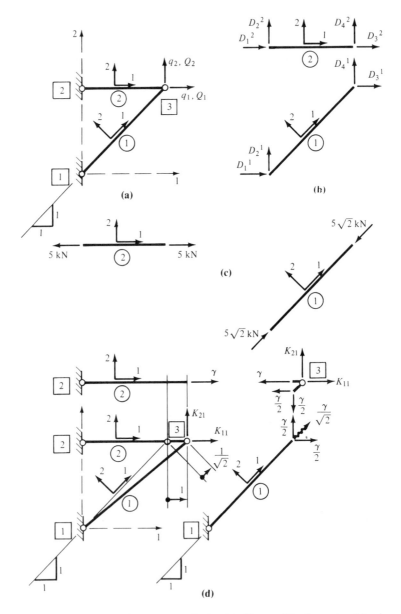

FIGURE 3.8 Matrix displacement analysis. **(a)** Truss; **(b)** elements; **(c)** element forces; **(d)** verification of K_{11}, K_{21}.

because for element 2 $c_1 = 1$, $c_2 = 0$. Hence,

$$\mathbf{K} = \sum_{i=1}^{2} \mathbf{K}^{(i)} = \frac{\gamma}{2}\begin{bmatrix} 3 & 1 \\ \text{sym.} & 1 \end{bmatrix} \tag{3.83}$$

and

$$\frac{\gamma}{2}\begin{bmatrix} 3 & 1 \\ 1 & 1 \end{bmatrix}\begin{bmatrix} q_1 \\ q_2 \end{bmatrix} = \begin{bmatrix} 0 \\ -5 \end{bmatrix} \tag{3.84}$$

To check by Eq. (3.35) the coefficients of the first column of the system stiffness matrix, we let $q_1 = 1$, $q_2 = 0$ as shown in Figure 3.8d. The resulting local element forces $f_b^1 = \gamma/\sqrt{2}$, $f_b^2 = \gamma$ follow from Eq. (3.76) for $d_a^1 = 0$, $d_b^1 = 1/\sqrt{2}$ and $d_a^2 = 0$, $d_b^2 = 1$. The projections of these forces onto the global axes are the global element forces shown. (Alternatively, the global element forces can be obtained directly from Eq. (3.19) with $\mathbf{K}^i$ defined by Eqs. (2.58), (3.78) and $\mathbf{D}^{1T} = \mathbf{D}^{2T} = \begin{bmatrix} 0 & 0 & 1 & 0 \end{bmatrix}$.) The conditions of joint equilibrium yield $K_{11} = 3\gamma/2$, $K_{21} = \gamma/2$.

These computations illuminate a contradiction of linear models (Section 1.5): The local element forces are determined from the deformed state, but equilibrium is formulated for the undeformed state. For example, by Eqs. (2.54) and (2.58)

$$\mathbf{F}_b^1 = \boldsymbol{\lambda}^{1T} f_b^1 = \frac{1}{\sqrt{2}}\begin{bmatrix} 1 \\ 1 \end{bmatrix}\frac{\gamma}{\sqrt{2}} = \frac{\gamma}{2}\begin{bmatrix} 1 \\ 1 \end{bmatrix} \tag{3.85}$$

In Eq. (3.85), $\boldsymbol{\lambda}^1$ is based on the initial orientation of element 1. Fortunately, the changes in configurations are usually so small that the resulting errors are negligible.

4. *Solution*

$$q_1 = \frac{5}{\gamma}, \quad q_2 = -\frac{15}{\gamma} \tag{3.86}$$

5. *Element Forces*
According to Eq. (3.81)

$$\mathbf{D}^1 = \mathbf{D}^2 = \begin{bmatrix} 0 & 0 \\ 0 & 0 \\ q_1 & 1 \\ q_2 & 2 \end{bmatrix} \begin{array}{l} \} \mathbf{D}_a \\ \} \mathbf{D}_b \end{array} \tag{3.87}$$

Eqs. (2.51), (2.58), (3.86), and (3.87) yield

$$\mathbf{d}_b^1 = \boldsymbol{\Lambda}^1 \mathbf{D}^1 = \frac{5\sqrt{2}}{\gamma}\begin{bmatrix} 0 \\ -1 \end{bmatrix}, \quad \mathbf{d}^2 = \boldsymbol{\Lambda}^2 \mathbf{D}^2 = \frac{5}{\gamma}\begin{bmatrix} 0 \\ 1 \end{bmatrix} \tag{3.88}$$

$$= \frac{1}{\sqrt{2}}\begin{bmatrix} 1 & 1 \end{bmatrix} \frac{5}{\gamma}\begin{bmatrix} 1 \\ -3 \end{bmatrix}$$

$$= \frac{-10}{\sqrt{2}\gamma} = -\frac{5\sqrt{2}}{\gamma}$$

The substitutions of Eqs. (3.88) into Eq. (3.80) result in the element forces (Figure 3.8c)

$$\mathbf{f}^1 = \begin{bmatrix} f_a^1 \\ f_b^1 \end{bmatrix} = \begin{bmatrix} 5\sqrt{2} \\ \hline -5\sqrt{2} \end{bmatrix}, \quad \mathbf{f}^2 = \begin{bmatrix} f_a^2 \\ f_b^2 \end{bmatrix} = \begin{bmatrix} -5 \\ \hline 5 \end{bmatrix} \tag{3.89}$$

6. *Joint Forces*

The joint forces are computed in Example 3 of Section 2.6.

3.6 ELEMENT ACTIONS

In the presence of element loads, the joint displacements no longer represent generalized displacements of the assemblage; that is, the joint displacements are not sufficient to define the configurations of elements subjected to loads.

However, the joint displacements, the element-end forces, and the joint forces can be computed by a procedure analogous to that used in the slope-deflection method (Appendix B). It includes load resolution, analysis of an equivalent joint load system, and response superposition.

Once the joint displacements and the element-end forces are known, we can solve the appropriate continuum models (Section 1.5) for the configurations and other response measures of elements subjected to loads.

Resolution of Actions

The load resolution used in the slope–deflection method to deal with element loads (Appendix B) is extended to incorporate element actions in the matrix displacement analysis. Element actions may consist of loads, temperature changes, element imperfections, prescribed displacements (such as support settlements), and the like. Specifically, the actions of the structure are transformed into equivalent joint loads by the resolution[2]

$$\overline{\mathscr{L}} = \hat{\mathscr{L}} + \mathscr{L} \tag{3.90}$$

where $\overline{\mathscr{L}}$ are the actions of the structure—they may consist of element and joint actions; $\hat{\mathscr{L}}$ are fixed-joint actions—they are composed of the *element actions* and *joint loads required to prevent joint displacements*; and $\mathscr{L}$ are equivalent joint loads. A specific load resolution is discussed in detail in Appendix B and illustrated in Figures B.3a–c.

Note that $\mathscr{L}$ contains only joint loads because the element actions are introduced in $\hat{\mathscr{L}}$ and by Eq. (3.90),

$$\mathscr{L} = \overline{\mathscr{L}} - \hat{\mathscr{L}} \tag{3.91}$$

---

[2] The principle of superposition is valid because our mathematical models are linear (Sections 1.5 and 1.6).

subtracted from $\bar{\mathscr{L}}$ to produce $\mathscr{L}$. Thus, the response to $\mathscr{L}$ can be determined by the matrix displacement method. In contrast, the elements of the structure subjected to $\hat{\mathscr{L}}$ can be analyzed independently because the fixed joints prevent their interaction.

In matrix form, the equivalent joint loads can be expressed as

$$\mathbf{Q} = \bar{\mathbf{Q}} - \hat{\mathbf{Q}} \qquad (3.92)$$

where $\bar{\mathbf{Q}}$ is the actual joint load vector, $\hat{\mathbf{Q}}$ the fixed-joint load vector, and $\mathbf{Q}$ the equivalent joint load vector. $\bar{\mathbf{Q}}$ is prescribed and $\hat{\mathbf{Q}}$ can be computed as follows:

1. For each element subjected to actions, the local fixed-end force vector, $\hat{\mathbf{f}}^i$, is computed and transformed into the global fixed-end force vector, $\hat{\mathbf{F}}^i$.
2. The generalized fixed-end force vector, $\hat{\mathbf{F}}^{(i)}$, is determined by the force transformation (Section 2.4)

$$\hat{\mathbf{F}}^i \overset{\mathbf{M}}{\to} \hat{\mathbf{F}}^{(i)} \qquad (3.93)$$

3. The condition of equilibrium, Eq. (2.13), is imposed to yield

$$\hat{\mathbf{Q}} = \sum_{i=1}^{NE} \hat{\mathbf{F}}^{(i)} \qquad (3.94)$$

Once the responses to $\hat{\mathscr{L}}$ and $\mathscr{L}$ are determined, they are superposed, Eq. (3.90), to yield the actual response. In particular, the actual element force vectors can be expressed, analogous to Eqs. (B.1a), as

$$\bar{\mathbf{f}}^i = \mathbf{f}^i + \hat{\mathbf{f}}^i$$

where   (3.95)

$$\mathbf{f}^i = \mathbf{k}^i \mathbf{d}^i$$

Then the actual joint force vector, $\bar{\mathbf{P}}_j$, can be computed by Eq. (2.29).

Note that the joint displacements corresponding to $\bar{\mathscr{L}}$ and $\mathscr{L}$ are identical because the joint displacements corresponding to $\hat{\mathscr{L}}$ are zero by definition; that is,

$$\bar{\mathbf{q}} = \mathbf{q} + \hat{\mathbf{q}} = \mathbf{q} \qquad (3.96)$$

because $\hat{\mathbf{q}} = \mathbf{0}$.

Matrix Analysis

The complete analysis process of structures with element actions is defined in Figure 3.9. It is illustrated next for structures with element loads. Other element actions are considered in Chapter 4.

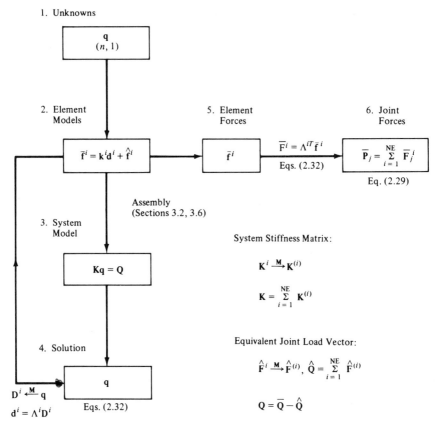

FIGURE 3.9 Matrix displacement method with element actions (n is the number of system degrees of freedom, m the number of element degrees of freedom, and NE the number of elements).

Example 1. To demonstrate once more the similarity between the slope-deflection method and the matrix displacement method, the continuous beams in Figure B.3a are analyzed on the basis of Figure 3.9. Observe that $\overline{\mathbf{Q}} = \mathbf{0}$. The analysis units are kilonewton (kN), meter (m), and radian (rad).

1. *Unknowns*

$$q_k, \quad k = 1, 2 \tag{3.97}$$

2. *Element Models*

$$\overline{\mathbf{f}}^i = \mathbf{f}^i + \hat{\mathbf{f}}^i, \quad i = 1, 2 \tag{3.98}$$

where

$$\mathbf{f}^i = \mathbf{k}\mathbf{d}^i$$

and according to Figure B.3d, the fixed-end force vectors are

$$\hat{\mathbf{f}}^1 = \begin{bmatrix} 11.2 \\ 14.0 \\ 11.2 \\ -14.0 \end{bmatrix}\begin{matrix} 0 \\ 0 \\ 0 \\ 1 \end{matrix}, \quad \hat{\mathbf{f}}^2 = \mathbf{0} \tag{3.99}$$

$\mathbf{M}$ is defined in Eq. (3.3)

3. System Model

System Stiffness Matrix
The system stiffness matrix is defined in Eq. (3.15).

Equivalent Joint Load Vector

$\bar{\mathscr{L}}$:

$$\bar{\mathbf{Q}} = \mathbf{0} \tag{3.100}$$

$\hat{\mathscr{L}}$:   By Eqs. (3.3) and (3.99)

$$\hat{\mathbf{f}}^1 \overset{\mathbf{M}}{\to} \hat{\mathbf{F}}^{(1)} = \begin{bmatrix} -14.0 \\ 0.0 \end{bmatrix}\begin{matrix} 1 \\ 2 \end{matrix}, \quad \hat{\mathbf{f}}^2 \overset{\mathbf{M}}{\to} \hat{\mathbf{F}}^{(2)} = \mathbf{0} \tag{3.101}$$

Thus,

$$\hat{\mathbf{Q}} = \sum_{i=1}^{2} \hat{\mathbf{F}}^{(i)} = \begin{bmatrix} -14.0 \\ 0.0 \end{bmatrix} \tag{3.102}$$

which agrees with Figure B.3b.

$\mathscr{L}$:

$$\mathbf{Q} = \bar{\mathbf{Q}} - \hat{\mathbf{Q}} = \begin{bmatrix} 14.0 \\ 0.0 \end{bmatrix} \tag{3.103}$$

which agrees with Figure B.3c. The matrix displacement analysis of Figure B.3c is presented in the introductory problem of Section 3.2. Hence, we can immediately compute the actual element forces.

5. Element Forces
By Eqs. (3.17), (3.98), and (3.99)

$$\bar{\mathbf{f}}^1 = \mathbf{f}^1 + \hat{\mathbf{f}}^1 = \begin{bmatrix} 13.6 \\ 18.0 \\ \hline 8.8 \\ -6.0 \end{bmatrix}, \quad \bar{\mathbf{f}}^2 = \mathbf{f}^2 = \begin{bmatrix} 1.2 \\ 6.0 \\ \hline -1.2 \\ 0.0 \end{bmatrix} \tag{3.104}$$

6. *Joint forces*

By Eqs. (2.29) and (3.104)

$$\bar{P}_1 = \bar{f}_a^1 = \begin{bmatrix} 13.6 \\ 18.0 \end{bmatrix}, \quad \bar{P}_3 = \bar{f}_b^2 = \begin{bmatrix} -1.2 \\ 0.0 \end{bmatrix} \tag{3.105}$$

$$\bar{P}_2 = \bar{f}_b^1 + \bar{f}_a^2 = \begin{bmatrix} 10.0 \\ 0.0 \end{bmatrix} \tag{3.106}$$

The actual element and joint forces are shown in Figure B.3e.

Example 2. The properties of the frame in Figure 3.10a are identical to those of the frame in Figure 3.7a. However, the frame is subjected to the joint loads $\bar{Q}_1 = \bar{Q}_2 = 0$, $\bar{Q}_3 = 14\,\text{k}\cdot\text{ft}$ and the uniformly distributed load on element 2. The analysis is based on the flow chart in Figure 3.9. The analysis units are kilopound (k), foot (ft), and radian (rad).

1. *Unknowns*

$$q_k, \quad k = 1, 2, 3 \tag{3.107}$$

2. *Element Models*

$$\bar{f}^i = f^i + \hat{f}^i, \quad i = 1, 2 \tag{3.108}$$

where

$$f^i = kd^i$$

k is defined in Eq. (3.59), and according to Figures 3.10b, d

$$\hat{f}^1 = 0, \quad \hat{f}^2 = \begin{bmatrix} 0.0 \\ -8.4 \\ -14.0 \\ 0.0 \\ -8.4 \\ 14.0 \end{bmatrix} \begin{matrix} 1 \\ 0 \\ 2 \\ 0 \\ 0 \\ 3 \end{matrix} \tag{3.109}$$

M is defined in Eq. (3.67).

3. *System Model*

System Stiffness Matrix

The system stiffness matrix is defined in Eq. (3.69).

Equivalent Joint Load Vector

$\bar{\mathscr{L}}$:

$$\bar{Q} = \begin{bmatrix} 0.0 \\ 0.0 \\ 14.0 \end{bmatrix} \tag{3.110}$$

$\hat{\mathscr{L}}$: By Eqs. (3.67), and (3.109)

$$\hat{F}^{(1)} = 0, \quad \hat{F}^2 = \hat{f}^2 \xrightarrow{M} \hat{F}^{(2)} = \begin{bmatrix} 0.0 \\ -14.0 \\ 14.0 \end{bmatrix} \begin{matrix} 1 \\ 2 \\ 3 \end{matrix} \tag{3.111}$$

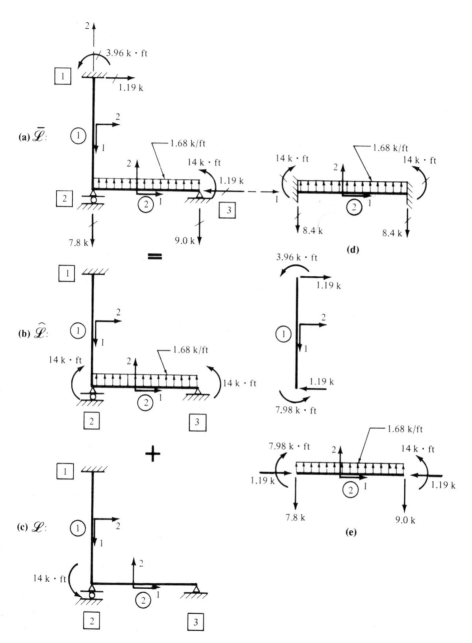

FIGURE 3.10 Matrix displacement analysis (Figure 3.9). **(a)** $\overline{\mathscr{L}}$; **(b)** $\widehat{\mathscr{L}}$; **(c)** $\mathscr{L}$; **(d)** fixed-end forces; **(e)** element forces.

Thus,

$$\hat{\mathbf{Q}} = \sum_{i=1}^{2} \hat{\mathbf{F}}^{(i)} = \begin{bmatrix} 0.0 \\ -14.0 \\ 14.0 \end{bmatrix} \qquad (3.112)$$

$\mathscr{L}$:

$$\mathbf{Q} = \bar{\mathbf{Q}} - \hat{\mathbf{Q}} = \begin{bmatrix} 0.0 \\ 14.0 \\ 0.0 \end{bmatrix} \qquad (3.113)$$

The resolution $\overline{\mathscr{L}} = \hat{\mathscr{L}} + \mathscr{L}$ is shown in Figures 3.10a–c. The analysis of Figure 3.10c is carried out in Example 1 of Section 3.4. Hence, we can proceed with the computation of the actual element forces.

5. *Element Forces*
By Eqs. (3.73), (3.108), and (3.109)

$$\bar{\mathbf{f}}^1 = \mathbf{f}^1 = \begin{bmatrix} 0.00 \\ 1.19 \\ 3.96 \\ \hline 0.00 \\ -1.19 \\ 7.98 \end{bmatrix}, \quad \bar{\mathbf{f}}^2 = \mathbf{f}^2 + \hat{\mathbf{f}}^2 = \begin{bmatrix} 1.19 \\ -7.80 \\ -7.98 \\ \hline -1.19 \\ -9.00 \\ 14.00 \end{bmatrix} \qquad (3.114)$$

the actual element forces are shown in Figure 3.10e.

6. *Joint Forces*
By Eqs. (2.29), (2.37), (2.47), and (3.114)

$$\bar{\mathbf{P}}_1 = \bar{\mathbf{F}}_a^1 = \boldsymbol{\lambda}^{1T}\bar{\mathbf{f}}_a^1 = \begin{bmatrix} 1.19 \\ 0.00 \\ 3.96 \end{bmatrix}, \quad \bar{\mathbf{P}}_3 = \bar{\mathbf{f}}_b^2 = \begin{bmatrix} -1.19 \\ -9.00 \\ 14.00 \end{bmatrix}$$

$$\bar{\mathbf{P}}_2 = \bar{\mathbf{F}}_b^1 + \bar{\mathbf{F}}_a^2 = \boldsymbol{\lambda}^{1T}\bar{\mathbf{f}}_b^1 + \bar{\mathbf{f}}_a^2 = \begin{bmatrix} 0.00 \\ -7.80 \\ 0.00 \end{bmatrix} \qquad (3.115)$$

The actual joint forces are shown in Figure 3.10a.

Element Response

The design of a structure is based on specific criteria that restrict the response to acceptable limits–for example, stress and displacement limits. Accordingly, we must be able to compute the measures of response used in the application of design criteria.

If the response computations of elements subjected to loads are to be automated (Chapter 7), we cannot rely on semigraphical procedures such as those used in the construction of shear and bending moment diagrams. Instead, we must use mathematical techniques that are suitable for computer implementation.

If the loads are continuous, we can integrate the differential load–response relations (Section 1.5) and express the responses by explicit functions. For discontinuous loads, such as concentrated loads and distributed loads with suddenly changing magnitudes, we can select a numerical integration procedure (Fenves, 1967; Godden, 1965) or use singularity functions specifically designed for this purpose (Crandall et al., 1978; Budynas, 1977; Pilkey and Pilkey, 1974; Shames, 1975).

Singularity functions are defined and illustrated in Appendix A. Here we present a simple illustration of their effectiveness.

Example 3. The beam ab of Figure B.3e (element 1) is reproduced in Figure 3.11a. The properties are $E = 200 \times 10^6 \text{ kN/m}^2$, $I = 20 \times 10^{-6} \text{ m}^4$, $EI = 4000 \text{ kN} \cdot \text{m}^2$; the cross section is doubly symmetric and the distance from the centroid to the extreme fiber is $c = 0.1 \text{ m}$. The shear force, $\overline{V}$, the bending moment, $\overline{M}$, the slope, $\overline{\theta}$, and the deflection, $\overline{v}$, are expressed as functions of x, and the extreme values of the normal stress and the deflection are computed. The analysis units are kilonewton (kN), meter (m), and radian (rad).

It follows from Figure 3.11a and the beam sign convention in Figure 1.8b that the boundary conditions at the a end of the beam are

$$\overline{V}_a = 13.6, \quad \overline{M}_a = -18.0, \quad \overline{\theta}_a = 0, \quad \overline{v}_a = 0 \qquad (3.116)$$

By means of the Dirac delta function, Eq. (A.36), the load can be expressed as

$$p(x) = -22.4\langle x - 2.5 \rangle^{-1} \qquad (3.117)$$

On the basis of Eqs. (3.116), (3.117), (A.38), (A.42), we obtain the response functions

$$\overline{V}(x) = \overline{V}_a + \int_0^x p(x)\,dx = 13.6 - 22.4\langle x - 2.5 \rangle^0 \qquad (3.118a)$$

$$\overline{M}(x) = \overline{M}_a + \int_0^x \overline{V}(x)\,dx = -18.0 + 13.6x - 22.4\langle x - 2.5 \rangle^1 \qquad (3.118b)$$

$$\overline{\theta}(x) = \overline{\theta}_a + \frac{1}{EI}\int_0^x \overline{M}(x)\,dx$$

$$\overline{\theta}(x) = \frac{1}{EI}(-18.0x + 6.8x^2 - 11.2\langle x - 2.5 \rangle^2) \qquad (3.118c)$$

$$\overline{v}(x) = \overline{v}_a + \int_0^x \overline{\theta}(x)\,dx$$

$$\overline{v}(x) = \frac{1}{EI}\left(-9.0x^2 + \frac{6.8}{3}x^3 - \frac{11.2}{3}\langle x - 2.5 \rangle^3\right) \qquad (3.118d)$$

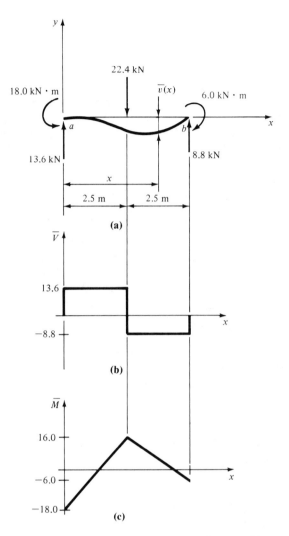

FIGURE 3.11 Element response. **(a)** Beam; **(b)** shear diagram; **(c)** moment diagram.

According to Eqs. (3.118a, b)

$$\overline{V}(x) = \begin{cases} 13.6, & 0 < x < 2.5 \\ -8.8, & 2.5 < x < 5.0 \end{cases} \qquad (3.119a)$$

and

$$\overline{M}(x) = \begin{cases} -18.0 + 13.6x, & 0 < x \le 2.5 \\ 38.0 - 8.8x, & 2.5 \le x < 5.0 \end{cases} \qquad (3.119b)$$

The resulting shear and bending moment diagrams are shown in Figure 3.11b and c. The extreme bending moment is $\overline{M}_a$. Thus,

$$\max \sigma = \frac{|\overline{M}_a|c}{I} = 90.0 \times 10^3 \text{ kN/m}^2 \tag{3.120}$$

The extreme value of the deflection occurs at $x = 2.66$ m because, by Eq. (3.118c), $\overline{\theta}(2.5) < 0$ and $\overline{\theta}(2.66) = 0$. Specifically, by Eq. (3.118d)

$$\min \overline{v} = \overline{v}(2.66) = -0.0053 \text{ m} = -5.3 \text{ mm} \tag{3.121}$$

3.7 FINITE ELEMENT FORMULATION

Finite element and system models are formulated by the principle of virtual work for skeletal structures with joint loads and element actions.

Element Models

The formulation of linearly elastic finite element models with element actions is described relative to the flexural deformation element in Figure 1.21c. The element actions consist of the element load, $p(x)$, shown in Figure 1.8, and the thermal strain, $\varepsilon_t(x, y)$, caused in the unconstrained element by temperature changes (Figure 3.12 and Section 4.8).

The formulation of the finite element model with element actions follows the procedure introduced in Section 1.9 for elements without actions:

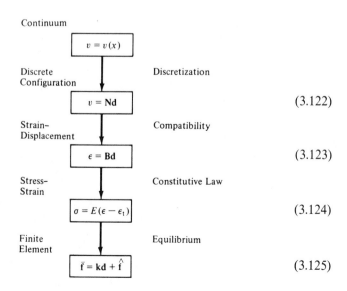

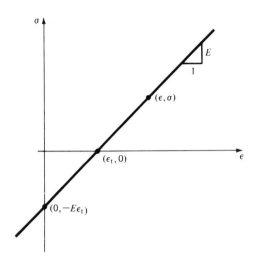

FIGURE 3.12 Stress–strain diagram.

where

$$\mathbf{f} = \mathbf{kd} \tag{3.126}$$

$$\mathbf{k} = \iiint_V \mathbf{B}^T E \mathbf{B} \, dV \tag{3.127}$$

$$\hat{\mathbf{f}} = \hat{\mathbf{f}}_p + \hat{\mathbf{f}}_t \tag{3.128}$$

$$\hat{\mathbf{f}}_p = -\int_0^L \mathbf{N}^T p \, dx \tag{3.129}$$

$$\hat{\mathbf{f}}_t = -\iiint_V \mathbf{B}^T E \varepsilon_t \, dV \tag{3.130}$$

Equations (3.125)–(3.130) indicate that in the absence of element actions

$$\bar{\mathbf{f}} = \mathbf{f} = \mathbf{kd} \tag{3.131}$$

and in the absence of nodal displacements

$$\bar{\mathbf{f}} = \hat{\mathbf{f}} \tag{3.132}$$

Thus, $\mathbf{f}$ is the nodal force vector caused by nodal displacements, $\hat{\mathbf{f}}$ is the fixed-end force vector, and $\bar{\mathbf{f}}$ is the total nodal force vector. This agrees with the formulation of Section 3.6.

PROOF. By Eq. (1.169), the finite element is in a configuration of equilibrium if

$$\delta W = \delta W_e - \delta U = 0 \tag{3.133}$$

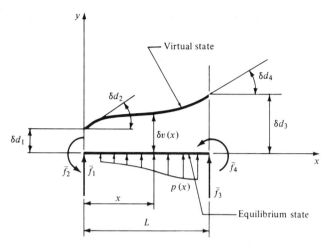

FIGURE 3.13 Beam element

for any virtual displacement. According to Figure 3.13, the external virtual work is

$$\delta W_e = \delta \mathbf{d}^T \bar{\mathbf{f}} + \int_0^L \delta v^T(x) p(x)\, dx \qquad (3.134)$$

By Eq. (3.122), $\delta v^T = \delta \mathbf{d}^T \mathbf{N}^T$; hence,

$$\delta W_e = \delta \mathbf{d}^T \bar{\mathbf{f}} + \delta \mathbf{d}^T \int_0^L \mathbf{N}^T p\, dx \qquad (3.135)$$

or

$$\delta W_e = \delta \mathbf{d}^T (\bar{\mathbf{f}} - \hat{\mathbf{f}}_p) \qquad (3.136)$$

where $\hat{\mathbf{f}}_p$ is defined in Eq. (3.129). From Eqs. (1.171), (3.123), and (3.124), we obtain

$$\delta U = \iiint_V \delta \varepsilon^T \sigma\, dV = \delta \mathbf{d}^T \iiint_V \mathbf{B}^T E(\varepsilon - \varepsilon_t)\, dV \qquad (3.137)$$

$$\delta U = \delta \mathbf{d}^T \left[\left(\iiint_V \mathbf{B}^T E \mathbf{B}\, dV \right) \mathbf{d} - \iiint_V \mathbf{B}^T E \varepsilon_t\, dV \right] \qquad (3.138)$$

or

$$\delta U = \delta \mathbf{d}^T (\mathbf{k}\mathbf{d} + \hat{\mathbf{f}}_t) \qquad (3.139)$$

where $\mathbf{k}$ and $\hat{\mathbf{f}}_t$ are defined in Eqs. (3.127) and (3.130). Equations (3.133), (3.136), and (3.139) yield

$$\delta W = \delta \mathbf{d}^T(\bar{\mathbf{f}} - \mathbf{k}\mathbf{d} - \hat{\mathbf{f}}_p - \hat{\mathbf{f}}_t) = 0 \tag{3.140}$$

Thus,

$$\bar{\mathbf{f}} = \mathbf{k}\mathbf{d} + \hat{\mathbf{f}}_p + \hat{\mathbf{f}}_t \tag{3.141}$$

which was to be shown.   □

The formulation of element stiffness matrices by Eq. (3.127) is illustrated in Section 1.9. The formulation of fixed-end forces by Eqs. (3.129) and (3.130) is illustrated next.

Example 1. The beam in Figure 3.14a is subjected to the uniformly distributed load $p(x) = -p_0$. It follows from Eq. (3.129) that the jth nodal force

$$\hat{f}_{pj} = -\int_0^L N_j p \, dx = p_0 L \int_0^1 N_j(\xi) \, d\xi \tag{3.142}$$

where $\xi = x/L$. If we select the prismatic beam functions in Eqs. (1.183) as interpolation functions, we obtain, for example,

$$\hat{f}_{p1} = p_0 L \int_0^1 N_1(\xi) \, d\xi = \tfrac{1}{2}p_0 L \tag{3.143}$$

FIGURE 3.14 Fixed-end forces

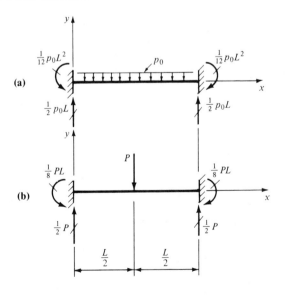

Similarly,

$$\hat{f}_{p2} = \tfrac{1}{12}p_0 L^2, \quad \hat{f}_{p3} = \tfrac{1}{2}p_0 L, \quad \hat{f}_{p4} = -\tfrac{1}{12}p_0 L^2 \tag{3.144}$$

The fixed-end forces (Figure 3.14a) are identical to those of the elementary beam theory.

Example 2. To compute the fixed-end forces caused by the concentrated load in Figure 3.14b, we express the load by the Dirac delta function (Appendix A) as

$$p(x) = -P\left\langle x - \frac{L}{2} \right\rangle^{-1} \tag{3.145}$$

Thus,

$$\int_0^L p(x)\, dx = -P \int_0^L \left\langle x - \frac{L}{2} \right\rangle^{-1} dx = -P \tag{3.146}$$

By Eqs. (3.129) and (3.145)

$$\hat{f}_{pj} = P \int_0^L N_j(x) \left\langle x - \frac{L}{2} \right\rangle^{-1} dx \tag{3.147a}$$

or

$$\hat{f}_{pj} = P \int_0^1 N_j(\xi) \langle \xi - \tfrac{1}{2} \rangle^{-1} d\xi = P N_j(\tfrac{1}{2}) \tag{3.147b}$$

Equations (1.183) and (3.147) yield the fixed-end forces (Figure 3.14b)

$$\hat{f}_{p1} = \tfrac{1}{2}P, \quad \hat{f}_{p2} = \tfrac{1}{8}PL, \quad \hat{f}_{p3} = \tfrac{1}{2}P, \quad \hat{f}_{p4} = -\tfrac{1}{8}PL \tag{3.148}$$

which agree with the elementary beam theory.

The fixed-end forces in Examples 1 and 2 are identical to those resulting from the exact solution of the nonhomogeneous differential equation, Eq. (1.53), even though the interpolation functions in Examples 1 and 2 represent only the homogeneous solution to Eq. (1.53). This surprising result can be explained by Betti's law.

Consider the fixed-end beam in Figure 3.15a with the load $p(x)$. Figure 3.15b depicts the configuration $N_1(x)$ that satisfies the homogeneous differential equation, Eq. (1.55a), and the boundary conditions $v(0) = 1$, $dv(0)/dx = 0$, $v(L) = 0$, $dv(L)/dx = 0$. The application of Betti's law to Figures 3.15a and b yields

$$\hat{f}_{p1} \cdot 1 + \int_0^L N_1(x)\, p(x)\, dx = 0 \tag{3.149a}$$

or

$$\hat{f}_{p1} = -\int_0^L N_1(x)\, p(x)\, dx \tag{3.149b}$$

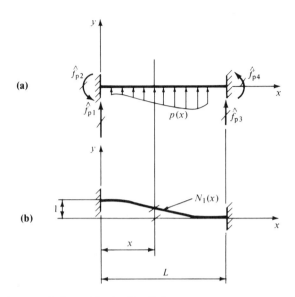

FIGURE 3.15 Formulation of $\mathbf{f}_{p1}$ by Betti's law

Similarly, by selecting appropriate configurations, we obtain

$$\hat{f}_{pj} = -\int_0^L N_j(x)\, p(x)\, dx, \quad j = 1, 2, 3, 4 \tag{3.150}$$

which is identical to Eq. (3.129). Accordingly, Eq. (3.129) yields the exact fixed-end forces if the interpolation functions satisfy the homogeneous differential equation.

Example 3. The concentrated axial load P in Figure 3.16 can be described, analogous to Example 2, by the Dirac delta function as

$$p(x) = -P\langle x - a \rangle^{-1} \tag{3.151}$$

FIGURE 3.16 Fixed-end forces

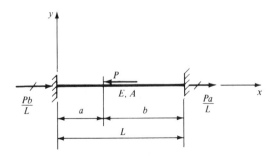

Thus, the fixed-end forces can be expressed as

$$\hat{f}_{pj} = -\int_0^L N_j p \, dx = P \int_0^1 N_j(\xi) \left\langle \xi - \frac{a}{L} \right\rangle^{-1} d\xi \qquad (3.152)$$

or

$$\hat{f}_{pj} = P N_j \left(\frac{a}{L} \right), \quad j = 1, 2 \qquad (3.153)$$

Equations (1.177) and (3.153) yield the exact fixed-end forces (Figure 3.16)

$$\hat{f}_{p1} = \frac{Pb}{L}, \quad \hat{f}_{p2} = \frac{Pa}{L} \qquad (3.154)$$

Example 4. The fixed-end forces induced by the temperature change (Figures 4.27a, b)

$$t(x, y) = t_0 + \frac{2}{h} \Delta t \, y \qquad (3.155)$$

are computed by Eq. (3.130). An alternative solution, by elementary tools, and physical interpretations are presented in Example 1 of Section 4.8. The resulting thermal strain is

$$\varepsilon_t = \varepsilon_0 + (-\phi_0 y) \qquad (3.156)$$

where

$$\varepsilon_0 = \alpha_t t_0, \quad \phi_0 = -\frac{2}{h} \alpha_t \Delta t \qquad (3.157)$$

Equation (3.156) indicates that the thermal strain consists of the axial strain, ε_0 and the flexural strain $-\phi_0 y$. It is convenient to compute the axial and flexural responses independently.

Axial Response (Figure 3.17b). By Eqs. (1.180) and (3.130)

$$\bar{f}_t = -\iiint_V \mathbf{B}^T E \varepsilon_0 \, dV = -\mathbf{B}^T E \varepsilon_0 AL = EA\varepsilon_0 \begin{bmatrix} 1 \\ -1 \end{bmatrix} \qquad (3.158)$$

Flexural Response (Figure 3.17c). By Eqs. (1.187), (1.188), and (3.130)

$$\hat{\mathbf{f}}_t = -\iiint_V \mathbf{B}^T E(-\phi_0 y) \, dA \, dx = -E\phi_0 \underbrace{\iint_A y^2 \, dA}_{I} \int_0^1 \frac{1}{L} \bar{\mathbf{B}}^T d\xi \qquad (3.159)$$

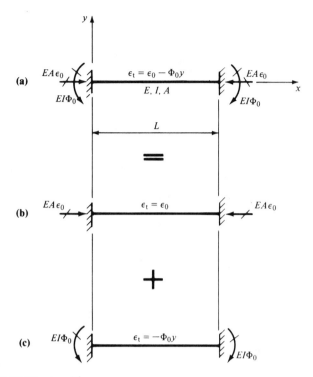

FIGURE 3.17 Fixed-end forces

or

$$\hat{\mathbf{f}}_t = -EI\phi_0 \int_0^1 \frac{1}{L}\,\bar{\mathbf{B}}^T\,d\xi = EI\phi_0 \begin{bmatrix} 0 \\ 1 \\ 0 \\ -1 \end{bmatrix} \tag{3.160}$$

The superposition of the axial and flexural responses yields the resultant fixed-end forces in Figure 3.17a.

System Model

The model of an assemblage of finite elements can be expressed as

$$\mathbf{Kq} = \mathbf{Q} \tag{3.161}$$

where

$$\mathbf{K} = \sum_{i=1}^{NE} \mathbf{K}^{(i)} \tag{3.162}$$

and

$$\mathbf{Q} = \bar{\mathbf{Q}} - \sum_{i=1}^{NE} \hat{\mathbf{F}}^{(i)} \tag{3.163}$$

It is identical to the system model formulated by the matrix displacement method [see Eqs. (3.23), (3.24), (3.92), and (3.94)].

PROOF. The proof of Eqs. (3.161)–(3.163) is presented with reference to the assemblage of beam finite elements in Figure 1.21. However, it is generally valid.

Instead of considering isolated elements, we apply the principle of virtual work to the assemblage of elements. For this purpose, we need to express the element configurations and the components of the model relative to the nodal displacements of the assemblage. Thus,

$$v^i = \mathbf{N}^{(i)}\mathbf{q} \tag{3.164}$$

$$\varepsilon^i = \mathbf{B}^{(i)}\mathbf{q} \tag{3.165}$$

$$\sigma^i = E^i(\varepsilon^i - \varepsilon_t^i) \tag{3.166}$$

where by Eqs. (1.162), (1.165), (2.32), and (2.63)

$$\mathbf{N}^{(i)} = \mathbf{N}^i \mathbf{\Lambda}^i \mathbf{A}^i, \quad \mathbf{B}^{(i)} = \mathbf{B}^i \mathbf{\Lambda}^i \mathbf{A}^i \tag{3.167}$$

The assemblage is in a configuration of equilibrium if

$$\delta W = \delta W_e - \delta U = 0 \tag{3.168}$$

for any virtual displacement. The virtual work of the external forces is

$$\delta W_e = \delta \mathbf{q}^T \overline{\mathbf{Q}} + \sum_{i=1}^{NE} \int_0^{L^i} \delta v^{iT} p^i \, dx \tag{3.169}$$

where

$$\int_0^{L^i} \delta v^{iT} p^i \, dx = \delta \mathbf{q}^T \int_0^{L^i} \mathbf{N}^{(i)T} p^i \, dx = -\delta \mathbf{q}^T \hat{\mathbf{F}}_p^{(i)} \tag{3.170}$$

Thus,

$$\delta W_e = \delta \mathbf{q}^T \left(\overline{\mathbf{Q}} - \sum_{i=1}^{NE} \hat{\mathbf{F}}_p^{(i)} \right) \tag{3.171}$$

where

$$\hat{\mathbf{F}}_p^{(i)} = -\int_0^{L^i} \mathbf{N}^{(i)T} p^i \, dx \tag{3.172}$$

The first variation of the strain energy is

$$\delta U = \sum_{i=1}^{NE} \delta U^i \tag{3.173}$$

where

$$\delta U^i = \iiint_{V^i} \delta \varepsilon^{iT} \sigma^i \, dV^i = \delta \mathbf{q}^T \left[\iiint_{V^i} \mathbf{B}^{(i)T} E^i (\mathbf{B}^{(i)}\mathbf{q} - \varepsilon_t^i) \, dV^i \right] \tag{3.174}$$

or

$$\delta U^i = \delta \mathbf{q}^T(\mathbf{K}^{(i)}\mathbf{q} + \hat{\mathbf{F}}_t^{(i)}) \tag{3.175}$$

where

$$\mathbf{K}^{(i)} = \iiint\limits_{V^i} \mathbf{B}^{(i)T} E^i \mathbf{B}^{(i)} \, dV^i \tag{3.176}$$

and

$$\hat{\mathbf{F}}_t^{(i)} = -\iiint\limits_{V^i} \mathbf{B}^{(i)T} E^i \varepsilon_t^i \, dV^i \tag{3.177}$$

From Eqs. (3.168), (3.171), and (3.175) we obtain

$$\delta W = \delta \mathbf{q}^T\left(\overline{\mathbf{Q}} - \sum_{i=1}^{NE}\hat{\mathbf{F}}^{(i)} - \sum_{i=1}^{NE}\mathbf{K}^{(i)}\mathbf{q}\right) = 0 \tag{3.178}$$

where

$$\hat{\mathbf{F}}^{(i)} = \hat{\mathbf{F}}_p^{(i)} + \hat{\mathbf{F}}_t^{(i)} \tag{3.179}$$

Thus,

$$\sum_{i=1}^{NE}\mathbf{K}^{(i)}\mathbf{q} = \overline{\mathbf{Q}} - \sum_{i=1}^{NE}\hat{\mathbf{F}}^{(i)} \tag{3.180}$$

which verifies Eqs. (3.161)–(3.163).   □

Strains and Stresses

It follows from Eqs. (3.123) and (3.124) that once the nodal displacements are determined by the finite element (matrix) displacement method (Figure 3.9), the strain and stress at any point of each finite element can be computed as

$$\varepsilon^i = \mathbf{B}^i \mathbf{d}^i \tag{3.181}$$

and

$$\sigma^i = \mathbf{S}^i \mathbf{d}^i + \sigma_t^i \tag{3.182}$$

where the stress matrix

$$\mathbf{S}^i = E^i \mathbf{B}^i \tag{3.183}$$

and

$$\sigma_t^i = -E^i \varepsilon_t^i \tag{3.184}$$

This formulation neglects the strain $\hat{\varepsilon}^i$ and the stress $E^i\hat{\varepsilon}^i$ associated with the fixed-node actions $\hat{\mathscr{L}}$ (Section 3.6). Specifically, according to Eq. (3.90), the actual strain is

$$\bar{\varepsilon}^i = \varepsilon^i + \hat{\varepsilon}^i \tag{3.185}$$

or

$$\bar{\varepsilon}^i = \mathbf{B}^i \mathbf{d}^i + \hat{\varepsilon}^i \tag{3.186}$$

The substitution of Eq. (3.186) into the actual stress relation [Eq. (4.235)]

$$\bar{\sigma}^i = E^i(\bar{\varepsilon}^i - \varepsilon_t^i) \tag{3.187}$$

yields

$$\bar{\sigma}^i = \mathbf{S}^i \mathbf{d}^i + \sigma_t^i + E^i \hat{\varepsilon}^i \tag{3.188}$$

If we neglect $\hat{\varepsilon}^i$, Eqs. (3.186) and (3.188) reduce to Eqs. (3.181) and (3.182), respectively.

This approximation, which may be regarded as a discretization error, is standard in finite elements that are *internally statically indeterminate*, that is, finite elements (such as plate elements) for which internal forces cannot be computed in terms of nodal forces and element loads solely by conditions of equilibrium.

Example 5. Equations (3.123) and (3.124) are used to compute the strains and stresses in the elements of Example 4 in Section 4.8.
By Eqs. (1.184), (1.186)–(1.188), (4.283), and (4.284)

$$\varepsilon^1 = \varepsilon^2 = -\phi_0 y\left(\frac{1}{2} - \frac{3x}{4L}\right) \tag{3.189}$$

Equations (3.124), (3.190), and (4.274) yield

$$\sigma^1 = -E\phi_0 y\left(\frac{1}{2} - \frac{3x}{4L}\right), \quad \sigma^2 = E\phi_0 y\left(\frac{1}{2} + \frac{3x}{4L}\right) \tag{3.190}$$

Equations (3.189) and (3.190) agree with Eqs. (4.289) and (4.288), respectively, because for the temperature distributions in Example 4 of Section 4.8, $\hat{\varepsilon}^i = 0$ (see Example 1 of Section 4.8).
The student may find it constructive to show that in Example 3 of Section 4.8, $\hat{\varepsilon}^2 \neq 0$ and, hence, $\bar{\varepsilon}^2 \neq \varepsilon^2$ and $\bar{\sigma}^2 \neq \sigma^2$.

PROBLEMS

3.1 Use the matrix displacement method to analyze the continuous beams on the basis of Figure 3.2. Check some of the system stiffness coefficients by Eq. (3.35). Draw the free-body diagrams of the elements and show the joint forces on each structure. It is convenient not to substitute for L until step 5. Accordingly, express the stiffness matrix for element 2 of Problem 3.1d as

$$\mathbf{k}^2 = \alpha \begin{bmatrix} 40.5 & 13.5L & -40.5 & 13.5L \\ & 6.0L^2 & -13.5L & 3.0L^2 \\ & & 40.5 & -13.5L \\ \text{sym.} & & & 6.0L^2 \end{bmatrix}, \quad \alpha = \frac{EI}{L^3}$$

Verify this matrix.

PROBLEM 3.1

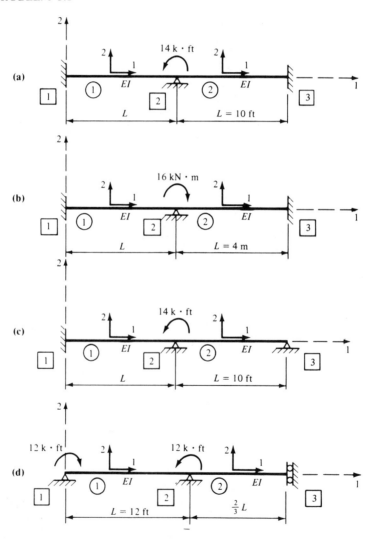

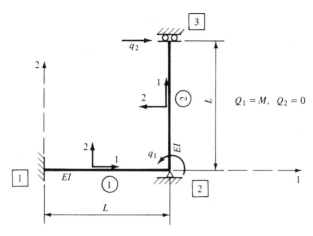

PROBLEM 3.2

3.2 Analyze the frame by the matrix displacement method (Figure 3.2). Verify the coefficients of the second column of the system stiffness matrix by Eq. (3.35). Draw the free-body diagrams of the elements and show the joint forces on the frame.

3.3 The elements of the frame have identical properties: E, I, A, L.
   a. Use the matrix displacement method to formulate the system stiffness matrix (steps 1–3 of Figure 3.2).
   b. Sketch the deformed state to which the joint force K_{43} is applied [see Eq. (3.35)]; use this state to compute $K^{\leftrightarrow}$. K_{43}

PROBLEM 3.3

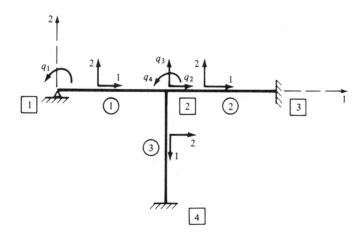

3.4   This problem tests your understanding of individual matrix operations introduced in Chapters 2 and 3:
   a. Compute the rotation matrix λ^1 and define the member code matrix $\mathbf{M}$.
   b. Compute the generalized force vectors $\mathbf{F}^{(1)}$, $\mathbf{F}^{(2)}$ if

$$\mathbf{f}^{1T} = [0 \quad -5 \quad -10 \quad 0 \quad 5 \quad -20]$$
$$\mathbf{f}^{2T} = [0 \quad -6 \quad 6 \quad 0 \quad 6 \quad -36]$$

   c. On the basis of (b), compute the *applied* joint force vector $\mathbf{Q}$, and draw the free-body diagram of joint 2.
   d. Formulate the generalized stiffness matrix of element 1, $\mathbf{K}^{(1)}$; the element properties are E, I, A, L.

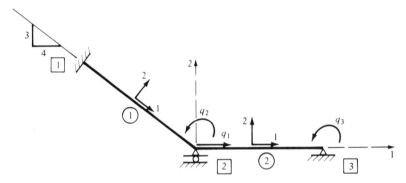

PROBLEM 3.4

3.5   Analyze the truss of Problem 2.10 by the matrix displacement method (Figure 3.2). The elements have the same extensional stiffness γ. Draw the free-body diagrams of the elements and show the joint forces on the truss.

3.6   The elements of the truss have the same properties: A, E, L. Analyze the truss by the matrix displacement method (Figure 3.2). Draw the free-body diagrams of the elements and show the joint forces on the truss.

PROBLEM 3.6

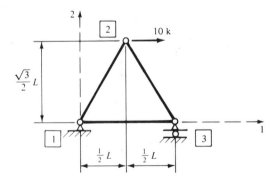

3.7   The continuous beams of Problems 3.1a–d are subjected to the following element loads; there are no joint loads. *Problems 3.1a and c*: The uniformly distributed load $p(x) = -1.68$ k/ft (see sign convention in Figure 1.8b) is applied to element 1. *Problem 3.1b*: The uniformly distributed load $p(x) = -12$ kN/m is applied to element 2. *Problem 3.1d*: The uniformly distributed load $p(x) = -1$ k/ft is applied to element 1. Perform the matrix displacement analysis (Figure 3.9): Sketch the component load systems $\hat{\mathscr{L}}$ and $\mathscr{L}$. Note that the equivalent joint loads are the joint loads of Problems 3.1a–d. Hence, you can use the solution for $\mathbf{f}^i$ from Problems 3.1a–d. Show the actual element forces in free-body diagrams of elements; show the actual joint forces on the continuous beams.

3.8   Element 1 of the frame in Problem 3.2 is subjected to a concentrated load at the midspan of value 8 k that acts in the negative direction of the local 2 axis; no joint forces are applied. $L = 10$ ft.
   a. Perform the matrix displacement analysis (Figure 3.9): Sketch the component load systems $\hat{\mathscr{L}}$ and $\mathscr{L}$; note that $\mathbf{f}^i$ can be obtained from Problem 3.2 by substituting for M and L. Show the actual element forces on element free-body diagrams and the actual joint forces on the frame.
   b. Check the element-end moments by the slope–deflection method.

3.9   The elements of the frame are prismatic. Compute $\hat{\mathbf{Q}}$ and $\mathbf{Q}$, and sketch $\hat{\mathscr{L}}$ and $\mathscr{L}$. Don't analyze the structure.

PROBLEM 3.9

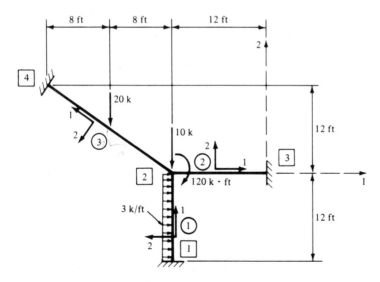

3.10   Use the finite element approach [Eq. (3.129)] to compute the fixed-end force vector $\hat{\mathbf{f}}$. Select the interpolation functions in Eqs. (1.177) and (1.183) to represent axial and flexural deformations, respectively. Show the fixed-end forces on each element.

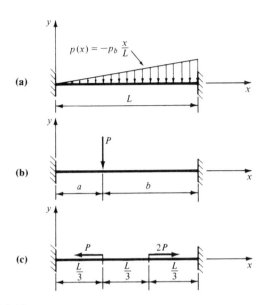

PROBLEM 3.10

3.11   Use the interpolation functions of Problem 3.10 to compute the fixed-end force vector $\hat{\mathbf{f}}$ [Eq. (3.130)] for the beam in Figure 4.27a, subject to the temperature change:

     a. $t(x, y) = t_b(x/L)$; t_b is a constant value of temperature.

     b. $t(x, y) = (2/h)\,\Delta t\,y(x/L)$; Δt is a constant value of temperature.

3.12   Consider the continuous beams in Figure 3.5a. Assume that the temperature in element 1 remains constant while element 2 experiences the temperature increase defined in Problem 3.11b. Analyze the structure on the basis of Figure 3.9 and compute the strains and stresses in the elements.

REFERENCES

Bathe, K. J., and E. L. Wilson. 1976. *Numerical Methods in Finite Element Analysis.* Prentice-Hall, Englewood Cliffs, NJ.

Budynas, R. G. 1977. *Advanced Strength and Applied Stress Analysis.* McGraw-Hill, New York.

Fenves, S. J. 1967. *Computer Methods in Civil Engineering.* Prentice-Hall, Englewood Cliffs, NJ.

Godden, W. G. 1965. *Numerical Analysis of Beam and Column Structures.* Prentice-Hall, Englewood Cliffs, NJ.

Hohn, F. E. 1972. *Elementary Matrix Algebra*, 3d ed. Macmillan, New York.

Irons, B., and S. Ahmad. 1980. *Techniques of Finite Elements.* Wiley, New York.

Langhaar, H. L. 1962. *Energy Methods in Applied Mechanics.* Wiley, New York.

Pilkey, W. D., and O. H. Pilkey. 1974. *Mechanics of Solids.* Quantum, New York.

Rubinstein, M. F. 1970. *Structural Systems—Statics, Dynamics and Stability.* Prentice-Hall, Englewood Cliffs, NJ.

Schnobrich, W. C., and D. A. W. Pecknold. 1973. *Introduction to the Finite Element Method,* Course Notes for CE 478, University of Illinois at Urbana-Champaign.

Shames, I. H. 1975. *Introduction to Solid Mechanics.* Prentice-Hall, Englewood Cliffs, NJ.

Turner, M. J., R. W. Clough, H. C. Martin, and L. J. Topp. 1956. "Stiffness and Deflection Analysis of Complex Structures," *Journal of the Aeronautical Sciences,* 25, 805–823, 854.

MATRIX DISPLACEMENT METHOD:
SPECIAL TOPICS

4.1 INTRODUCTION

This chapter illustrates how special structural features can be incorporated in the matrix displacement method formulated in Chapter 3. In particular, they can be implemented directly or with minor modifications in a computer program based on Figure 3.9.

The topics considered fall into three groups:

1. The *reduction of the degrees of freedom* of an assemblage by utilizing structural symmetry (Sections 4.2 and 4.9), by introducing internal constraints (Section 4.3), and by condensation (Section 4.6).

2. The representation of various disturbances as *element actions*. Considered are geometric imperfections of elements and support locations (Section 4.7), temperature changes (Section 4.8), and unit displacements at releases imposed in the construction of influence lines by the Müller–Breslau principle (Section 4.10).

3. The *formulation of assemblages* with internal releases (Section 4.4 and 4.6), distinct elements (4.5), and distinct joint reference frames (Section 4.9).

The sections are primarily based on Chapter 3 and can be studied in any order. However, some sections are linked through formulas (Sections 4.7, 4.8) and examples (Sections 4.2, 4.3). Although the presentation is confined to plane structures, it can be extended directly to space structures.

4.2 SYMMETRIC STRUCTURES

Symmetry is evident in many natural and man-made objects. It may be defined as "exact correspondence of form and constituent configuration on opposite sides of a dividing line or plane or about a center or axis" (American

Heritage, 1971). The points, lines, and planes of symmetry are called *elements of symmetry*. The analysis of symmetric structures can be simplified by confining it to substructures isolated along elements of symmetry.

Although this definition may conform to our intuitive notion of symmetry, it lacks mathematical precision to serve as a working definition. Consequently, rigorous definitions of concepts and principles required in the analysis of symmetric structures are introduced. The presentation is based in part on the work by Glockner (Glockner, 1973) and confined to mirror (bilateral) symmetry of plane structures.

Concepts and Principles

Reflection

Consider Figure 4.1. Point P' is a reflection of point P about the y axis. P' is obtained by replacing the coordinate x by $-x$. Alternatively, P' can be obtained by rotating P about the y axis through π radians.

Symmetric Structures

A structure is symmetric if it is equivalent to its reflection about an element of symmetry. Equivalence means pointwise correspondence of properties such as the configuration, the boundary conditions, and the materials.

For example, the reflection of all material points of the frame in Figure 4.2a about the y axis yields the frame in Figure 4.2b. Alternatively, the frame in Figure 4.2b can be obtained by rotating the frame in Figure 4.2a about the y axis through π radians. By definition, the frame in Figure 4.2a is symmetric with respect to the y axis if it is equivalent to the frame in Figure 4.2b. Note that symmetry requires, for example, identical boundary conditions at joints 1 and 2'.

Symmetric Actions and Responses

An action or response is symmetric if it is equivalent to its reflection about an element of symmetry. Equivalence means exact correspondence in the

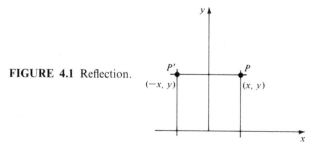

FIGURE 4.1 Reflection.

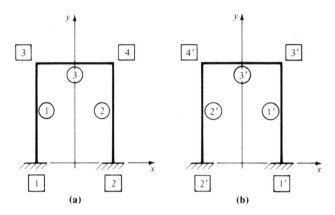

FIGURE 4.2 Symmetric frame. **(a)** Frame; **(b)** reflection of (a).

parmaters that define the action or response (Section 1.2), such as the point of application, the direction, and the magnitude of a force.

For example, the structure, the load, and the configuration in Figure 4.3a are symmetric with respect to the y axis. Their equivalent reflections are shown in Figure 4.3b.

Antisymmetric Actions and Responses

An action or response is antisymmetric if it is equivalent to its reflection modified by a sign change. Recall that a sign change reverses the direction of a force, transforms a compressive strain into a tensile strain, and changes a temperature increase into a temperature decrease.

The structure in Figure 4.3c is symmetric while the load and the deformed configuration are antisymmetric with respect to the y axis. The reflection of Figure 4.3c is shown in Figure 4.3d, and the reflection modified by a sign change is shown in Figure 4.3e. Figures 4.3c and e are equivalent.

Note that symmetric and antisymmetric actions and responses are characterized mathematically by even and odd functions, respectively. For example, for symmetric deflections (Figure 4.3a), $v(x) = v(-x)$, while for antisymmetric deflections (Figure 4.3c), $v(x) = -v(-x)$.

Decomposition of Actions

Any action can be decomposed into a symmetric and antisymmetric component with respect to the element of symmetry of a structure: The symmetric component consists of one half of the action plus the reflection of one half of the action about the element of symmetry. The antisymmetric component is the difference between the action and the symmetric component.

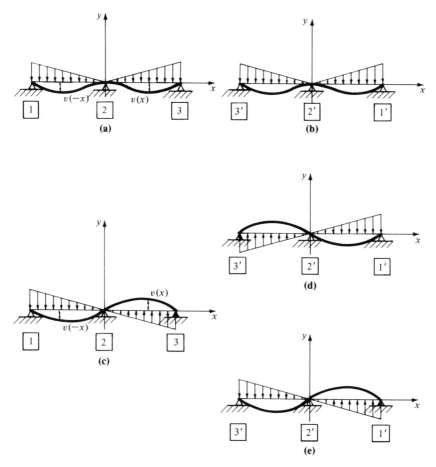

FIGURE 4.3 Symmetric structures. **(a)** Symmetric action and response; **(b)** reflection of (a); **(c)** antisymmetric action and response; **(d)** reflection of (c); **(e)** sign change of (d).

The forces in Figure 4.4a are resolved into the symmetric components (Figure 4.4b) and the antisymmetric components (Figure 4.4c) with respect to the y axis. The symmetric components consist of one half of the forces (Figure 4.4d) and the reflection of one half of the forces (Figure 4.4e).

Symmetry Axiom

On symmetric structures, symmetric actions cause symmetric responses and antisymmetric actions cause antisymmetric responses (Hoff, 1956).

Thus we may divide a symmetric structure along elements of symmetry and analyze one of the resulting substructures. The response of the complete

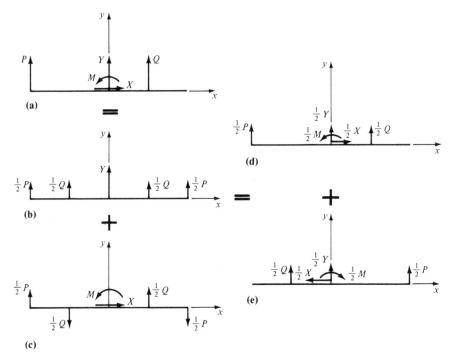

FIGURE 4.4 Decomposition of forces. **(a)** Asymmetric forces; **(b)** symmetric components; **(c)** antisymmetric components; **(d)** one half of (a); **(e)** reflection of (d).

structure can be determined from the response of the substructure by reflection for symmetric action and by reflection and sign change for antisymmetric action.

Boundary Conditions Along Elements of Symmetry

The boundary conditions of substructures along elements of symmetry can be determined by applying the definitions of symmetry or antisymmetry.

Symmetric Truss

Consider the truss in Figure 4.5a, which is symmetric with respect to the y axis. Let us determine the possible loads and deflections at the axis of symmetry for symmetric and antisymmetric responses.

The forces and deflections of the joint on the axis of symmetry in Figure 4.5b are defined relative to the x–y coordinate axes. The applied forces and deflections are denoted by X, Y and u, v, respectively; the subscripts L, R, and C identify the element forces at the left, right, and center, respectively,

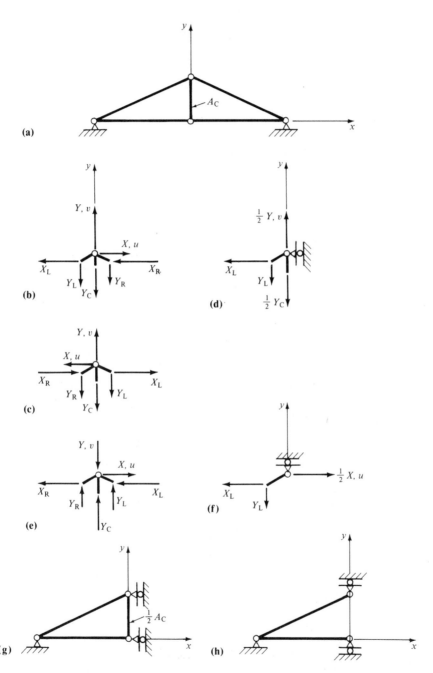

FIGURE 4.5 Symmetric and antisymmetric truss response. **(a)** Symmetric truss; **(b)** joint forces and deflections; **(c)** reflection of (b); **(d)** symmetric response; **(e)** sign change of (c); **(f)** antisymmetric response; **(g)** subtruss for symmetric response; **(h)** subtruss for antisymmetric response.

of the joint; the cross-sectional area of the center element is denoted by A_C. The conditions of equilibrium of the joint in Figure 4.5b are

$$Y = Y_L + Y_R + Y_C$$
$$X = X_L + X_R \qquad\qquad (4.1)$$

Symmetric Response. The reflection of the joint in Figure 4.5b is shown in Figure 4.5c. By definition, the forces and deflections of the joint are symmetric if Figures 4.5b and c are equivalent. If we supplement this requirement with Eqs. (4.1), we obtain the *conditions of symmetry* for the balanced joint:

$$X = 0, \quad u = 0, \quad X_L = -X_R \qquad\qquad (4.2a)$$

$$Y_L = Y_R = \tfrac{1}{2}Y - \tfrac{1}{2}Y_C \qquad\qquad (4.2b)$$

Thus, for symmetric response, a joint on the axis of symmetry may not be subjected to a load normal to the axis of symmetry nor deflect normal to the axis of symmetry. These conditions are reflected by the joint in Figure 4.5d and the subtruss in Figure 4.5g, separated at the axis of symmetry. The reduction of Y_C to $\tfrac{1}{2}Y_C$ [Eq. (4.2b) and Figure 4.5d] is accomplished in the subtruss by reducing A_C to $\tfrac{1}{2}A_C$.

Antisymmetric Response. The reflection of the joint in Figure 4.5b modified by a sign change is shown in Figure 4.5e. By definition, the forces and deflections of the joint are antisymmetric if Figures 4.5b and e are equivalent. This requirement subject to Eqs. (4.1) yields the *conditions of antisymmetry* for the balanced joint:

$$X_L = X_R = \tfrac{1}{2}X \qquad\qquad (4.3a)$$

$$Y = 0, \quad Y_C = 0, \quad v = 0, \quad Y_L = -Y_R \qquad\qquad (4.3b)$$

Thus, for antisymmetric response, a joint on the axis of symmetry may not be subjected to forces in the direction of the axis of symmetry nor deflect in the direction of the axis of symmetry. A single joint separated at the axis of symmetry is shown in Figure 4.5f, and the subtruss suitable for the analysis of antisymmetric response is shown in Figure 4.5h. Since $Y_C = 0$ [Eqs. (4.3b)], the corresponding element is eliminated in Figures 4.5f and h.

Symmetric Frame

The frame in Figure 4.6a is symmetric with respect to the y axis. Analogous to the symmetric truss, conditions for symmetric and antisymmetric response can be determined by applying the definitions of symmetry and antisymmetry to the joint on the axis of symmetry in Figure 4.6b (Problem 4.1). All forces and displacements are defined relative to the right-handed x–y–z coordinate system. The applied forces and deflections are denoted by X, Y, M and u, v, θ, respectively; the element forces at the left, right, and center of the joint are identified by the subscripts L, R, and C, respectively; the cross-sectional area

167

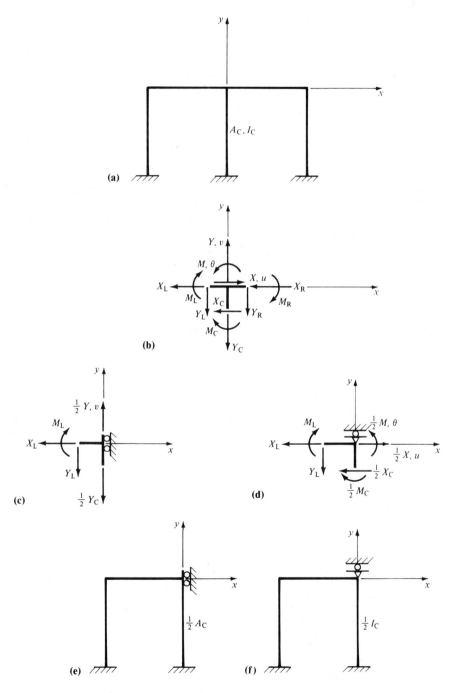

FIGURE 4.6 Symmetric and antisymmetric frame response. **(a)** Symmetric frame; **(b)** joint forces and displacements; **(c)** symmetric response; **(d)** antisymmetric response; **(e)** subframe for symmetric response; **(f)** subframe for antisymmetric response.

and moment of inertia of the center element are denoted by A_C and I_C, respectively.

Symmetric Response. The conditions of symmetric response for the balanced joint in Figure 4.6b are

$$X = 0, \quad X_C = 0, \quad u = 0, \quad X_L = -X_R \tag{4.4a}$$

$$Y_L = Y_R = \tfrac{1}{2}Y - \tfrac{1}{2}Y_C \tag{4.4b}$$

$$M = 0, \quad M_C = 0, \quad \theta = 0, \quad M_L = -M_R \tag{4.4c}$$

They are reflected by the joint and the subframe in Figures 4.6c and e, which are separated at the axis of symmetry. Equation (4.4b) is satisfied by reducing A_C to $\tfrac{1}{2}A_C$ in the subframe. Thus, a rigid joint on the axis of symmetry experiencing symmetric response does not rotate nor translate normal to the axis of symmetry.

Antisymmetric Response. The conditions of antisymmetric response for the balanced joint in Figure 4.6b are

$$X_L = X_R = \tfrac{1}{2}X - \tfrac{1}{2}X_C \tag{4.5a}$$

$$Y = 0, \quad Y_C = 0, \quad v = 0, \quad Y_L = -Y_R \tag{4.5b}$$

$$M_L = M_R = \tfrac{1}{2}M - \tfrac{1}{2}M_C \tag{4.5c}$$

They are reflected by the joint and the subframe in Figures 4.6d and f, which are separated at the axis of symmetry. Equations (4.5a) and (4.5c) are satisfied by reducing I_C to $\tfrac{1}{2}I_C$ in the subframe. Thus, a rigid joint on the axis of symmetry experiencing antisymmetric response does not translate in the direction of the axis of symmetry.

Analysis of Symmetric Structures

Symmetric or Antisymmetric Actions

For symmetric or antisymmetric actions, the behavior of the symmetric structure is completely determined by the behavior of the substructure separated along the element of symmetry. Hence, the analysis can be confined to the substructure.

Example 1. The continuous beams and the load in Figure 4.7a are symmetric with respect to the y axis. The left-hand portion of the structure is isolated and the symmetry conditions, Eqs. (4.4), are imposed at the line of symmetry. The resulting substructure in Figure 4.7b is analyzed by the matrix displacement method (Figure 3.9). The analysis units are kilopound (k), foot (ft), and radian (rad).

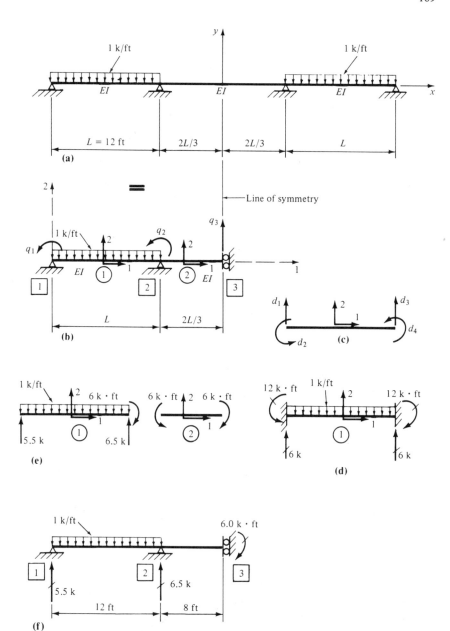

FIGURE 4.7 Symmetric response. **(a)** Continuous beams; **(b)** substructure; **(c)** elements; **(d)** fixed-end forces; **(e)** element forces; **(f)** joint forces.

1. *Unknowns*

$$q_k, \quad k = 1, 2, 3 \tag{4.6}$$

2. *Element Forces*

$$\bar{\mathbf{f}}^i = \mathbf{k}^i \mathbf{d}^i + \hat{\mathbf{f}}^i, \quad i = 1, 2 \tag{4.7}$$

$\mathbf{k}^1$ is defined in Eq. (3.37), and by substituting $2L/3$ for L in Eq. (3.37) we obtain

$$\mathbf{k}^2 = \alpha \begin{bmatrix} 40.5 & 13.5L & -40.5 & 13.5L \\ & 6L^2 & -13.5L & 3L^2 \\ & & 40.5 & -13.5L \\ \text{sym.} & & & 6L^2 \end{bmatrix}, \quad \alpha = \frac{EI}{L^3} \tag{4.8}$$

The fixed-end force vectors are (Figure 4.7d)

$$\hat{\mathbf{f}}^1 = \begin{bmatrix} 6 \\ 12 \\ 6 \\ -12 \end{bmatrix} \begin{matrix} 0 \\ 1 \\ 0 \\ 2 \end{matrix}, \quad \hat{\mathbf{f}}_2 = \mathbf{0} \tag{4.9}$$

3. *System Model*
According to Figure 4.7b, c and Eq. (2.5),

$$\mathbf{M}^T = \begin{bmatrix} 0 & 1 & 0 & 2 \\ 0 & 2 & 3 & 0 \end{bmatrix} \tag{4.10}$$

System Stiffness Matrix

$$\mathbf{K}^1 = \mathbf{k}^1 = \alpha \begin{bmatrix} & 0 & 1 & 0 & 2 \\ & & & & 0 \\ & & 4L2 & & 2L^2 \\ & & & & \\ & & & & \\ \text{sym.} & & & 4L^2 \end{bmatrix} \begin{matrix} 0 \\ 1 \\ 0 \\ 2 \end{matrix} \xrightarrow{\mathbf{M}} \mathbf{K}^{(1)} = \alpha \begin{bmatrix} 1 & 2 & 3 \\ 4L^2 & 2L^2 & 0 \\ & 4L^2 & 0 \\ \text{sym.} & & 0 \end{bmatrix} \begin{matrix} 1 \\ 2 \\ 3 \end{matrix} \tag{4.11a}$$

$$\mathbf{K}^2 = \mathbf{k}^2 = \alpha \begin{bmatrix} 0 & 2 & 3 & 0 \\ & & & & 0 \\ & 6L^2 & -13.5L & & 2 \\ & & 40.5 & & 3 \\ \text{sym.} & & & & 0 \end{bmatrix} \xrightarrow{\mathbf{M}} \mathbf{K}^{(2)}$$

$$= \alpha \begin{bmatrix} 1 & 2 & 3 \\ 0 & 0 & 0 \\ & 6L^2 & -13.5L \\ \text{sym.} & & 40.5 \end{bmatrix} \begin{matrix} 1 \\ 2 \\ 3 \end{matrix} \tag{4.11b}$$

$$\mathbf{K} = \sum_{i=1}^{2} \mathbf{K}^{(i)} = \alpha \begin{bmatrix} 4L^2 & 2L^2 & 0 \\ & 10L^2 & -13.5L \\ \text{sym.} & & 40.5 \end{bmatrix} \quad (4.12)$$

Equivalent Joint Load Vector

$\overline{\mathcal{L}}$:

$$\overline{\mathbf{Q}} = \mathbf{0}$$

$\hat{\mathcal{L}}$:

$$\hat{\mathbf{f}}^1 \overset{M}{\to} \hat{\mathbf{F}}^{(1)} = \begin{bmatrix} 12 & 1 \\ -12 & 2, \\ 0 & 3 \end{bmatrix} \quad \hat{\mathbf{F}}^{(2)} = \mathbf{0} \quad (4.13)$$

$$\hat{\mathbf{Q}} = \sum_{i=1}^{2} \hat{\mathbf{F}}^{(i)} = \hat{\mathbf{F}}^{(1)}$$

$\mathcal{L}$:

$$\mathbf{Q} = \overline{\mathbf{Q}} - \hat{\mathbf{Q}} = \begin{bmatrix} -12 \\ 12 \\ 0 \end{bmatrix} \quad (4.14)$$

System Model

$$\alpha \begin{bmatrix} 4L^2 & 2L^2 & 0 \\ 2L^2 & 10L^2 & -13.5L \\ 0 & -13.5L & 40.5 \end{bmatrix} \begin{bmatrix} q_1 \\ q_2 \\ q_3 \end{bmatrix} = \begin{bmatrix} -12 \\ 12 \\ 0 \end{bmatrix} \quad (4.15)$$

4. *Solution*

$$q_1 = -\frac{5}{\alpha L^2}, \quad q_2 = \frac{4}{\alpha L^2}, \quad q_3 = \frac{4}{3\alpha L} \quad (4.16)$$

5. *Element Forces* (Figure 4.7e)

$$\mathbf{d}^1 = \mathbf{D}^1 = \begin{bmatrix} 0 & 0 \\ q_1 & 1 \\ 0 & 0 \\ q_2 & 2 \end{bmatrix}, \quad \mathbf{d}^2 = \mathbf{D}^2 = \begin{bmatrix} 0 & 0 \\ q_2 & 2 \\ q_3 & 3 \\ 0 & 0 \end{bmatrix} \quad (4.17)$$

By Eqs. (3.37), (4.7)–(4.9), (4.16), and (4.17),

$$\mathbf{\bar{f}}^1 = \begin{bmatrix} 5.5 \\ 0 \\ \hline 6.5 \\ -6 \end{bmatrix} = \begin{bmatrix} \mathbf{\bar{f}}_a^1 \\ \hline \mathbf{\bar{f}}_b^1 \end{bmatrix}, \quad \mathbf{\bar{f}}^2 = \begin{bmatrix} 0 \\ 6 \\ \hline 0 \\ -6 \end{bmatrix} = \begin{bmatrix} \mathbf{\bar{f}}_a^2 \\ \hline \mathbf{\bar{f}}_b^2 \end{bmatrix} \quad (4.18)$$

6. *Joint Forces* (Figure 4.7f)

$$\bar{\mathbf{P}}_1 = \bar{\mathbf{f}}_a^1 = \begin{bmatrix} 5.5 \\ 0 \end{bmatrix}, \quad \bar{\mathbf{P}}_2 = \bar{\mathbf{f}}_b^1 + \bar{\mathbf{f}}_a^2 = \begin{bmatrix} 6.5 \\ 0 \end{bmatrix}, \quad \bar{\mathbf{P}}_3 = \bar{\mathbf{f}}_b^2 = \begin{bmatrix} 0 \\ -6 \end{bmatrix} \quad (4.19)$$

Asymmetric Actions

Although an asymmetric action on a symmetric structure causes an asymmetric response, we can still take advantage of structural symmetry as follows:

a. Decompose the asymmetric action into symmetric and antisymmetric components.
b. Analyze the substructure corresponding to the symmetric component, and reflect the response about the axis of symmetry to define the symmetric response for the complete structure.
c. Analyze the substructure corresponding to the antisymmetric component, and reflect the response modified by a sign change about the axis of symmetry to define the antisymmetric response for the complete structure.
d. Superpose the symmetric and antisymmetric responses to obtain the complete response of the structure.

To facilitate the computation of the symmetric and antisymmetric responses of the complete structure (steps b and c), the substructures together with their local and global reference frames are reflected about the axis of symmetry. The numbers of reflected joints, elements, and coordinate axes are marked with primes (Figures 4.8 and 4.9). Accordingly the force vectors for reflected elements and joints are defined as follows. For symmetric element forces (Figures 4.8a, b)

$$\mathbf{f}^{i'} = \mathbf{f}^i \qquad (4.20)$$

For antisymmetric element forces (Figures 4.8c, d)

$$\mathbf{f}^{i'} = -\mathbf{f}^i \qquad (4.21)$$

For symmetric joint forces (Figures 4.8e, f)

$$\mathbf{P}_{j'} = \mathbf{P}_j \qquad (4.22)$$

and for antisymmetric joint forces (Figures 4.8g, h)

$$\mathbf{P}_{j'} = -\mathbf{P}_j \qquad (4.23)$$

Joints on the axis of symmetry are defined relative to the global (unreflected) reference frame (Figure 4.9). For symmetric joint forces, only the force in the direction of the axis of symmetry can be nonzero; its value is equal to twice the value of the corresponding joint force of the substructure [Eqs. (4.4) and Figures 4.6b and c]. For antisymmetric joint forces, the force in the direction

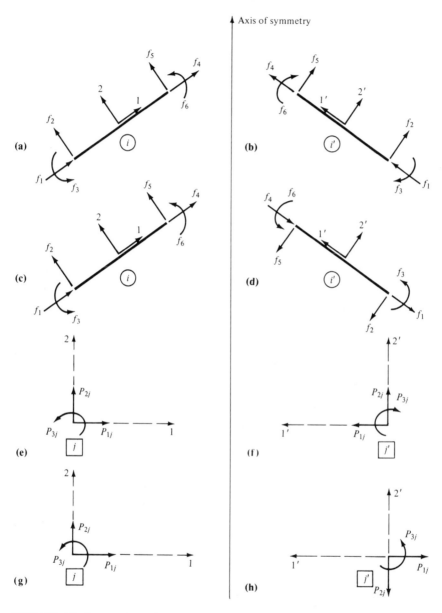

FIGURE 4.8 Symmetric and antisymmetric forces. **(a, b)** Symmetric element forces; **(c, d)** antisymmetric element forces; **(e, f)** symmetric joint forces; **(g, h)** antisymmetric joint forces.

174

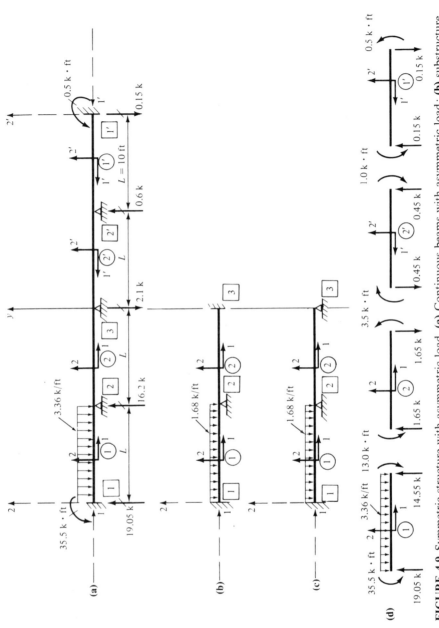

FIGURE 4.9 Symmetric structure with asymmetric load. (a) Continuous beams with asymmetric load; (b) substructure for symmetric response; (c) substructure for antisymmetric response; (d) response of structure.

of the axis of symmetry is zero, and the values of the force normal to the axis of symmetry and the moment are equal to twice the values of the corresponding joint forces of the substructure [Eqs. (4.5) and Figures 4.6b and d].

In general, the joint force vectors on the axis of symmetry are defined as follows: The symmetric joint force vector is equal to the sum of the corresponding force vector of the substructure and its reflection. The antisymmetric joint force vector is equal to the sum of the corresponding force vector of the substructure plus its reflection modified by a sign change.

Example 2. The continuous beams in Figure 4.9a with identical properties are symmetric about the y axis. The load is resolved into symmetric and antisymmetric components. The corresponding substructures are shown in Figures 4.9b and c. The analysis units are kilopound, foot, and radian.

Response of Substructure (Figure 4.9b)

$$\bar{\mathbf{f}}^1 = \begin{bmatrix} 9.45 \\ 17.50 \\ \hline 7.35 \\ -7.00 \end{bmatrix}, \quad \bar{\mathbf{f}}^2 = \begin{bmatrix} 1.05 \\ 7.00 \\ \hline -1.05 \\ 3.50 \end{bmatrix} \tag{4.24}$$

$$\bar{\mathbf{P}}_1 = \begin{bmatrix} 9.45 \\ 17.50 \end{bmatrix}, \quad \bar{\mathbf{P}}_2 = \begin{bmatrix} 8.40 \\ 0.00 \end{bmatrix}, \quad \bar{\mathbf{P}}_3 = \begin{bmatrix} -1.05 \\ 3.50 \end{bmatrix} \tag{4.25}$$

Response of Substructure (Figure 4.9c)

$$\bar{\mathbf{f}}^1 = \begin{bmatrix} 9.60 \\ 18.00 \\ \hline 7.20 \\ -6.00 \end{bmatrix}, \quad \bar{\mathbf{f}}^2 = \begin{bmatrix} 0.60 \\ 6.00 \\ \hline -0.60 \\ 0.00 \end{bmatrix} \tag{4.26}$$

$$\bar{\mathbf{P}}_1 = \begin{bmatrix} 9.60 \\ 18.00 \end{bmatrix}, \quad \bar{\mathbf{P}}_2 = \begin{bmatrix} 7.80 \\ 0.00 \end{bmatrix}, \quad \bar{\mathbf{P}}_3 = \begin{bmatrix} -0.60 \\ 0.00 \end{bmatrix} \tag{4.27}$$

Symmetric Response of Structure
By Eqs. (4.20), (4.22), (4.24), and (4.25),

$$\bar{\mathbf{f}}^{1'} = \bar{\mathbf{f}}^1 = \begin{bmatrix} 9.45 \\ 17.50 \\ \hline 7.35 \\ -7.00 \end{bmatrix}, \quad \bar{\mathbf{f}}^{2'} = \bar{\mathbf{f}}^2 = \begin{bmatrix} 1.05 \\ 7.00 \\ \hline -1.05 \\ 3.50 \end{bmatrix} \tag{4.28}$$

$$\bar{\mathbf{P}}_{1'} = \bar{\mathbf{P}}_1 = \begin{bmatrix} 9.45 \\ 17.50 \end{bmatrix}, \quad \bar{\mathbf{P}}_{2'} = \bar{\mathbf{P}}_2 = \begin{bmatrix} 8.40 \\ 0.00 \end{bmatrix}, \quad \bar{\mathbf{P}}_3 = \begin{bmatrix} -2.10 \\ 0.00 \end{bmatrix} \tag{4.29}$$

Antisymmetric Response of Structure

By Eqs. (4.21), (4.23), (4.26), and (4.27),

$$-\bar{\mathbf{f}}^{1'} = \bar{\mathbf{f}}^1 = \begin{bmatrix} 9.60 \\ 18.00 \\ \hline 7.20 \\ -6.00 \end{bmatrix}, \quad -\bar{\mathbf{f}}^{2'} = \bar{\mathbf{f}}^2 = \begin{bmatrix} 0.60 \\ 6.00 \\ \hline -0.60 \\ 0.00 \end{bmatrix} \quad (4.30)$$

$$-\bar{\mathbf{P}}_{1'} = \bar{\mathbf{P}}_1 = \begin{bmatrix} 9.60 \\ 18.00 \end{bmatrix}, \quad -\bar{\mathbf{P}}_{2'} = \bar{\mathbf{P}}_2 = \begin{bmatrix} 7.80 \\ 0.00 \end{bmatrix}, \quad \bar{\mathbf{P}}_3 = \begin{bmatrix} 0.00 \\ 0.00 \end{bmatrix} \quad (4.31)$$

Complete Response of Structure

The addition of Eqs. (4.28) and (4.30) yields (Figure 4.9d)

$$\bar{\mathbf{f}}^1 = \begin{bmatrix} 19.05 \\ 35.50 \\ \hline 14.55 \\ -13.00 \end{bmatrix}, \quad \bar{\mathbf{f}}^2 = \begin{bmatrix} 1.65 \\ 13.00 \\ \hline -1.65 \\ 3.50 \end{bmatrix} \quad (4.32a)$$

$$\bar{\mathbf{f}}^{1'} = \begin{bmatrix} -0.15 \\ -0.50 \\ \hline 0.15 \\ -1.00 \end{bmatrix}, \quad \bar{\mathbf{f}}^{2'} = \begin{bmatrix} 0.45 \\ 1.00 \\ \hline -0.45 \\ 3.50 \end{bmatrix} \quad (4.32b)$$

The addition of Eqs. (4.29) and (4.31) yields (Figure 4.9a)

$$\bar{\mathbf{P}}_1 = \begin{bmatrix} 19.05 \\ 35.50 \end{bmatrix}, \quad \bar{\mathbf{P}}_2 = \begin{bmatrix} 16.20 \\ 0.00 \end{bmatrix}, \quad \bar{\mathbf{P}}_3 = \begin{bmatrix} -2.10 \\ 0.00 \end{bmatrix}$$

$$\bar{\mathbf{P}}_{1'} = \begin{bmatrix} -0.15 \\ -0.50 \end{bmatrix}, \quad \bar{\mathbf{P}}_{2'} = \begin{bmatrix} 0.60 \\ 0.00 \end{bmatrix} \quad (4.33)$$

4.3 ORTHOGONAL FRAMES WITH INTERNAL CONSTRAINTS

In the slope-deflection method (and the moment-distribution method) the centroidal axes of the elements are assumed to be inextensible. This assumption reduces the number of degrees of freedom of a frame by the number of elements with externally unconstrained centroidal axes. Yet, the resulting effect on the response of the frame is frequently negligible (see Example 1 in Section 3.3). In any case, this approximation may be useful in preliminary design because it is economical.

Orthogonal frames with inextensible centroidal axes can be analyzed directly by the matrix displacement method defined in Figure 3.2 or 3.9. All we need to do is represent the frame as an assemblage of beam elements and select the local frames of reference to express the conditions of compatibility in local coordinates. However, since axial deformations are neglected, we cannot compute axial forces or compute joint forces by Eq. (2.29). Thus, step 6 in Figures 3.2 or 3.9 must be omitted from the analysis.

Consider the frame in Figure 4.10a. If we assume the centroidal axes of the elements to be inextensible,

$$q_2 = 0, \quad q_5 = 0, \quad q_4 - q_1 = 0 \tag{4.34}$$

These constraints reduce the six degree-of-freedom frame to the three degree-of-freedom frame in Figure 4.10b. Note that as a consequence of the inextensibility constraint, the beam in Figure 4.10b moves as a rigid body in the direction of the local 1 axis; that is, every point of the beam experiences the same axial deflection q_1. This is reflected in Figure 4.10d for the configuration $q_1 = 1, q_2 = q_3 = 0$.

The conditions of compatibility are (Figures 4.10b, c):

$$\mathbf{d}^1 = \begin{bmatrix} d_1^1 \\ d_2^1 \\ d_3^1 \\ d_4^1 \end{bmatrix} = \begin{bmatrix} q_1 \\ q_2 \\ 0 \\ 0 \end{bmatrix}, \quad \mathbf{d}^2 = \begin{bmatrix} d_1^2 \\ d_2^2 \\ d_3^2 \\ d_4^2 \end{bmatrix} = \begin{bmatrix} q_1 \\ q_3 \\ 0 \\ 0 \end{bmatrix}, \quad \mathbf{d}^3 = \begin{bmatrix} d_1^3 \\ d_2^3 \\ d_3^3 \\ d_4^3 \end{bmatrix} = \begin{bmatrix} 0 \\ q_2 \\ 0 \\ q_3 \end{bmatrix} \tag{4.35}$$

In general, if the local reference frames are oriented as shown in Figure 4.10b, that is, $0 = -\pi/2$ for columns and $0 = 0$ for beams (Figure C.2a), the member code matrix can be defined in terms of the local element displacements:

$$M_{li} = \begin{cases} k & \text{if } d_l^i = q_k \\ 0 & \text{otherwise} \end{cases} \tag{4.36}$$

Thus, Eqs. (4.35) can be expressed as

$$\mathbf{M}^T = \begin{bmatrix} 1 & 2 & 0 & 0 \\ 1 & 3 & 0 & 0 \\ 0 & 2 & 0 & 3 \end{bmatrix} \tag{4.37}$$

By Eq. (4.36), the local force vector and the local stiffness matrix of each element can be transformed directly into the generalized force vector and the generalized stiffness matrix, respectively; that is,

$$\hat{\mathbf{f}}^i \overset{\mathbf{M}}{\to} \hat{\mathbf{F}}^{(i)}$$
$$\mathbf{k}^i \overset{\mathbf{M}}{\to} \mathbf{K}^{(i)} \tag{4.38}$$

It was noted above that Eq. (2.29) is not applicable. However, we can perform a check on the solution accuracy (an equilibrium check) by computing the applied joint forces as follows:

$$\bar{\mathbf{f}}^i \overset{\mathbf{M}}{\to} \bar{\mathbf{F}}^{(i)}, \quad \bar{\mathbf{Q}} = \sum_{i=1}^{NE} \bar{\mathbf{F}}^{(i)} \tag{4.39}$$

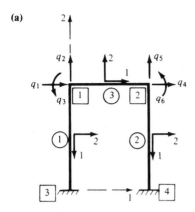

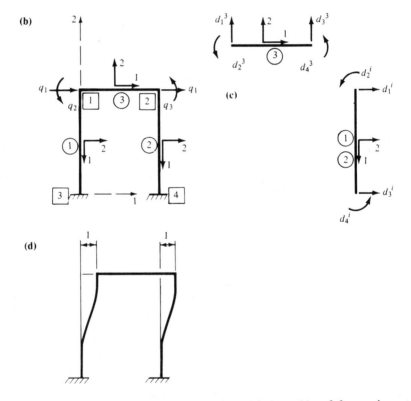

FIGURE 4.10 Frame with internal constraints. **(a)** Assembly of frame elements; **(b)** assembly of beam elements; **(c)** beam elements; **(d)** configuration of (b): $q_1 = 1$, $q_2 = q_3 = 0$.

Example 1. The element centroidal axes of the frame in Figure 3.10a are assumed to be inextensible. The resulting two degree-of-freedom frame (Figure 4.11a) is analyzed on the basis of Figure 3.9. The joint loads are $\bar{Q}_1 = 0$, $\bar{Q}_2 = 14 \text{ k} \cdot \text{ft}$. The analysis units are kilopound, foot, and radian.

1. *Unknowns*

$$q^k, \quad k = 1, 2 \tag{4.40}$$

FIGURE 4.11 Frame with internal constraints. **(a)** Assembly of beam elements; **(b)** beam elements; **(c)** free-body diagrams.

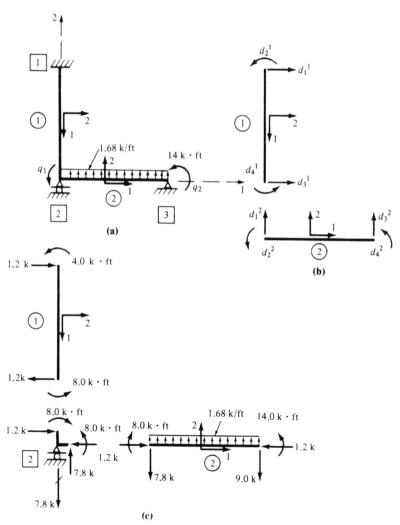

2. *Element Models*

$$\bar{\mathbf{f}}^i = \mathbf{k}\mathbf{d}^i + \hat{\mathbf{f}}^i, \quad i = 1, 2 \tag{4.41}$$

where

$$\mathbf{k} = \alpha \begin{bmatrix} 12 & 6L & -12 & 6L \\ & 4L^2 & -6L & 2L^2 \\ & & 12 & -6L \\ \text{sym.} & & & 4L^2 \end{bmatrix} \tag{4.42}$$

and from Figures 3.10d and 4.11b

$$\hat{\mathbf{f}}^1 = \mathbf{0}, \quad \hat{\mathbf{f}}^2 = \begin{bmatrix} -8.4 & 0 \\ -14.0 & 1 \\ -8.4 & 0 \\ 14.0 & 2 \end{bmatrix} \tag{4.43}$$

Figures 4.11a, b and Eq. (4.36) yield

$$\mathbf{M}^T = \begin{bmatrix} 0 & 0 & 0 & 1 \\ 0 & 1 & 0 & 2 \end{bmatrix} \tag{4.44}$$

3. *System Model*

System Stiffness Matrix

$$\mathbf{k}^1 = \alpha \begin{matrix} 0 & 0 & 0 & 1 \\ \begin{bmatrix} & & & \\ & & & \\ & & & \\ \text{sym.} & & & 4L^2 \end{bmatrix} \begin{matrix} 0 \\ 0 \\ 0 \\ 1 \end{matrix} \end{matrix} \xrightarrow{\mathbf{M}} \mathbf{K}^{(1)} = \alpha \begin{matrix} 1 & 2 \\ \begin{bmatrix} 4L^2 & 0 \\ \text{sym.} & 0 \end{bmatrix} \begin{matrix} 1 \\ 2 \end{matrix} \end{matrix}$$

$$\tag{4.45}$$

$$\mathbf{k}^2 = \alpha \begin{matrix} 0 & 1 & 0 & 2 \\ \begin{bmatrix} & 4L^2 & & 2L^2 \\ & & & \\ & & & \\ \text{sym.} & & & 4L^2 \end{bmatrix} \begin{matrix} 0 \\ 1 \\ 0 \\ 2 \end{matrix} \end{matrix} \xrightarrow{\mathbf{M}} \mathbf{K}^{(2)} = \alpha \begin{matrix} 1 & 2 \\ \begin{bmatrix} 4L^2 & 2L^2 \\ \text{sym.} & 4L^2 \end{bmatrix} \begin{matrix} 1 \\ 2 \end{matrix} \end{matrix}$$

Thus,

$$\mathbf{K} = \sum_{i=1}^{2} \mathbf{K}^{(i)} = \alpha \begin{bmatrix} 8L^2 & 2L^2 \\ \text{sym.} & 4L^2 \end{bmatrix} \tag{4.46}$$

Equivalent Joint Load Vector

$\mathscr{L}$:

$$\bar{Q} = \begin{bmatrix} 0 \\ 14.0 \end{bmatrix} \tag{4.47}$$

$\mathscr{L}$: By Eqs. (4.43) and (4.44)

$$\hat{\mathbf{F}}^{(1)} = \mathbf{0}, \quad \hat{\mathbf{f}}^2 \overset{M}{\to} \hat{\mathbf{F}}^{(2)} = \begin{bmatrix} -14.0 \\ 14.0 \end{bmatrix} \begin{matrix} 1 \\ 2 \end{matrix} \tag{4.48}$$

Thus,

$$\hat{\mathbf{Q}} = \sum_{i=1}^{2} \hat{\mathbf{F}}^{(i)} = \hat{\mathbf{F}}^{(2)} \tag{4.49}$$

$\mathscr{L}$:

$$\mathbf{Q} = \bar{\mathbf{Q}} - \hat{\mathbf{Q}} = \begin{bmatrix} 14.0 \\ 0 \end{bmatrix} \tag{4.50}$$

System Model

$$\alpha \begin{bmatrix} 8L^2 & 2L^2 \\ 2L^2 & 4L^2 \end{bmatrix} \begin{bmatrix} q_1 \\ q_2 \end{bmatrix} = \begin{bmatrix} 14.0 \\ 0 \end{bmatrix} \tag{4.51}$$

4. *Solution*

$$q_1 = \frac{2}{\alpha L^2}, \quad q_2 = -\frac{1}{\alpha L^2} \tag{4.52}$$

5. *Element Forces*
According to Eqs. (4.36) and (4.44)

$$\mathbf{d}^1 = \begin{bmatrix} 0 & 0 \\ 0 & 0 \\ 0 & 0 \\ q_1 & 1 \end{bmatrix}, \quad \mathbf{d}^2 = \begin{bmatrix} 0 & 0 \\ q_1 & 1 \\ 0 & 0 \\ q_2 & 2 \end{bmatrix} \tag{4.53}$$

By Eqs. (4.41)–(4.43), (4.52) and (4.53),

$$\bar{\mathbf{f}}^1 = \begin{bmatrix} 1.2 & 0 \\ 4.0 & 0 \\ -1.2 & 0 \\ 8.0 & 1 \end{bmatrix}, \quad \bar{\mathbf{f}}^2 = \begin{bmatrix} -7.8 & 0 \\ -8.0 & 1 \\ -9.0 & 0 \\ 14.0 & 2 \end{bmatrix} \tag{4.54}$$

The free-body diagrams of the elements and joint 2 are shown in Figure 4.11c. Observe that the shear force of element 1 causes the axial force in element 2. A comparison of Figures 3.10e and 4.11c indicates that the effect of axial deformation is negligible in this case.

By Eqs. (4.39) we obtain a check on the applied joint forces:

$$\bar{\mathbf{f}}^1 \overset{M}{\to} \bar{\mathbf{F}}^{(1)} = \begin{bmatrix} 8.0 \\ 0.0 \end{bmatrix} \begin{matrix} 1 \\ 2 \end{matrix}, \quad \bar{\mathbf{f}}^2 \overset{M}{\to} \bar{\mathbf{F}}^{(2)} = \begin{bmatrix} -8.0 \\ 14.0 \end{bmatrix} \tag{4.55}$$

$$\bar{\mathbf{Q}} = \sum_{i=1}^{2} \bar{\mathbf{F}}^{(i)} = \begin{bmatrix} 0.0 \\ 14.0 \end{bmatrix} \tag{4.56}$$

Example 2. The frame in Figure 4.12a has nine degrees of freedom. The elements have identical properties. If we assume the centroidal axes of the elements to be inextensible and take advantage of symmetry (Section 4.2), we can confine the analysis to the one degree-of-freedom frame in Figure 4.12b. The frame is analyzed by the matrix displacement method (Figure 3.9). The analysis units are kilopound, foot, and radian.

1. *Unknowns*

$$q_k, \quad k = 1 \tag{4.57}$$

2. *Element Models*

$$\bar{\mathbf{f}}^i = \mathbf{k} \mathbf{d}^i + \hat{\mathbf{f}}^i, \quad i = 1, 2 \tag{4.58}$$

where **k** is defined in Eq. (4.42),

$$\hat{\mathbf{f}}^1 = \mathbf{0}, \quad \hat{\mathbf{f}}^2 = \begin{bmatrix} 12.0 & 0 \\ 24.0 & 1 \\ 12.0 & 0 \\ -24.0 & 0 \end{bmatrix} \tag{4.59}$$

and according to Figures 4.12b, c and Eq. (4.36)

$$\mathbf{M}^T = \begin{bmatrix} 0 & 1 & 0 & 0 \\ 0 & 1 & 0 & 0 \end{bmatrix} \tag{4.60}$$

3. *System Model*

System Stiffness Matrix

$$\mathbf{k}^1 = \mathbf{k}^2 = \alpha \begin{array}{cccc} 0 & 1 & 0 & 0 \\ \begin{bmatrix} & & & 0 \\ & 4L^2 & & 1 \\ & & & 0 \\ \text{sym.} & & & 0 \end{bmatrix} \end{array} \xrightarrow{\mathbf{M}} K^{(1)} = K^{(2)} = \alpha 4L^2 \tag{4.61}$$

$$K = \sum_{i=1}^{2} K^{(i)} = \alpha 8L^2 \tag{4.62}$$

Equivalent Joint Load Vector
$\bar{\mathscr{L}}$:

$$\bar{Q} = 0 \tag{4.63}$$

$\mathscr{L}$:

$$\hat{F}^{(1)} = 0, \quad \hat{\mathbf{f}}^2 \xrightarrow{\mathbf{M}} \hat{F}^{(2)} = 24.0 \tag{4.64}$$

$$\hat{Q} = \sum_{i=1}^{2} \hat{F}^{(i)} = 24.0 \tag{4.65}$$

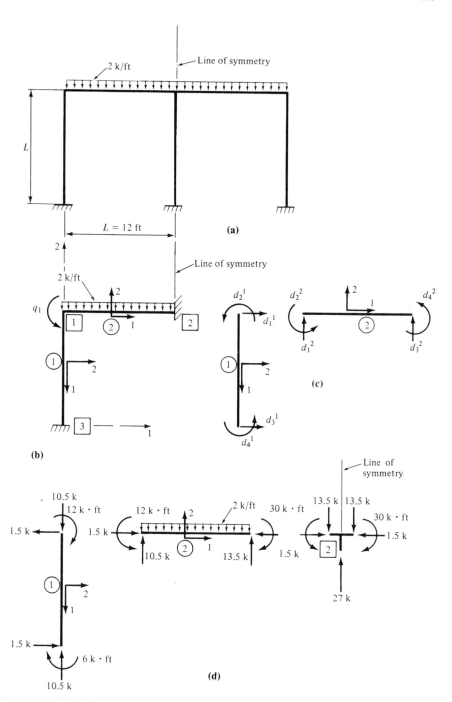

FIGURE 4.12 Symmetric frame with internal constraints. **(a)** Symmetric frame; **(b)** subassembly of beam elements; **(c)** beam elements; **(d)** free-body diagrams.

$\mathcal{L}$:

$$Q = \bar{Q} - \hat{Q} = -24.0 \tag{4.66}$$

System Model

$$\alpha 8 L^2 q_1 = -24 \tag{4.67}$$

4. *Solution*

$$q_1 = -\frac{3}{\alpha L^2} \tag{4.68}$$

5. *Element Forces*
From Eq. (4.60)

$$\mathbf{d}^1 = \mathbf{d}^2 = \begin{bmatrix} 0 & 0 \\ q_1 & 1 \\ 0 & 0 \\ 0 & 0 \end{bmatrix} \tag{4.69}$$

By Eqs. (4.42), (4.58), (4.59), (4.68), (4.69)

$$\bar{\mathbf{f}}^1 = \begin{bmatrix} -1.5 \\ -12.0 \\ 1.5 \\ -6.0 \end{bmatrix}, \quad \bar{\mathbf{f}}^2 = \begin{bmatrix} 10.5 \\ 12.0 \\ 13.5 \\ -30.0 \end{bmatrix} \tag{4.70}$$

The free-body diagrams of elements 1, 2 and joint 2 are shown in Figure 4.12d. Since the axial forces of the elements are not included in the analysis, they are computed from conditions of element and joint equilibrium.

4.4 INTERNAL RELEASES

To introduce a release in a structure means to remove a continuity constraint. For example, the insertion of a hinge, a moment release, removes the slope continuity constraint at the point of the hinge.

Internal releases can be represented as joint releases or element releases. The advantage of a joint release is that the element models remain unchanged; the disadvantage is that it increases the degrees of freedom of the system model. Joint releases of plane structures are considered in this section. Element releases are treated in Section 4.6.

Consider the two element ends in Figure 4.13a. If they are rigidly connected, as shown in Figure 4.13b, the joint has three degrees of freedom and the continuity constraints are

$$D_1^i = D_1^j = q_1 \tag{4.71a}$$

$$D_2^i = D_2^j = q_2 \tag{4.71b}$$

$$D_3^i = D_3^j = q_3 \tag{4.71c}$$

This assures continuity in deflections and slope at the joint (Figure 4.13c).

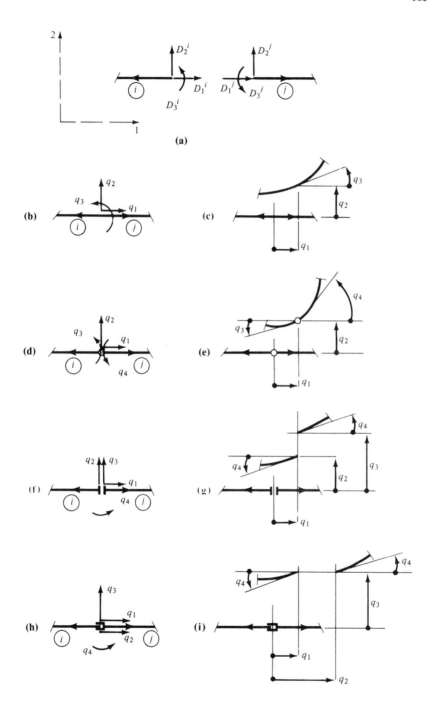

FIGURE 4.13 Joint releases. **(a)** Element ends; **(b, c)** rigid joint; **(d, e)** moment release; **(f, g)** shear release, **(h, i)** axial force release.

If we insert a frictionless hinge, the joint has four degrees of freedom (Figure 4.13d), and Eq. (4.71c) is replaced by the conditions

$$D_3^i = q_3, \quad D_3^j = q_4 \qquad (4.72)$$

The resulting slope discontinuity is shown in Figure 4.13e.

Two alternative releases, a shear release and an axial force release, are shown in Figures 4.13f and h. The corresponding deflection discontinuities are depicted in Figures 4.13g and i. The conditions of compatibility at the joints in Figures 4.13f and h are, respectively,

$$D_1^i = D_1^j = q_1$$
$$D_2^i = q_2, \quad D_2^j = q_3 \qquad (4.73)$$
$$D_3^i = D_3^j = q_4$$

and

$$D_1^i = q_1, \quad D_1^j = q_2$$
$$D_2^i = D_2^j = q_3 \qquad (4.74)$$
$$D_3^i = D_3^j = q_4$$

If more than two elements are incident to a joint, a joint release may not affect all elements. For example, consider the three element ends in Figure 4.14a. In Figure 4.14b, elements i and j are rigidly interconnected, while element k is pinned to the joint. Accordingly, the joint has four degrees of freedom and the conditions of compatibility are

$$D_1^i = D_1^j = D_1^k = q_1 \qquad (4.75a)$$

$$D_2^i = D_2^j = D_2^k = q_2 \qquad (4.75b)$$

$$D_3^i = D_3^j = q_3, \quad D_3^k = q_4 \qquad (4.75c)$$

However, in Figure 4.14c, all elements are joined by a hinge. Thus, the joint has five degrees of freedom and Eqs. (4.75c) become

$$D_3^i = q_3, \quad D_3^j = q_4, \quad D_3^k = q_5 \qquad (4.76)$$

Structures with joint releases can be analyzed directly by the matrix displacement method defined in Figures 3.9 and 3.2.

Example. The elements in Figure 4.15a are connected by a frictionless hinge, $\bar{Q} = 0$. The analysis is based on Figure 3.9. The analysis units are kilopound, foot, and radian.

1. *Unknowns*

$$q_k, \quad k = 1, 2, 3 \qquad (4.77)$$

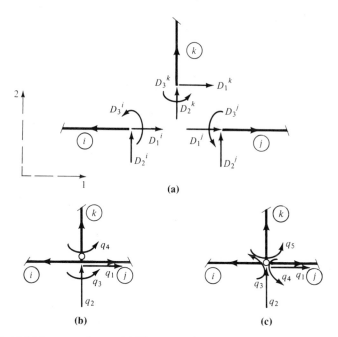

(a)

(b)   **(c)**

FIGURE 4.14 Moment releases. **(a)** Element ends; **(b)** element release; **(c)** joint release.

2. *Element Models*

$$\bar{\mathbf{f}}^i = \mathbf{k}\mathbf{d}^i + \hat{\mathbf{f}}^i \tag{4.78}$$

where

$$\mathbf{k} = \alpha \begin{bmatrix} 12 & 6L & -12 & 6L \\ & 4L^2 & -6L & 2L^2 \\ & & 12 & -6L \\ \text{sym.} & & & 4L^2 \end{bmatrix} \tag{4.79}$$

From Figures 4.15b, e

$$\hat{\mathbf{f}}^1 = \begin{bmatrix} 4 & 0 \\ 12 & 0 \\ 4 & 1 \\ -12 & 2 \end{bmatrix}, \quad \hat{\mathbf{f}}^2 = \mathbf{0} \tag{4.80}$$

and by Figures 4.15a and d

$$\mathbf{M}^T = \begin{bmatrix} 0 & 0 & 1 & 2 \\ 1 & 3 & 0 & 0 \end{bmatrix} \tag{4.81}$$

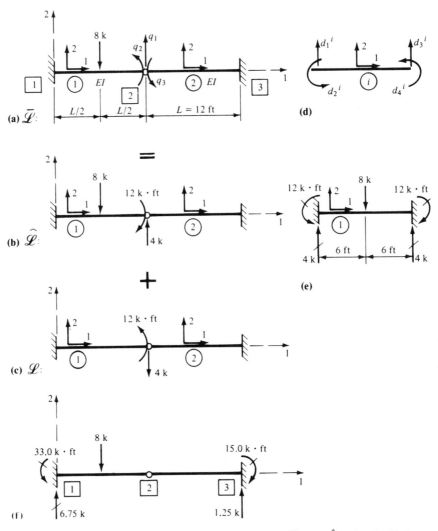

FIGURE 4.15 Beam assembly with joint release. **(a)** $\overline{\mathscr{L}}$; **(b)** $\hat{\mathscr{L}}$; **(c)** $\mathscr{L}$; **(d)** beam elements; **(e)** fixed-end forces; **(f)** joint forces.

3. *System Model*
System Stiffness Matrix

$$
\mathbf{k}^1 = \alpha
\begin{matrix}
0 & \;\;0 & \;\;1 & \;\;2 \\
\begin{bmatrix}
 & & & \\
 & & & \\
 & & 12 & -6L \\
\text{sym.} & & & 4L^2
\end{bmatrix}
\begin{matrix} 0 \\ 0 \\ 1 \\ 2 \end{matrix}
\end{matrix}
\;\xrightarrow{\text{M}}\;
\mathbf{K}^{(1)} = \alpha
\begin{matrix}
1 & \;\;2 & \;\;3 \\
\begin{bmatrix}
12 & -6L & 0 \\
 & 4L^2 & 0 \\
\text{sym.} & & 0
\end{bmatrix}
\begin{matrix} 1 \\ 2 \\ 3 \end{matrix}
\end{matrix}
$$

$$(4.82)$$

$$\mathbf{k}^2 = \alpha \begin{bmatrix} \overset{1}{12} & \overset{3}{6L} & \overset{0}{} & \overset{0}{} \\ & 4L^2 & & \\ & & & \\ \text{sym.} & & & \end{bmatrix} \begin{matrix} 1 \\ 3 \\ 0 \\ 0 \end{matrix} \overset{\mathbf{M}}{\to} \mathbf{K}^{(2)} = \alpha \begin{bmatrix} \overset{1}{12} & \overset{2}{0} & \overset{3}{6L} \\ & 0 & 0 \\ \text{sym.} & & 4L^2 \end{bmatrix} \begin{matrix} 1 \\ 2 \\ 3 \end{matrix}$$

$$\mathbf{K} = \sum_{i=1}^{2} \mathbf{K}^{(i)} = \alpha \begin{bmatrix} 24 & -6L & 6L \\ & 4L^2 & 0 \\ \text{sym.} & & 4L^2 \end{bmatrix} \tag{4.83}$$

Equivalent Joint Load Vector

$\overline{\mathscr{L}}$:

$$\overline{\mathbf{Q}} = \mathbf{0} \tag{4.84}$$

$\hat{\mathscr{L}}$:

$$\hat{\mathbf{f}}^1 \overset{\mathbf{M}}{\to} \hat{\mathbf{F}}^{(1)} = \begin{bmatrix} 4 \\ -12 \\ 0 \end{bmatrix} \begin{matrix} 1 \\ 2, \\ 3 \end{matrix} \quad \hat{\mathbf{F}}^{(2)} = \mathbf{0} \tag{4.85}$$

$$\hat{\mathbf{Q}} = \sum_{i=1}^{2} \hat{\mathbf{F}}^{(i)} = \hat{\mathbf{F}}^{(1)} \tag{4.86}$$

$\mathscr{L}$:

$$\mathbf{Q} = \overline{\mathbf{Q}} - \hat{\mathbf{Q}} = \begin{bmatrix} -4 \\ 12 \\ 0 \end{bmatrix} \tag{4.87}$$

The resolution of $\overline{\mathscr{L}}$ into $\hat{\mathscr{L}}$ and $\mathscr{L}$ is shown in Figures 4.15a–c.

System Model

$$\alpha \begin{bmatrix} 24 & -6L & 6L \\ -6L & 4L^2 & 0 \\ 6L & 0 & 4L^2 \end{bmatrix} \begin{bmatrix} q_1 \\ q_2 \\ q_3 \end{bmatrix} = \begin{bmatrix} -4 \\ 12 \\ 0 \end{bmatrix} \tag{4.88}$$

4. *Solution*

$$q_1 = -\frac{5}{12\alpha}, \quad q_2 = -\frac{3}{8\alpha L}, \quad q_3 = \frac{5}{8\alpha L} \tag{4.89}$$

5. *Element Forces*

$$\mathbf{d}^1 = \begin{bmatrix} 0 \\ 0 \\ q_1 \\ q_2 \end{bmatrix} \begin{matrix} 0 \\ 0 \\ 1, \\ 2 \end{matrix} \quad \mathbf{d}^2 = \begin{bmatrix} q_1 \\ q_3 \\ 0 \\ 0 \end{bmatrix} \begin{matrix} 1 \\ 3 \\ 0 \\ 0 \end{matrix} \tag{4.90}$$

From Eqs. (4.78)–(4.80), (4.89), and (4.90)

$$\bar{\mathbf{f}}^1 = \begin{bmatrix} 6.75 \\ 33.00 \\ \hdashline 1.25 \\ 0.00 \end{bmatrix} = \begin{bmatrix} \bar{\mathbf{f}}_a^1 \\ \hdashline \bar{\mathbf{f}}_b^1 \end{bmatrix}, \quad \bar{\mathbf{f}}^2 = \begin{bmatrix} -1.25 \\ 0.00 \\ \hdashline 1.25 \\ -15.00 \end{bmatrix} = \begin{bmatrix} \bar{\mathbf{f}}_a^2 \\ \hdashline \bar{\mathbf{f}}_b^2 \end{bmatrix} \qquad (4.91)$$

6. *Joint Forces* (Figure 4.15f)

$$\bar{\mathbf{P}}_1 = \bar{\mathbf{f}}_a^1 = \begin{bmatrix} 6.75 \\ 33.00 \end{bmatrix}, \quad \bar{\mathbf{P}}_2 = \bar{\mathbf{f}}_b^1 + \bar{\mathbf{f}}_a^2 = \begin{bmatrix} 0 \\ 0 \end{bmatrix}$$

$$\bar{\mathbf{P}}_3 = \bar{\mathbf{f}}_b^2 = \begin{bmatrix} 1.25 \\ -15.00 \end{bmatrix} \qquad (4.92)$$

Note that in the computation of joint forces by Eq. (2.29), the moments at joint 2 are represented by their sum. Clearly, Eq. (2.29) cannot be used to compute applied forces that correspond to distinct degrees of freedom at a joint. In this case, the applied joint forces can only be computed (to check the solution accuracy) by the condition of equilibrium for the assemblage

$$\bar{\mathbf{Q}} = \sum_{i=1}^{NE} \bar{\mathbf{F}}^{(i)} \qquad (4.93)$$

However, if the internal releases are represented as element releases (Section 4.6) rather than joint releases, Eq. (2.29) is applicable.

4.5 ASSEMBLAGES OF DISTINCT ELEMENTS

Skeletal structures may be composed of elements with distinct character-istics and degrees of freedom. For example, braced frames consist of members that resist axial and flexural deformations and braces that are frequently designed to transmit only axial forces. Similarly, a flexible joint (Livesley, 1964, 1975) that transmits a moment proportional to the relative joint rotation may be represented by a rotational spring to which the frame members are connected.

Assemblages of distinct elements can be analyzed by the matrix displace-ment method defined in Figures 3.9 and 3.2. In general, however, Eq. (2.29) cannot be used to compute joint forces.

Example 1. The beams in Figure 4.16a are joined by an infinitesimal rotational spring whose moment–rotation relation follows from Figure 4.16b:

$$M = \gamma(\theta_{\mathbf{R}} - \theta_{\mathbf{L}}) \qquad (4.94)$$

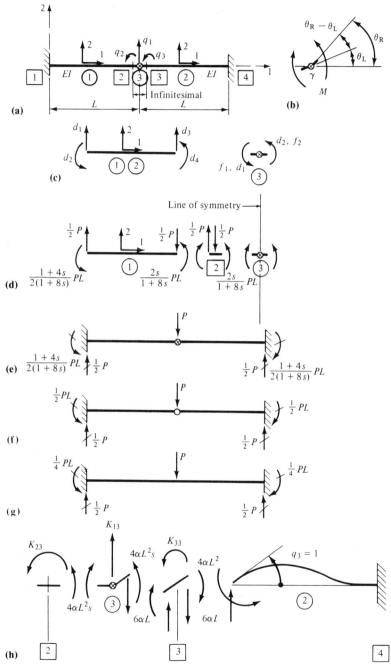

FIGURE 4.16 Beam–spring assembly. **(a)** Beam–spring assembly; **(b)** rotational spring; **(c)** elements; **(d)** free-body diagrams; **(e)** spring stiffness s; **(f)** $s = 0$; **(g)** $s = \infty$; **(h)** joint equilibrium for $q_3 = 1$, $q_1 = q_2 = 0$.

where the spring stiffness γ has units of moment per unit radian. The substitutions

$$-f_1 = M = f_2, \quad d_1 = \theta_L, \quad d_2 = \theta_R \tag{4.95}$$

into Eq. (4.94) yield the spring model (Figure 4.16c)

$$\begin{bmatrix} f_1 \\ f_2 \end{bmatrix} = \gamma \begin{bmatrix} 1 & -1 \\ -1 & 1 \end{bmatrix} \begin{bmatrix} d_1 \\ d_2 \end{bmatrix} \tag{4.96}$$

It is convenient to express γ as

$$\gamma = 4L^2 \alpha s, \quad \alpha = \frac{EI}{L^3} \tag{4.97}$$

where s is a nondimensional spring stiffness.

The structure in Figure 4.16a is analyzed by the matrix displacement method (Figure 3.2). The response is specialized for $s = 0$, which corresponds to a frictionless hinge, and $s = \infty$, which corresponds to a rigid joint. The joint loads are $Q_1 = -P$, $Q_2 = Q_3 = 0$.

1. *Unknowns*

$$q_k, \quad k = 1, 2, 3 \tag{4.98}$$

Since the length of the spring is assumed to be infinitesimal, joints 2 and 3 have the same deflection q_1. Thus, the corresponding generalized external force Q_1 (Section 1.4) is equal to the sum of the forces applied at joints 2 and 3 in the direction of q_1.

2. *Element Models*

$$\mathbf{f}^i = \mathbf{k}^i \mathbf{d}^i, \quad i = 1, 2, 3 \tag{4.99}$$

where $\mathbf{k}^1$ and $\mathbf{k}^2$ are defined in Eq. (4.79), and $\mathbf{k}^3$ is defined in Eq. (4.96).

3. *System Model*

According to Figures 4.16a and c, the member codes are

$$\mathbf{M}^{1T} = \begin{bmatrix} 0 & 0 & 1 & 2 \end{bmatrix}$$

$$\mathbf{M}^{2T} = \begin{bmatrix} 1 & 3 & 0 & 0 \end{bmatrix} \tag{4.100}$$

$$\mathbf{M}^{3T} = \begin{bmatrix} 2 & 3 \end{bmatrix}$$

System Stiffness Matrix

$$
\mathbf{k}^1 = \alpha \begin{array}{c} \begin{matrix} 0 & & 0 & & 1 & & 2 \end{matrix} \\ \begin{bmatrix} & & & \\ & & & \\ & & 12 & -6L \\ \text{sym.} & & & 4L^2 \end{bmatrix} \begin{matrix} 0 \\ 0 \\ 1 \\ 2 \end{matrix} \end{array}
\xrightarrow{\text{M}}
\mathbf{K}^{(1)} = \alpha \begin{array}{c} \begin{matrix} 1 & & 2 & & 3 \end{matrix} \\ \begin{bmatrix} 12 & -6L & 0 \\ & 4L^2 & 0 \\ \text{sym.} & & 0 \end{bmatrix} \begin{matrix} 1 \\ 2 \\ 3 \end{matrix} \end{array}
$$

$$\tag{4.101a}$$

$$\mathbf{k}^2 = \alpha \begin{array}{cc} \begin{array}{cccc} 1 & 3 & 0 & 0 \end{array} \\ \begin{bmatrix} 12 & 6L & & \\ & 4L^2 & & \\ & & & \\ \text{sym.} & & \end{bmatrix} \begin{array}{c} 1 \\ 3 \\ 0 \\ 0 \end{array} \end{array} \xrightarrow{\mathbf{M}} \mathbf{K}^{(2)} = \alpha \begin{array}{c} \begin{array}{ccc} 1 & 2 & 3 \end{array} \\ \begin{bmatrix} 12 & 0 & 6L \\ & 0 & 0 \\ \text{sym.} & & 4L^2 \end{bmatrix} \begin{array}{c} 1 \\ 2 \\ 3 \end{array} \end{array} \quad (4.101b)$$

$$\mathbf{k}^3 = \alpha \begin{array}{c} \begin{array}{cc} 2 & 3 \end{array} \\ \begin{bmatrix} 4L^2s & -4L^2s \\ \text{sym.} & 4L^2s \end{bmatrix} \begin{array}{c} 2 \\ 3 \end{array} \end{array} \xrightarrow{\mathbf{M}} \mathbf{K}^{(3)} = \alpha \begin{array}{c} \begin{array}{ccc} 1 & 2 & 3 \end{array} \\ \begin{bmatrix} 0 & 0 & 0 \\ & 4L^2s & -4L^2s \\ \text{sym.} & & 4L^2s \end{bmatrix} \begin{array}{c} 1 \\ 2 \\ 3 \end{array} \end{array} \quad (4.101c)$$

Thus,

$$\mathbf{K} = \sum_{i=1}^{3} \mathbf{K}^{(i)} = \alpha \begin{bmatrix} 24 & -6L & 6L \\ & 4L^2(1+s) & -4L^2s \\ \text{sym.} & & 4L^2(1+s) \end{bmatrix} \quad (4.102)$$

System Model

$$\alpha \begin{bmatrix} 24 & -6L & 6L \\ -6L & 4L^2(1+s) & -4L^2s \\ 6L & -4L^2s & 4L^2(1+s) \end{bmatrix} \begin{bmatrix} q_1 \\ q_2 \\ q_3 \end{bmatrix} = \begin{bmatrix} -P \\ 0 \\ 0 \end{bmatrix} \quad (4.103)$$

4. *Solution*

$$q_1 = \frac{-(1+2s)P}{\alpha 6(1+8s)}, \quad q_2 = \frac{3}{2L(1+2s)} q_1, \quad q_3 = -q_2 \quad (4.104)$$

5. *Element Forces*
Since the response is symmetric with respect to the rotational spring, only
$\mathbf{f}^1$ and $\mathbf{f}^2$ are computed:

$$\mathbf{d}^1 = \begin{bmatrix} 0 \\ 0 \\ q_1 \\ q_2 \end{bmatrix} \begin{array}{c} 0 \\ 0 \\ 1 \\ 2 \end{array}, \quad \mathbf{d}^3 = \begin{bmatrix} q_2 \\ q_3 \end{bmatrix} \begin{array}{c} 2 \\ 3 \end{array} \quad (4.105)$$

From Eqs. (4.79), (4.96), (4.99), (4.104), and (4.105)

$$
\mathbf{f}^1 = \begin{bmatrix} \frac{1}{2}P \\ \frac{1+4s}{2(1+8s)}\,PL \\ \hline -\frac{1}{2}P \\ \frac{2s}{1+8s}\,PL \end{bmatrix}, \quad
\mathbf{f}^3 = \begin{bmatrix} \frac{-2s}{1+8s}\,PL \\ \hline \frac{2s}{1+8s}\,PL \end{bmatrix} \tag{4.106}
$$

The free-body diagrams of elements 1, 3 and joint 2 are shown in Figure 4.16d, and the joint forces are shown in Figure 4.16e.

Special Cases

If we specialize $\mathbf{f}^1$ in Eqs. (4.106) for $s = 0$ and $s = \infty$, we obtain, respectively,

$$
\mathbf{f}^1 = \begin{bmatrix} \frac{1}{2}P \\ \frac{1}{2}PL \\ \hline -\frac{1}{2}P \\ 0 \end{bmatrix} \quad \text{and} \quad
\mathbf{f}^1 = \begin{bmatrix} \frac{1}{2}P \\ \frac{1}{4}PL \\ \hline -\frac{1}{2}P \\ \frac{1}{4}PL \end{bmatrix} \tag{4.107}
$$

The corresponding joint forces are shown in Figures 4.16f and g.

Comments

Joint Forces. Joint forces can only be computed by Eq. (2.29) if there is a one-to-one correspondence between element-end forces and joint forces. Since in this example the spring element need not transmit shear forces (Figure 4.16d), a suitable spring model that permits the computation of joint forces is

$$
\begin{bmatrix} f_1 \\ f_2 \\ f_3 \\ f_4 \end{bmatrix} = \gamma
\begin{bmatrix} 0 & 0 & 0 & 0 \\ 0 & 1 & 0 & -1 \\ 0 & 0 & 0 & 0 \\ 0 & -1 & 0 & 1 \end{bmatrix}
\begin{bmatrix} d_1 \\ d_2 \\ d_3 \\ d_4 \end{bmatrix} \tag{4.108}
$$

With this model, the matrix displacement analysis yields the element force vector

$$
\mathbf{f}^3 = \begin{bmatrix} 0 \\ \frac{-2s}{1+8s}\,PL \\ \hline 0 \\ \frac{2s}{1+8s}\,PL \end{bmatrix} \tag{4.109}
$$

and the joint force vector (Figures 4.16a and d)

$$\mathbf{P}_2 = \mathbf{f}_b^1 + \mathbf{f}_a^3 = \begin{bmatrix} -\frac{1}{2}P \\ 0 \end{bmatrix} \tag{4.110}$$

A general approach is to incorporate the spring element into a complex element whose end forces match the joint forces (Section 4.6).

Stiffness Coefficients. It is instructive to verify stiffness coefficients of unusual structures by Eq. (3.35). For example, the coefficients of the third column of $\mathbf{K}$ in Eq. (4.102) follow from Figure 4.16h. Specifically, the conditions of equilibrium yield $K_{13} = 6\alpha L$, $K_{23} = -4\alpha L^2 s$, $K_{33} = 4\alpha L^2(1 + s)$.

Example 2. Consider the braced frame in Figure 4.17a. The braces consist of cables whose cross-sectional areas are much smaller than those of the frame members. Accordingly, let us assume that the centroidal axes of the frame members are inextensible (Section 4.3) and construct the system stiffness matrix. The frame members have identical properties E, I, L and the extensional stiffness of a cable is $\gamma = EA/(\sqrt{2}L) = \alpha\bar{\beta}$, $\bar{\beta} = AL^2/(\sqrt{2}I)$.

System Stiffness Matrix
The member code matrix for the frame members is defined in Eq. (4.37), and the member code for a cable is

$$\mathbf{M}^{4T} = \begin{bmatrix} 1 & 0 & 0 & 0 \end{bmatrix} \tag{4.111}$$

Since only one cable can contribute to the stiffness of the frame at one time,

FIGURE 4.17 Braced frame. **(a)** Frame–truss assembly; **(b)** truss element.

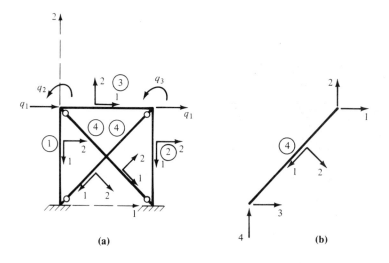

(a)   (b)

the cable in tension, both cables are assigned the same number. The generalized element stiffness matrices are

$$\mathbf{k}^1 = \alpha \begin{bmatrix} 1 & 2 & 0 & 0 \\ 12 & 6L^2 & & \\ & 4L^2 & & \\ & & & \\ \text{sym.} & & & \end{bmatrix} \begin{matrix} 1 \\ 2 \\ 0 \\ 0 \end{matrix} \xrightarrow{\mathbf{M}} \mathbf{K}^{(1)} = \alpha \begin{bmatrix} 1 & 2 & 3 \\ 12 & 6L_2 & 0 \\ & 4L^2 & 0 \\ \text{sym.} & & 0 \end{bmatrix} \begin{matrix} 1 \\ 2 \\ 3 \end{matrix} \qquad (4.112a)$$

$$\mathbf{k}^2 = \alpha \begin{bmatrix} 1 & 3 & 0 & 0 \\ 12 & 6L & & \\ & 4L^2 & & \\ & & & \\ \text{sym.} & & & \end{bmatrix} \begin{matrix} 1 \\ 3 \\ 0 \\ 0 \end{matrix} \xrightarrow{\mathbf{M}} \mathbf{K}^{(2)} = \alpha \begin{bmatrix} 1 & 2 & 3 \\ 12 & 0 & 6L \\ & 0 & 0 \\ \text{sym.} & & 4L^2 \end{bmatrix} \begin{matrix} 1 \\ 2 \\ 3 \end{matrix} \qquad (4.112b)$$

$$\mathbf{k}^3 = \alpha \begin{bmatrix} 0 & 2 & 0 & 3 \\ & 4L^2 & & 2L^2 \\ & & & \\ \text{sym.} & & & 4L^2 \end{bmatrix} \begin{matrix} 0 \\ 2 \\ 0 \\ 3 \end{matrix} \xrightarrow{\mathbf{M}} \mathbf{K}^{(3)} = \alpha \begin{bmatrix} 1 & 2 & 3 \\ 0 & 0 & 0 \\ & 4L^2 & 2L^2 \\ \text{sym.} & & 4L^2 \end{bmatrix} \begin{matrix} 1 \\ 2 \\ 3 \end{matrix} \qquad (4.112c)$$

$$\mathbf{K}^4 = \alpha \begin{bmatrix} 1 & 0 & 0 & 0 \\ \frac{1}{2}\bar{\beta} & & & \\ & 0 & & \\ & & 0 & \\ \text{sym.} & & & 0 \end{bmatrix} \begin{matrix} 1 \\ 0 \\ 0 \\ 0 \end{matrix} \xrightarrow{\mathbf{M}} \mathbf{K}^{(4)} = \alpha \begin{bmatrix} 1 & 2 & 3 \\ \frac{1}{2}\bar{\beta} & 0 & 0 \\ & 0 & 0 \\ \text{sym.} & & 0 \end{bmatrix} \begin{matrix} 1 \\ 2 \\ 3 \end{matrix} \qquad (4.112d)$$

$\mathbf{K}^4$ follows from Eq. (3.78) for $c_1^2 = \frac{1}{2}$. Thus, the system stiffness matrix is

$$\mathbf{K} = \sum_{i=1}^{4} \mathbf{K}^{(i)} = \alpha \begin{bmatrix} 24 + \frac{1}{2}\bar{\beta} & 6L & 6L \\ & 8L^2 & 2L^2 \\ \text{sym.} & & 8L^2 \end{bmatrix} \qquad (4.113)$$

4.6 CONDENSATION AND SUBSTRUCTURING

This section deals with complex elements, also called substructures or super-elements, that are assembled from simple elements (Livesley, 1975; McGuire and Gallagher, 1979; Tong and Rossettos, 1977; Zienkiewicz, 1977). Since the *internal degrees of freedom* of complex elements are not shared with other elements or joints, they can be regarded as *dependent variables* and eliminated from the element model. Thus, the model of the complex element can be *condensed* to an explicit relation that involves only the *external degrees of freedom*.[1]

---

[1] To condense a model means to reduce the number of independent variables and, thus, to reduce the number of degrees of freedom (Section 1.4).

For example, it was shown in Section 4.5 that a flexible joint can be represented as an infinitesimal rotational spring. The analysis of a frame with flexible joints can be simplified by incorporating them into a complex element as shown in Figure 4.18a. Moreover, by eliminating the internal displacement d_5 and d_6, we can express the complex element in the standard form a beam element (Figure 1.14b). The axial deformations (Figure 1.14a) are not included in Figure 4.18a because they are not affected by the rotational springs.

FIGURE 4.18 Complex element. **(a)** Beam–spring assembly; **(b)** elements; **(c)** fixed-end forces; **(d)** special case: $s_a = 0$, $s_b = \infty$.

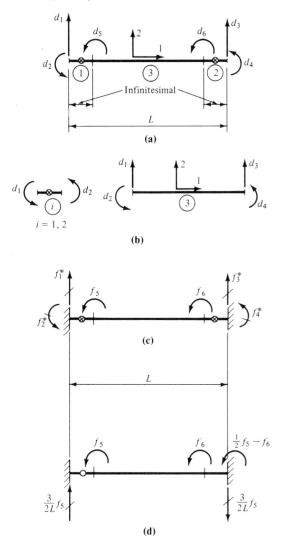

The formulation and condensation of complex elements is useful if a structure consists of many identical complex elements or if the size of the structure exceeds the computer core storage.

Condensation

Symbolically, the condensation proceeds as follows: The complex element model is expressed in partitioned form as[2]

$$\begin{bmatrix} \bar{\mathbf{f}}_e \\ \mathbf{f}_i \end{bmatrix} = \begin{bmatrix} \mathbf{k}_{ee} & \mathbf{k}_{ei} \\ \mathbf{k}_{ie} & \mathbf{k}_{ii} \end{bmatrix} \begin{bmatrix} \mathbf{d}_e \\ \mathbf{d}_i \end{bmatrix} \tag{4.114}$$

where the subscript e identifies the *external displacements and forces* that are involved in the assembly process, and i identifies the *internal displacements and forces* that are unique to the element. Equation (4.114) yields

$$\bar{\mathbf{f}}_e = \mathbf{k}_{ee}\mathbf{d}_e + \mathbf{k}_{ei}\mathbf{d}_i \tag{4.115a}$$

$$\mathbf{f}_i = \mathbf{k}_{ie}\mathbf{d}_e + \mathbf{k}_{ii}\mathbf{d}_i \tag{4.115b}$$

Equation (4.115b) is solved for

$$\mathbf{d}_i = -\mathbf{k}_{ii}^{-1}\mathbf{k}_{ie}\mathbf{d}_e + \mathbf{k}_{ii}^{-1}\mathbf{f}_i \tag{4.116}$$

and Eq. (4.116) is substituted into Eq. (4.115a) to yield the condensed model

$$\bar{\mathbf{f}}_e = \mathbf{f}_e + \mathbf{f}_e^* \tag{4.117}$$
$$\mathbf{f}_e = \mathbf{k}_e\mathbf{d}_e$$

where

$$\mathbf{k}_e = \mathbf{k}_{ee} - \mathbf{k}_{ei}\mathbf{k}_{ii}^{-1}\mathbf{k}_{ie} \tag{4.118}$$

and

$$\mathbf{f}_e^* = \mathbf{k}_{ei}\mathbf{k}_{ii}^{-1}\mathbf{f}_i \tag{4.119}$$

Equations (4.117) and (4.119) indicate that $\bar{\mathbf{f}}_e = \mathbf{f}_e^*$ if $\mathbf{d}_e = \mathbf{0}$, and $\bar{\mathbf{f}}_e = \mathbf{f}_e$ if $\mathbf{f}_i = \mathbf{0}$. Thus, $\mathbf{f}_e^*$ is the external force vector induced by the element loads $\mathbf{f}_i$ in the absence of external displacements. For the element in Figure 4.18a, $\mathbf{f}_e^*$ is the fixed-end force vector. Accordingly, Eqs. (4.117) are analogous to Eqs. (3.95). The only difference between the "fixed-end" force vectors $\hat{\mathbf{f}}$ and $\mathbf{f}^*$ is that $\hat{\mathbf{f}}$ is due to element actions in general whereas $\mathbf{f}^*$ is caused specifically by the element loads $\mathbf{f}_i$.

Note that once the external element displacements are determined in the matrix displacement analysis (Figure 3.9), the internal element displacements can be computed by Eq. (4.116).

Example 1. The condensed model for the complex element in Figure 4.18a is formulated. The properties of the beam element are E, I, L, and the rotational spring stiffnesses are defined according to Eq. (4.97) as

$$\gamma_a = 4\alpha L^2 s_a, \quad \gamma_b = 4\alpha L^2 s_b \tag{4.120}$$

---

[2] The bar over $\bar{\mathbf{f}}_e$ is introduced to make the notation consistent with that in Section 3.6.

where s_a and s_b are nondimensional spring stiffnesses at the a and b end, respectively.

System Model (Figure 3.2)
According to Figures 4.18a and b, the member codes are (see Section 4.5 for assembly of distinct elements)

$$\mathbf{M}^{1T} = [2 \quad 5]$$

$$\mathbf{M}^{2T} = [6 \quad 4] \tag{4.121}$$

$$\mathbf{M}^{3T} = [1 \quad 5 \quad 3 \quad 6]$$

The degrees of freedom are labeled in Figure 4.18a to separate the external and internal degrees of freedom as in Eq. (4.114). The element stiffness matrices are [Eqs. (3.37), (4.96), and (4.97)]

$$\mathbf{k}^1 = \alpha \begin{bmatrix} 4L^2s_a & -4L^2s_a \\ \text{sym.} & 4L^2s_a \end{bmatrix} \begin{matrix} 2 \\ 5 \end{matrix} \qquad \begin{matrix} 2 & \quad & 5 \end{matrix} \tag{4.122a}$$

$$\mathbf{k}^2 = \alpha \begin{bmatrix} 4L^2s_b & -4L^2s_b \\ \text{sym.} & 4L^2s_b \end{bmatrix} \begin{matrix} 6 \\ 4 \end{matrix} \qquad \begin{matrix} 6 & \quad & 4 \end{matrix} \tag{4.122b}$$

$$\mathbf{k}^3 = \alpha \begin{bmatrix} 12 & 6L & -12 & 6L \\ & 4L^2 & -6L & 2L^2 \\ & & 12 & -6L^2 \\ & & & 4L^2 \end{bmatrix} \begin{matrix} 1 \\ 5 \\ 3 \\ 6 \end{matrix} \qquad \begin{matrix} 1 & 5 & 3 & 6 \end{matrix} \tag{4.122c}$$

Equations (4.122) are transformed into generalized element stiffness matrices and added to yield the stiffness matrix of the complex element

$$\mathbf{k} = \alpha \left[\begin{array}{cccc:cc} 12 & 0 & -12 & 0 & 6L & 6L \\ 0 & 4L^2s_a & 0 & 0 & -4L^2s_a & 0 \\ -12 & 0 & 12 & 0 & -6L & -6L \\ 0 & 0 & 0 & 4L^2s_b & 0 & -4L^2s_b \\ \hdashline 6L & -4L^2s_a & -6L & 0 & 4L^2(1+s_a) & 2L^2 \\ 6L & 0 & -6L & -4L^2s_b & 2L^2 & 4L^2(1+s_b) \end{array} \right] \begin{matrix} 1 \\ 2 \\ 3 \\ 4 \\ 5 \\ 6 \end{matrix}$$

with column headers $1\ 2\ 3\ 4\ 5\ 6$, where the block $\frac{1}{\alpha}\mathbf{k}_{ee}$ spans columns 1–4, the block $\frac{1}{\alpha}\mathbf{k}_{ei}$ spans columns 5–6, the block $\frac{1}{\alpha}\mathbf{k}_{ie}$ spans the lower-left, and $\frac{1}{\alpha}\mathbf{k}_{ii}$ spans the lower-right.

$$\tag{4.123}$$

which is partitioned consistent with Eq. (4.114). Note that the upper triangular portion of Eq. (4.122c) does not map into the upper triangular
portion of Eq. (4.123) because all nonzero integers of the member code do
not increase from top to bottom. However, by invoking symmetry (for
example, $k_{35} = k_{53}$), we can map Eq. (4.122c) into the upper triangular
portion of Eq. (4.123).

The inverse of $\mathbf{k}_{ii}$ [Eq. (4.123)] can be expressed as

$$\mathbf{k}_{ii}^{-1} = \frac{1}{\alpha 4 L^2 s} \begin{bmatrix} 2(1 + s_b) & -1 \\ -1 & 2(1 + s_a) \end{bmatrix} \tag{4.124}$$

where

$$s = 2(1 + s_a)(1 + s_b) - \tfrac{1}{2} \tag{4.125}$$

By performing the matrix operations defined in Eqs. (4.118) and (4.119) on
the submatrices in Eqs. (4.123) and (4.124) we obtain the condensed complex
element model

$$\bar{\mathbf{f}}_e = \mathbf{k}_e \mathbf{d}_e + \mathbf{f}_e^* \tag{4.126}$$

where

$$\mathbf{k}_e = \alpha \begin{bmatrix} 12s_1 & 6Ls_2 & -12s_1 & 6Ls_4 \\ & 4L^2 s_3 & -6Ls_2 & 2L^2 s_5 \\ & & 12s_1 & -6Ls_4 \\ \text{sym.} & & & 4L^2 s_6 \end{bmatrix} \tag{4.127}$$

$$s_1 = \frac{1}{2s}(s_a + s_b + 4 s_a s_b)$$

$$s_2 = \frac{s_a}{s}(1 + 2s_b)$$

$$s_3 = \frac{s_a}{2s}(3 + 4s_b)$$

$$\tag{4.128}$$

$$s_4 = \frac{s_b}{s}(1 + 2s_a)$$

$$s_5 = \frac{2}{s}(s_a s_b)$$

$$s_6 = \frac{s_b}{2s}(3 + 4s_a)$$

and

$$
\mathbf{f}_e^* =
\begin{bmatrix} f_1^* \\ f_2^* \\ f_3^* \\ f_4^* \end{bmatrix}
=
\begin{bmatrix}
\dfrac{3}{2Ls}[(1 + 2s_b)f_5 + (1 + 2s_a)f_6] \\[2ex]
\dfrac{s_a}{s}[-2(1 + s_b)f_5 + f_6] \\[2ex]
\dfrac{-3}{2Ls}[(1 + 2s_b)f_5 + (1 + 2s_a)f_6] \\[2ex]
\dfrac{s_b}{s}[f_5 - 2(1 + s_a)f_6]
\end{bmatrix}
\tag{4.129}
$$

The internal forces and the fixed-end forces defined by Eq. (4.129) are shown in Figure 4.18c.

Special Case

Equations (4.127)–(4.129) are specialized for the spring stiffnesses $s_a = 0$, $s_b = \infty$, which characterize the element in Figure 1.18 with a frictionless hinge near the a end and the slope continuity restored near the b end. The limits of the coefficients in Eqs. (4.128) and (4.129) with the indeterminate form ∞/∞ can be obtained by L'Hôpital's rule (Thomas, 1956). For example,

$$
\lim_{\substack{s_b \to \infty \\ s_a \to 0}} s_1 = \lim_{s_a \to 0} \frac{1 + 4s_a}{4(1 + s_a)} = \frac{1}{4}
\tag{4.130}
$$

The resulting stiffness matrix

$$
\mathbf{k}_e = \alpha
\begin{bmatrix}
3 & 0 & -3 & 3L \\
 & 0 & 0 & 0 \\
 & & 3 & -3L \\
\text{sym.} & & & 3L^2
\end{bmatrix}
\tag{4.131}
$$

agrees with the flexural coefficients in Eq. (1.149), and the resulting fixed-end force vector, whose components are shown in Figure 4.18d, is

$$
\mathbf{f}_e^* =
\begin{bmatrix}
\dfrac{3}{2L}f_5 \\[2ex]
0 \\[2ex]
-\dfrac{3}{2L}f_5 \\[2ex]
\dfrac{1}{2}f_5 - f_6
\end{bmatrix}
\tag{4.132}
$$

Example 2. The fixed-end forces of the element in Figure 4.19a are computed with the aid of Eq. (4.132). Accordingly, the load is resolved into the component loads of Figures 4.19b and c. In Figure 4.19b, fixed-end moments are

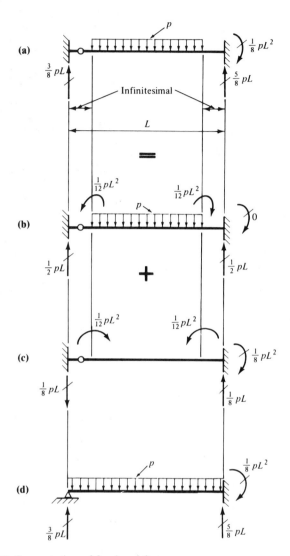

FIGURE 4.19 Computation of fixed-end forces.

applied to the internal beam element to prevent end rotations, that is, to cause $d_5 = d_6 = 0$. The corresponding shear forces are transmitted through the infinitesimal springs to the supports. The substitution for $f_6 = (1/12)pL^2 = -f_5$ (Figure 4.19c) in Eq. (4.132) yields the fixed-end forces in Figure 4.19c

$$\mathbf{f}_e^* = \begin{bmatrix} -\frac{1}{8}pL \\ 0 \\ \frac{1}{8}pL \\ -\frac{1}{8}pL^2 \end{bmatrix} \tag{4.133}$$

Finally, the superposition of the fixed-end forces in Figures 4.19b and c yields the fixed-end forces in Figure 4.19a

$$\hat{\mathbf{f}}_e = \begin{bmatrix} \frac{3}{8}pL \\ 0 \\ \frac{5}{8}pL \\ -\frac{1}{8}pL^2 \end{bmatrix} \tag{4.134}$$

Note that Figures 4.19a and d are equivalent.

Substructure Analysis

Since condensed complex element models assume the form of Eqs. (3.95), assemblages of complex elements can be analyzed by the matrix displacement method defined in Figure 3.9. However, it is constructive to reformulate the system model by starting with the condensed complex element model in global coordinates

$$\overline{\mathbf{F}}_e^i = \mathbf{K}_e^i \mathbf{D}_e^i + \hat{\mathbf{F}}_e^i \tag{4.135}$$

where

$$\hat{\mathbf{F}}_e^i = \mathbf{F}_e^{*i} \tag{4.136}$$

if the complex element actions consist solely of the load vector $\mathbf{F}_i$ [see Eqs. (4.117)–(4.119)].

We assemble the complex elements by imposing conditions of compatibility and equilibrium (Section 2.4). The conditions of compatibility permit us to transform the global element model, Eq. (4.135), into the generalized element model

$$\overline{\mathbf{F}}^{(i)} = \mathbf{K}^{(i)}\mathbf{q} + \hat{\mathbf{F}}^{(i)} \tag{4.137}$$

where

$$\mathbf{K}_e^i \xrightarrow{\mathbf{M}^i} \mathbf{K}^{(i)}$$
$$\overline{\mathbf{F}}_e^i \xrightarrow{\mathbf{M}^i} \hat{\mathbf{F}}^{(i)} \tag{4.138}$$

$\mathbf{M}^i$ is the member code of element i. Note that the condensed complex elements may have distinct degrees of freedom (Section 4.5). This is illustrated in Figure 4.20, where the truss in Figure 4.20a is represented by the three condensed complex elements in Figure 4.20b with four, eight, and four degrees of freedom, respectively. The assembly of the condensed complex elements is shown in Figure 4.20c. Accordingly the member codes are

$$\mathbf{M}^{1T} = \begin{bmatrix} 1 & 2 & 3 & 4 \end{bmatrix}$$
$$\mathbf{M}^{2T} = \begin{bmatrix} 1 & 2 & 3 & 4 & 5 & 6 & 7 & 8 \end{bmatrix} \tag{4.139}$$
$$\mathbf{M}^{3T} = \begin{bmatrix} 5 & 6 & 7 & 8 \end{bmatrix}$$

204

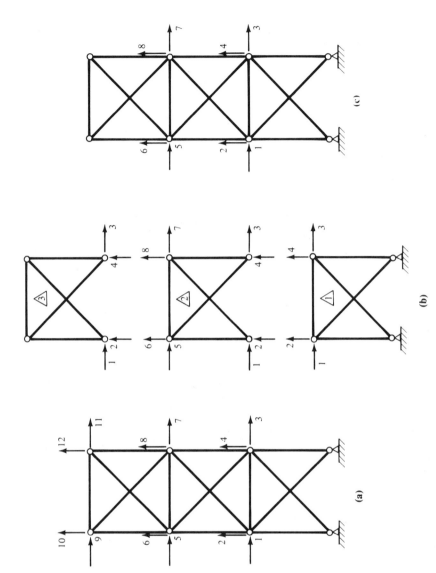

FIGURE 4.20 Substructuring. **(a)** Truss; **(b)** complex elements; **(c)** assembly of complex elements.

The condition of equilibrium of the assemblage is

$$\bar{Q} = \sum_{i=1}^{NE} \bar{F}^{(i)} \qquad (4.140)$$

The substitution of Eq. (4.137) into Eq. (4.140) yields the system model

$$Q = Kq \qquad (4.141)$$

where

$$K = \sum_{i=1}^{NE} K^{(i)}$$

$$Q = \bar{Q} - \hat{Q} \qquad (4.142)$$

$$\hat{Q} = \sum_{i=1}^{NE} \hat{F}^{(i)}$$

Example 3. The structure in Figure 4.21a consists of a standard beam element and an element with a frictionless hinge near the a end. Alternatively, the hinge could be placed into joint 2 as in Figure 4.15a. No joint loads are applied. The analysis is based on Figure 3.9.

1. *Unknowns*

$$q_k, \quad k = 1, 2 \qquad (4.143)$$

2. *Element Models*

$$\bar{f}^i = f^i + \hat{f}^i \qquad (4.144)$$

where

$$f^i = k^i d^i \qquad (4.145)$$

k^1 is defined in Eq. (3.37), k^2 is defined in Eq. (4.131),

$$\hat{f}^1 = pL \begin{bmatrix} \dfrac{1}{2} & 0 \\[2mm] \dfrac{L}{12} & 0 \\[2mm] \dfrac{1}{2} & 1 \\[2mm] -\dfrac{L}{12} & 2 \end{bmatrix}, \quad \hat{f}^2 = pL \begin{bmatrix} \dfrac{3}{8} & 1 \\[2mm] 0 & 2 \\[2mm] \dfrac{5}{8} & 0 \\[2mm] -\dfrac{L}{8} & 0 \end{bmatrix} \qquad (4.146)$$

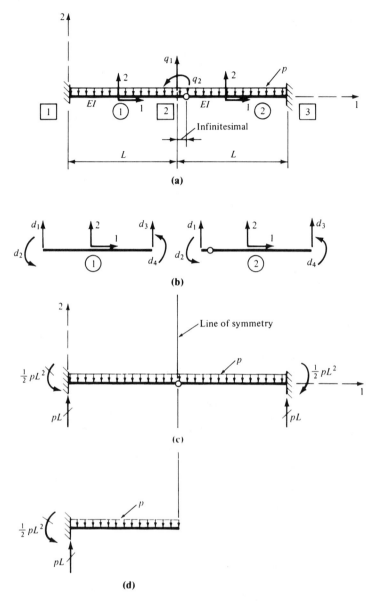

FIGURE 4.21 Beam assembly with element release. **(a)** Beam assembly; **(b)** elements; **(c)** joint forces; **(d)** cantilever.

and according to Figures 4.21a, b

$$\mathbf{M}^T = \begin{bmatrix} 0 & 0 & 1 & 2 \\ 1 & 2 & 0 & 0 \end{bmatrix} \tag{4.147}$$

In Eqs. (4.146), $\hat{\mathbf{f}}^1$ is the standard fixed-end force vector and $\hat{\mathbf{f}}^2$ follows from Eq. (4.134).

3. *System Model*

System Stiffness Matrix

$$\mathbf{k}^1 = \alpha \begin{array}{cccc} & 0 \quad\quad 0 \quad\quad 1 \quad\quad 2 \\ \begin{bmatrix} & & & \\ & & & \\ & & 12 & -6L \\ \text{sym.} & & & 4L^2 \end{bmatrix} \begin{array}{l} 0 \\ 0 \\ 1 \\ 2 \end{array} \end{array} \tag{4.148a}$$

$$\mathbf{k}^2 = \alpha \begin{array}{cccc} & 1 \quad\quad 2 \quad\; 0 \quad 0 \\ \begin{bmatrix} 3 & 0 & & \\ & 0 & & \\ & & & \\ \text{sym.} & & & \end{bmatrix} \begin{array}{l} 1 \\ 2 \\ 0 \\ 0 \end{array} \end{array} \tag{4.148b}$$

By Eqs. (4.148) we obtain the system stiffness matrix

$$\mathbf{K} = \alpha \begin{array}{cc} & 1 \quad\;\; 2 \\ \begin{bmatrix} 15 & -6L \\ \text{sym.} & 4L^2 \end{bmatrix} \begin{array}{l} 1 \\ 2 \end{array} \end{array} \tag{4.149}$$

Equivalent Joint Load Vector

$\overline{\mathscr{L}}$:

$$\overline{\mathbf{Q}} = \mathbf{0} \tag{4.150}$$

$\hat{\mathscr{L}}$:

$$\hat{\mathbf{f}}^1 \overset{\mathbf{M}}{\to} \hat{\mathbf{F}}^{(1)} = pL \begin{bmatrix} \dfrac{1}{2} \\[2mm] -\dfrac{L}{12} \end{bmatrix} \begin{array}{l} 1 \\ 2 \end{array}, \quad \hat{\mathbf{f}}^2 \overset{\mathbf{M}}{\to} \hat{\mathbf{F}}^{(2)} = pL \begin{bmatrix} \dfrac{3}{8} \\[2mm] 0 \end{bmatrix} \begin{array}{l} 1 \\ 2 \end{array} \tag{4.151}$$

$$\hat{\mathbf{Q}} = \sum_{i=1}^{2} \hat{\mathbf{F}}^{(i)} = pL \begin{bmatrix} \dfrac{7}{8} \\[2mm] -\dfrac{L}{12} \end{bmatrix} \tag{4.152}$$

$\mathscr{L}$:

$$Q = \bar{Q} - \hat{Q} = pL \begin{bmatrix} -\dfrac{7}{8} \\[2mm] \dfrac{L}{12} \end{bmatrix} \tag{4.153}$$

System Model

$$\alpha \begin{bmatrix} 15 & -6L \\ -6L & 4L^2 \end{bmatrix} \begin{bmatrix} q_1 \\ q_2 \end{bmatrix} = pL \begin{bmatrix} -\dfrac{7}{8} \\[2mm] \dfrac{L}{22} \end{bmatrix} \tag{4.154}$$

4. *Solution*

$$q_1 = -\frac{pL}{8\alpha}, \quad q_2 = -\frac{p}{6\alpha} \tag{4.155}$$

5. *Element Forces*

By Eqs. (4.145), (4.147), and (4.155)

$$\mathbf{d}^1 = \begin{bmatrix} 0 \\ 0 \\ q_1 \\ q_2 \end{bmatrix} \begin{matrix} 0 \\ 0 \\ 1 \\ 2 \end{matrix}, \quad \mathbf{d}^2 = \begin{bmatrix} q_1 \\ q_2 \\ 0 \\ 0 \end{bmatrix} \begin{matrix} 1 \\ 2 \\ 0 \\ 0 \end{matrix} \tag{4.156a}$$

$$\mathbf{f}^1 = pL \begin{bmatrix} \dfrac{1}{2} \\[2mm] \dfrac{5L}{12} \\[2mm] -\dfrac{1}{2} \\[2mm] \dfrac{L}{12} \end{bmatrix}, \quad \mathbf{f}^2 = pL \begin{bmatrix} -\dfrac{3}{8} \\[2mm] 0 \\[2mm] \dfrac{3}{8} \\[2mm] -\dfrac{3L}{8} \end{bmatrix} \tag{4-156b}$$

Equations (4.144), (4.146), and (4.156) yield

$$\bar{\mathbf{f}}^1 = \begin{bmatrix} pL \\ \frac{1}{2}pL^2 \\ \hdashline 0 \\ 0 \end{bmatrix}, \quad \bar{\mathbf{f}}^2 = \begin{bmatrix} 0 \\ 0 \\ \hdashline pL \\ -\frac{1}{2}pL^2 \end{bmatrix} \tag{4.157}$$

6. *Joint Forces*

$$\bar{P}_1 = \bar{f}_a^1, \quad \bar{P}_2 = \bar{f}_b^1 + \bar{f}_a^2 = \begin{bmatrix} 0 \\ 0 \end{bmatrix}, \quad \bar{P}_3 = \bar{f}_b^2 \qquad (4.158)$$

The joint forces are shown in Figure 4.21c. The results can easily be checked because by symmetry (Section 4.2) the shear force at the hinge is zero. Thus, each half of the structure acts as a cantilever (Figure 4.21d).

Example 4. The truss in Figure 4.22a is regarded as an assemblage of two complex elements (Figures 4.22b). The elements in Figure 4.22a have identical properties A, E, L. The analysis is based on Figure 3.9. The analysis units are kilonewton and meter.

Complex Elements (Figures 4.22c, d)
The condensed models of the complex elements are formulated and expressed as [Eq. (4.135)]

$$\bar{F}_e = K_e D_e + \hat{F}_e \qquad (4.159)$$

where according to Eqs. (4.116), (4.118), and (4.119)

$$K_c = K_{ee} - K_{ei} K_{ii}^{-1} K_{ie} \qquad (4.160)$$

$$\hat{F}_e = K_{ei} K_{ii}^{-1} F_i \qquad (4.161)$$

$$D_i = -K_{ii}^{-1} K_{ie} D_e + K_{ii}^{-1} F_i \qquad (4.162)$$

It follows from Figures 4.22a–d that

$$D_e = \begin{bmatrix} D_1 \\ D_2 \end{bmatrix}, \quad D_i = \begin{bmatrix} D_3 \\ D_4 \end{bmatrix}, \quad \text{and} \quad F_i = \begin{bmatrix} 0 \\ -4\sqrt{3} \end{bmatrix} \qquad (4.163)$$

for each complex element.

Element 1 (Figure 4.22c)

$$\lambda^1 = [-1 \quad 0], \quad \lambda^2 = \tfrac{1}{2}[1 \quad -\sqrt{3}], \quad \lambda^3 = \tfrac{1}{2}[-1 \quad -\sqrt{3}] \quad (4.164)$$

$$M^T = \begin{bmatrix} 1 & 2 & 3 & 4 \\ 3 & 4 & 0 & 0 \\ 1 & 2 & 0 & 0 \end{bmatrix} \qquad (4.165)$$

210

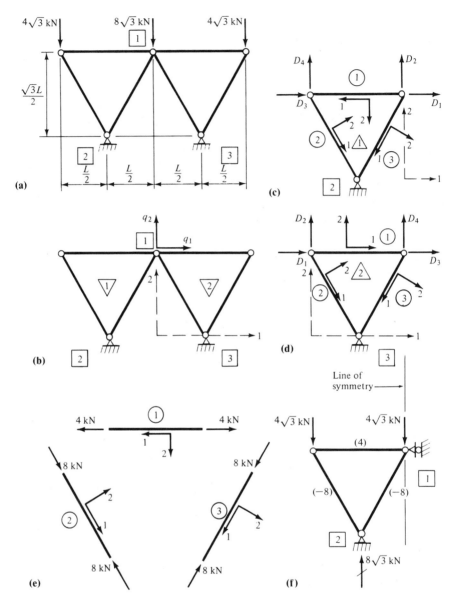

FIGURE 4.22 Substructure analysis. **(a)** Truss; **(b)** assembly of complex elements; **(c, d)** complex elements; **(e)** element forces; **(f)** joint forces.

By Eqs. (3.78), (4.164), and (4.165)

$$\mathbf{K}^1 = \frac{\gamma}{4} \begin{array}{cccc} ^1 & ^2 & ^3 & ^4 \\ \left[\begin{array}{cccc} 4 & 0 & -4 & 0 \\ & 0 & 0 & 0 \\ & & 4 & 0 \\ \text{sym.} & & & 0 \end{array}\right] & \begin{array}{c} 1 \\ 2 \\ 3 \\ 4 \end{array} \end{array} \tag{4.166a}$$

$$\mathbf{K}^2 = \frac{\gamma}{4} \begin{array}{cccc} ^3 & ^4 & ^0 & ^0 \\ \left[\begin{array}{cccc} 1 & -\sqrt{3} & & \\ & 3 & & \\ & & & \\ \text{sym.} & & & \end{array}\right] & \begin{array}{c} 3 \\ 4 \\ 0 \\ 0 \end{array} \end{array} \tag{4.166b}$$

$$\mathbf{K}^3 = \frac{\gamma}{4} \begin{array}{cccc} ^1 & ^2 & ^0 & ^0 \\ \left[\begin{array}{cccc} 1 & \sqrt{3} & & \\ & 3 & & \\ & & & \\ \text{sym.} & & & \end{array}\right] & \begin{array}{c} 1 \\ 2 \\ 0 \\ 0 \end{array} \end{array} \tag{4.166c}$$

From Eqs. (4.166) we obtain the stiffness matrix of complex element 1

$$\mathbf{K} = \frac{\gamma}{4} \left[\begin{array}{cc:cc} 5 & \sqrt{3} & -4 & 0 \\ \sqrt{3} & 3 & 0 & 0 \\ \hdashline -4 & 0 & 5 & -\sqrt{3} \\ 0 & 0 & -\sqrt{3} & 3 \end{array}\right] \tag{4.167}$$

with upper partitions labeled $\frac{4}{\gamma}\mathbf{K}_{ee}$ and $\frac{4}{\gamma}\mathbf{K}_{ei}$, and lower partitions labeled $\frac{4}{\gamma}\mathbf{K}_{ie}$ and $\frac{4}{\gamma}\mathbf{K}_{ii}$.

which is partitioned consistent with Eqs. (4.114) and (4.163). Equations (4.160)–(4.163) and (4.167) yield

$$\mathbf{K}_{ii}^{-1} = \frac{1}{\gamma} \left[\begin{array}{cc} 1 & \dfrac{\sqrt{3}}{3} \\ \dfrac{\sqrt{3}}{3} & \dfrac{5}{3} \end{array}\right] \tag{4.168}$$

$$\mathbf{K}_e = \frac{\gamma}{4}\begin{bmatrix} 1 & \sqrt{3} \\ \sqrt{3} & 3 \end{bmatrix}, \quad \hat{\mathbf{F}}_e = \begin{bmatrix} 4 \\ 0 \end{bmatrix} \tag{4.169}$$

$$\mathbf{D}_i = \begin{bmatrix} 1 & 0 \\ \dfrac{\sqrt{3}}{3} & 0 \end{bmatrix} \mathbf{D}_e + \frac{1}{\gamma}\begin{bmatrix} -4 \\ -\dfrac{20\sqrt{3}}{3} \end{bmatrix} \tag{4.170}$$

Element 2 (Figure 4.22d)
Similarly, we obtain for the condensed complex element 2

$$\mathbf{K}_e = \frac{\gamma}{4}\begin{bmatrix} 1 & -\sqrt{3} \\ -\sqrt{3} & 3 \end{bmatrix}, \quad \hat{\mathbf{F}}_e = \begin{bmatrix} -4 \\ 0 \end{bmatrix} \tag{4.171}$$

Analysis of Truss (Figure 4.22b)

1. *Unknowns*

$$q_k, \quad k = 1, 2 \tag{4.172}$$

2. *Element Models*

$$\bar{\mathbf{F}}^i = \mathbf{K}_e^i \mathbf{D}_e^i + \hat{\mathbf{F}}_e^i, \quad i = 1, 2 \tag{4.173}$$

where $\mathbf{K}_e^1, \hat{\mathbf{F}}_e^1$ and $\mathbf{K}_e^2, \hat{\mathbf{F}}_e^2$ are defined in Eqs. (4.169) and (4.171), respectively. The member codes for the condensed complex elements are (Figures 4.22b–d)

$$\mathbf{M}^{iT} = [1 \quad 2], \quad i = 1, 2 \tag{4.174}$$

Thus,

$$\mathbf{D}_e^i = \mathbf{q}, \quad \hat{\mathbf{F}}_e^i = \hat{\mathbf{F}}^{(i)}, \quad \mathbf{K}_e^i = \mathbf{K}^{(i)} \tag{4.175}$$

3. *System Model*
System Stiffness Matrix

$$\mathbf{K} = \sum_{i=1}^{2} \mathbf{K}^{(i)} = \frac{\gamma}{2}\begin{bmatrix} 1 & 0 \\ 0 & 3 \end{bmatrix} \tag{4.176}$$

The diagonal form of the system stiffness matrix signifies that the deflections are uncoupled; that is, a horizontal (vertical) force at joint 1 causes only a horizontal (vertical) deflection.

Equivalent Joint Load Vector
$\mathscr{L}$: According to Figures 4.22a and b

$$\bar{\mathbf{Q}} = \begin{bmatrix} 0 \\ -8\sqrt{3} \end{bmatrix} \tag{4.177}$$

$\hat{\mathscr{L}}$:

$$\hat{\mathbf{Q}} = \sum_{i=1}^{2} \hat{\mathbf{F}}^{(i)} = \mathbf{0} \tag{4.178}$$

$\mathscr{L}$:

$$\mathbf{Q} = \bar{\mathbf{Q}} - \hat{\mathbf{Q}} = \bar{\mathbf{Q}} \qquad (4.179)$$

System Model

$$\frac{\gamma}{2}\begin{bmatrix} 1 & 0 \\ 0 & 3 \end{bmatrix}\begin{bmatrix} q_1 \\ q_2 \end{bmatrix} = \begin{bmatrix} 0 \\ -8\sqrt{3} \end{bmatrix} \qquad (4.180)$$

4. *Solution*

$$q_1 = 0, \quad q_2 = -\frac{16\sqrt{3}}{3\gamma} \qquad (4.181)$$

5. *Element Forces.* The computations are confined to the *complex element* 1 (Figure 4.22c): It follows from Eqs. (4.170), (4.175), and (4.181) that

$$\mathbf{D}_e^1 = \frac{1}{\gamma}\begin{bmatrix} 0 \\ -\dfrac{16\sqrt{3}}{3} \end{bmatrix}, \quad \mathbf{D}_i^1 = \frac{1}{\gamma}\begin{bmatrix} -4 \\ -\dfrac{20\sqrt{3}}{3} \end{bmatrix} \qquad (4.182)$$

By Eqs. (4.164), (4.165), and (4.182)

$$d_a^1 = \lambda^1\mathbf{D}_e^1 = 0, \quad d_b^1 = \lambda^1\mathbf{D}_i^1 = \frac{4}{\gamma} \qquad (4.183)$$

Thus

$$\mathbf{f}^1 = \mathbf{k}^1\mathbf{d}^1 = \gamma\begin{bmatrix} 1 & -1 \\ -1 & 1 \end{bmatrix}\begin{bmatrix} 0 \\ \dfrac{4}{\gamma} \end{bmatrix} = \begin{bmatrix} -4 \\ 4 \end{bmatrix} \qquad (4.184)$$

Similarly, we obtain

$$\mathbf{f}^2 = \mathbf{f}^3 = \begin{bmatrix} 8 \\ -8 \end{bmatrix} \qquad (4.185)$$

The free-body diagrams of these elements are shown in Figure 4.22e. These results can easily be checked by statics because as a result of symmetry (Section 4.2), the trusses in Figures 4.22a and f are equivalent.

6. *Joint Forces*
By Eqs. (4.164) and (4.185)

$$\mathbf{P}_2 = \mathbf{F}_b^2 + \mathbf{F}_b^3 = \lambda^{2T}f_b^2 + \lambda^{3T}f_b^3 = \begin{bmatrix} 0 \\ 8\sqrt{3} \end{bmatrix} \qquad (4.186)$$

This agrees with Figure 4.22f.

The formulation of condensed models by Eqs. (4.117)–(4.119) and the computation of the internal displacements by Eq. (4.116) are not efficient. A practical tool for condensation is Gaussian elimination. This is illustrated in Section 6.3.

4.7 GEOMETRIC IMPERFECTIONS

Aside from unavoidable fabrication and construction tolerances, imperfections in the geometry of a structure may be due to support settlements and fabrication errors. In a statically determinate structure, deviations in support locations and member dimensions cause changes in the configuration of the structure but no stresses. However, in statically indeterminate structures, significant erection stresses may arise as a result of geometric imperfections.

The effects of geometric imperfections can be determined with the matrix displacement method (Figure 3.9) by transforming the geometric imperfections into equivalent joint loads (Section 3.6).

Support Settlements

Consider the continuous beams in Figure 4.23a. The settlement of joint 3 can be specified in terms of the element displacement

$$d_3^2 = -\Delta \tag{4.187}$$

and regarded as an element action. Thus, we can use Eq. (3.90) to transform the geometrically *imperfect structure* (Figure 4.23a) into a geometrically *perfect structure* (Figure 4.23c) with joint loads. Note that in Figure 4.23b the unknown joint displacements are constrained, that is, $\hat{q}_1 = \hat{q}_2 = 0$, but the prescribed joint displacement is retained. Accordingly,

$$\hat{\mathbf{d}}^1 = \mathbf{0}, \quad \hat{\mathbf{d}}^2 = \begin{bmatrix} 0 \\ 0 \\ -\Delta \\ 0 \end{bmatrix} \tag{4.188}$$

and the corresponding "fixed-end" force vectors are

$$\hat{\mathbf{f}}^1 = \mathbf{0}, \quad \hat{\mathbf{f}}^2 = \mathbf{k}^2 \hat{\mathbf{d}}^2 \tag{4.189}$$

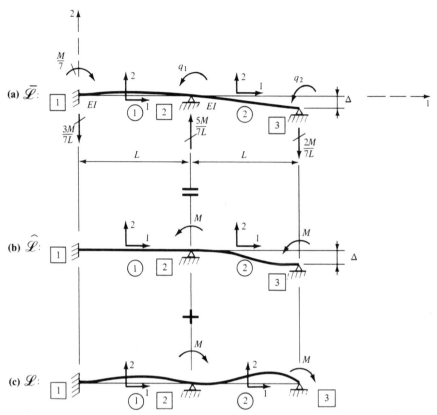

FIGURE 4.23 Support settlement analysis. (a) $\overline{\mathscr{L}}$; (b) $\widehat{\mathscr{L}}$; (c) $\mathscr{L}$.

In general, if the support displacements are specified in global coordinates, the global and local "fixed-end" force vectors can be computed as

$$\hat{\mathbf{F}}^i = \mathbf{K}^i\hat{\mathbf{D}}^i, \quad \hat{\mathbf{f}}^i = \mathbf{\Lambda}^i\hat{\mathbf{F}}^i \tag{4.190}$$

or

$$\hat{\mathbf{d}}^i = \mathbf{\Lambda}^i\hat{\mathbf{D}}^i, \quad \hat{\mathbf{f}}^i = \mathbf{k}^i\hat{\mathbf{d}}^i, \quad \hat{\mathbf{F}}^i = \mathbf{\Lambda}^{iT}\hat{\mathbf{f}}^i \tag{4.191}$$

Example 1. The continuous beams in Figure 4.23a are analyzed by the matrix displacement method (Figure 3.9)

1. *Unknowns*

$$q_k, \quad k = 1, 2 \tag{4.192}$$

2. *Element Models*

$$\bar{\mathbf{f}}^i = \mathbf{k}\mathbf{d}^i + \hat{\mathbf{f}}^i \tag{4.193}$$

where **k** is defined in Eq. (3.37), and by Eqs. (4.188) and (4.189)

$$\hat{\mathbf{f}}^1 = \mathbf{0}, \quad \hat{\mathbf{f}}^2 = M \begin{bmatrix} \dfrac{2}{L} & 0 \\[2mm] 1 & 1 \\[2mm] -\dfrac{2}{L} & 0 \\[2mm] 1 & 2 \end{bmatrix}, \quad M = 6\alpha L\Delta \qquad (4.194)$$

The member code matrix is defined in Eq. (3.3).

3. *System Model*

System Stiffness Matrix
The system stiffness matrix is defined in Eq. (3.15).

Equivalent Joint Load Vector (Figures 4.23a–c)
$\overline{\mathcal{L}}$:

$$\overline{\mathbf{Q}} = \mathbf{0} \qquad (4.195)$$

$\hat{\mathcal{L}}$: By Eqs. (4.194)

$$\hat{\mathbf{F}}^{(1)} = \mathbf{0}, \quad \hat{\mathbf{F}}^{(2)} = \begin{bmatrix} M \\ M \end{bmatrix} \qquad (4.196)$$

Thus,

$$\hat{\mathbf{Q}} = \sum_{i=1}^{2} \hat{\mathbf{F}}^{(i)} = \begin{bmatrix} M \\ M \end{bmatrix} \qquad (4.197)$$

$\mathcal{L}$:

$$\mathbf{Q} = \overline{\mathbf{Q}} - \hat{\mathbf{Q}} = \begin{bmatrix} -M \\ -M \end{bmatrix} \qquad (4.198)$$

System Model

$$\alpha \begin{bmatrix} 8L^2 & 2L^2 \\ 2L^2 & 4L^2 \end{bmatrix} \begin{bmatrix} q_1 \\ q_2 \end{bmatrix} = -M \begin{bmatrix} 1 \\ 1 \end{bmatrix} \qquad (4.199)$$

4. *Solution*

$$q_1 = -\frac{M}{\alpha 14L^2}, \quad q_2 = 3q_1 \qquad (4.200)$$

5. *Element Forces.*
By Eqs. (3.4), (4.193), (4.194), and (4.200),

$$\bar{\mathbf{f}}^1 = \frac{M}{7} \begin{bmatrix} -\dfrac{3}{L} \\ -1 \\ \hline \dfrac{3}{L} \\ -2 \end{bmatrix}, \quad \bar{\mathbf{f}}^2 = \frac{M}{7} \begin{bmatrix} \dfrac{2}{L} \\ 2 \\ \hline -\dfrac{2}{L} \\ 0 \end{bmatrix} \tag{4.201}$$

6. *Joint Forces* (Figure 4.23a)

$$\bar{\mathbf{P}}_1 = \bar{\mathbf{f}}_a^1 = \frac{M}{7} \begin{bmatrix} -\dfrac{3}{L} \\ -1 \end{bmatrix}, \quad \bar{\mathbf{P}}_3 = \bar{\mathbf{f}}_b^2 = \frac{M}{7} \begin{bmatrix} -\dfrac{2}{L} \\ 0 \end{bmatrix}$$

$$\bar{\mathbf{P}}_2 = \bar{\mathbf{f}}_b^1 + \bar{\mathbf{f}}_a^2 = \frac{M}{7} \begin{bmatrix} \dfrac{5}{L} \\ 0 \end{bmatrix} \tag{4.202}$$

Element Imperfections

Consider the imperfect element in Figure 4.24b whose deviation from the exact configuration is specified by the *deformation vector* **e** (Figure 1.17b). The corresponding fixed-end force vectors (Figure 4.24a) can be computed by the force method on the basis of Figures 4.24a–c (see references to textbooks on elementary structural analysis in Chapter 1): The condition of compatibility is

$$\hat{\mathbf{d}}_b = \mathbf{e} + \mathbf{e}_b = 0 \tag{4.203}$$

or

$$\mathbf{e}_b = -\mathbf{e} \tag{4.204}$$

The principle of superposition yields

$$\hat{\mathbf{f}}_a = \mathbf{f}_a, \quad \hat{\mathbf{f}}_b = \mathbf{f}_b \tag{4.205}$$

where by Eqs. (1.122), (1.124), and (1.142)

$$\mathbf{f}_a = -\mathbf{T}_{ba}^T \mathbf{f}_b, \quad \mathbf{f}_b = \mathbf{k}_{bb} \mathbf{e}_b \tag{4.206}$$

Thus, the fixed-end force vectors can be expressed in terms of the deformation vector as

$$\hat{\mathbf{f}}_b = -\mathbf{k}_{bb} \mathbf{e}, \quad \hat{\mathbf{f}}_a = -\mathbf{T}_{ba}^T \hat{\mathbf{f}}_b \tag{4.207}$$

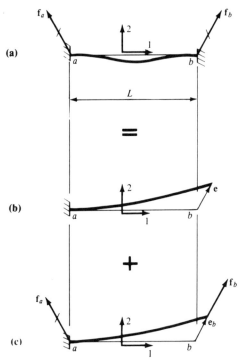

(a)

(b)

(c)

FIGURE 4.24 Fixed-end forces of imperfect element.

Alternatively, the fixed-end force vector can be defined as

$$\hat{\mathbf{f}} = \mathbf{kd} \qquad (4.208)$$

where

$$\mathbf{d} = \begin{bmatrix} \mathbf{d}_a \\ \mathbf{d}_b \end{bmatrix} = \begin{bmatrix} \mathbf{0} \\ -\mathbf{e} \end{bmatrix} \qquad (4.209)$$

Example 2. Consider a crooked beam element with the deformation vector

$$\mathbf{e} = \begin{bmatrix} e_1 \\ e_2 \end{bmatrix} = \begin{bmatrix} 0 \\ \theta \end{bmatrix} \qquad (4.210)$$

where $\theta = \frac{1}{4}$ deg $= \pi/720$ rad. The properties of the beam are defined in Example 1 of Section 3.4; in addition, the section modulus $S = 12.5$ in^3. The fixed-end forces and the maximum bending stress are computed. The analysis units are kilopound, inch, radian.

From Eqs. (1.98b) and (1.123), we obtain

$$\mathbf{k}_{bb} = \alpha \begin{bmatrix} 12 & -6L \\ -6L & 4L^2 \end{bmatrix}, \quad \mathbf{T}_{ba} = \begin{bmatrix} 1 & L \\ 0 & 1 \end{bmatrix} \qquad (4.211)$$

Equations (4.207), (4.210), and (4.211) yield the fixed-end force vectors

$$\hat{\mathbf{f}}_b = \begin{bmatrix} -2.7 \\ 218.2 \end{bmatrix}, \quad \hat{\mathbf{f}}_a = \begin{bmatrix} 2.7 \\ 109.1 \end{bmatrix} \tag{4.212}$$

Thus, the maximum bending stress

$$\max \sigma = \frac{218.2}{12.5} = 17.5 \text{ k/in}^2 \tag{4.213}$$

Example 3. The elements of the truss in Figure 4.25a have the identical extensional stiffness $\gamma = 100$ k/in. Element 3 is 0.5 inch too short, that is,

FIGURE 4.25 Analysis of imperfect truss. **(a)** $\overline{\mathscr{L}}$; **(b)** $\hat{\mathscr{L}}$; **(c)** $\mathscr{L}$; **(d)** joint forces.

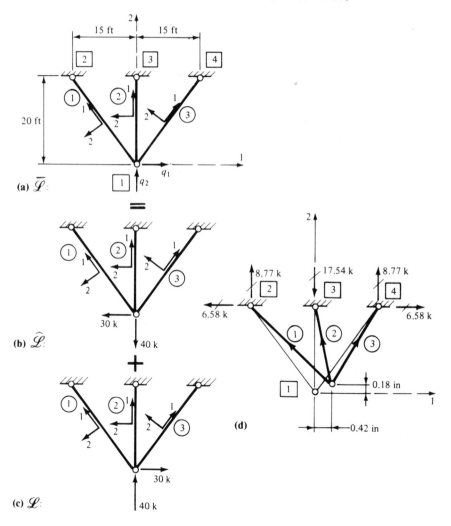

$e = -0.5$. The truss is analyzed by the matrix displacement method (Figure 3.9). The analysis units are kilopound, inch.

1. *Unknowns*

$$q_k, \quad k = 1, 2 \tag{4.214}$$

2. *Element Models*

$$\bar{\mathbf{f}}^i = \mathbf{f}^i + \hat{\mathbf{f}}^i, \quad i = 1, 2, 3 \tag{4.215}$$

$$\mathbf{f}^i = \mathbf{k}\mathbf{d}^i \tag{4.216}$$

where

$$\mathbf{k} = \gamma \begin{bmatrix} 1 & -1 \\ -1 & 1 \end{bmatrix}, \quad \gamma = 100 \tag{4.217}$$

$$\hat{\mathbf{f}}^1 = \hat{\mathbf{f}}^2 = \mathbf{0}, \quad \hat{\mathbf{f}}^3 = \begin{bmatrix} -50 \\ 50 \end{bmatrix} \tag{4.218}$$

$\hat{\mathbf{f}}^3$ is obtained by specializing Eqs. (4.207) for a truss element (Figure 1.14a): $\hat{\mathbf{f}}_b^3 = -\gamma e = 50 = -\hat{\mathbf{f}}_a^3$.

3. *System Model*
The rotation matrices are

$$\boldsymbol{\lambda}^1 = [-\tfrac{3}{5} \quad \tfrac{4}{5}], \quad \boldsymbol{\lambda}^2 = [0 \quad 1], \quad \boldsymbol{\lambda}^3 = [\tfrac{3}{5} \quad \tfrac{4}{5}] \tag{4.219}$$

and the member code matrix is

$$\mathbf{M}^T = \begin{bmatrix} 1 & 2 & 0 & 0 \\ 1 & 2 & 0 & 0 \\ 1 & 2 & 0 & 0 \end{bmatrix} \tag{4.220}$$

System Stiffness Matrix
By Eqs. (3.78), (4.219), and (4.220)

$$\mathbf{K}^1 = \frac{\gamma}{25} \begin{bmatrix} \overset{1}{9} & \overset{2}{-12} & \overset{0}{} & \overset{0}{} \\ & 16 & & \\ & & & \\ \text{sym.} & & & \end{bmatrix} \begin{matrix} 1 \\ 2 \\ 0 \\ 0 \end{matrix} \tag{4.221a}$$

$$\mathbf{K}^2 = \frac{\gamma}{25} \begin{bmatrix} \overset{1}{0} & \overset{2}{0} & \overset{0}{} & \overset{0}{} \\ & 25 & & \\ & & & \\ \text{sym.} & & & \end{bmatrix} \begin{matrix} 1 \\ 2 \\ 0 \\ 0 \end{matrix} \tag{4.221b}$$

$$
\mathbf{K}^3 = \frac{\gamma}{25}
\begin{bmatrix}
1 & 2 & 0 & 0 \\
9 & 12 & & \\
 & 16 & & \\
 & & & \\
\text{sym.} & & &
\end{bmatrix}
\begin{matrix}
\\ 1 \\ 2 \\ 0 \\ 0
\end{matrix}
\tag{4.221c}
$$

From Eqs. (4.221), we obtain the system stiffness matrix

$$
\mathbf{K} = \frac{\gamma}{25}
\begin{bmatrix}
18 & 0 \\
0 & 57
\end{bmatrix}
\tag{4.222}
$$

Equivalent Joint Load Vector (Figures 4.25a–c)

$\overline{\mathscr{L}}$:

$$
\overline{\mathbf{Q}} = \mathbf{0}
\tag{4.223}
$$

$\hat{\mathscr{L}}$:

$$
\hat{\mathbf{F}}^{(1)} = \hat{\mathbf{F}}^{(2)} = \mathbf{0}
\tag{4.224}
$$

$$
\hat{\mathbf{F}}^3 = \mathbf{\Lambda}^{3T}\hat{\mathbf{f}}^3 =
\begin{bmatrix}
-30 \\ -40 \\ 30 \\ 40
\end{bmatrix}
\begin{matrix}
1 \\ 2 \\ 0 \\ 0
\end{matrix}
\xrightarrow{\mathbf{M}} \hat{\mathbf{F}}^{(3)} =
\begin{bmatrix}
-30 \\ -40
\end{bmatrix}
\begin{matrix}
1 \\ 2
\end{matrix}
\tag{4.225}
$$

$$
\hat{\mathbf{Q}} = \sum_{i=1}^{3} \hat{\mathbf{F}}^{(i)} = \hat{\mathbf{F}}^{(3)}
\tag{4.226}
$$

$\mathscr{L}$:

$$
\mathbf{Q} = \overline{\mathbf{Q}} - \hat{\mathbf{Q}} =
\begin{bmatrix}
30 \\ 40
\end{bmatrix}
\tag{4.227}
$$

System Model

$$
\frac{\gamma}{25}
\begin{bmatrix}
18 & 0 \\
0 & 57
\end{bmatrix}
\begin{bmatrix}
q_1 \\ q_2
\end{bmatrix}
=
\begin{bmatrix}
30 \\ 40
\end{bmatrix}
\tag{4.228}
$$

4. *Solution*

$$
\mathbf{q} =
\begin{bmatrix}
q_1 \\ q_2
\end{bmatrix}
= \frac{1}{\gamma}
\begin{bmatrix}
41.67 \\ 17.54
\end{bmatrix}
=
\begin{bmatrix}
0.42 \\ 0.18
\end{bmatrix}
\tag{4.229}
$$

5. *Element Forces*
According to Eq. (4.220)

$$
\mathbf{D}^i =
\begin{bmatrix}
\mathbf{D}^i_a \\
\hline
\mathbf{D}^i_b
\end{bmatrix}
=
\begin{bmatrix}
q_1 \\ q_2 \\ \hline 0 \\ 0
\end{bmatrix}
\begin{matrix}
1 \\ 2 \\ 0 \\ 0
\end{matrix}, \quad i = 1, 2, 3
\tag{4.230}
$$

Hence,

$$d_a^i = \lambda^i \mathbf{q} \quad \text{and} \quad d_b^i = 0 \tag{4.231}$$

Equations (4.215)–(4.220), (4.229), and (4.231) yield

$$\mathbf{f}^1 = \begin{bmatrix} -10.96 \\ 10.96 \end{bmatrix}, \quad \mathbf{f}^2 = \begin{bmatrix} 17.54 \\ -17.54 \end{bmatrix}, \quad \mathbf{f}^3 = \begin{bmatrix} 39.03 \\ -39.03 \end{bmatrix}$$

$$\bar{\mathbf{f}}^1 = \mathbf{f}^1, \qquad \bar{\mathbf{f}}^2 = \mathbf{f}^2, \qquad \bar{\mathbf{f}}^3 = \bar{\mathbf{f}}^1 \tag{4.232}$$

6. *Joint Forces* (Figure 4.25d)

$$\bar{\mathbf{P}}_1 = \sum_{i=1}^{3} \bar{\mathbf{F}}_a^i = \sum_{i=1}^{3} \lambda^{iT} \bar{f}_a^i = \mathbf{0}$$

$$\bar{\mathbf{P}}_2 = \bar{\mathbf{F}}_b^1 = \lambda^{1T} \bar{f}_b^1 = \begin{bmatrix} -6.58 \\ 8.77 \end{bmatrix} \tag{4.233a}$$

$$\bar{\mathbf{P}}_3 = \bar{\mathbf{F}}_b^2 = \lambda^{2T} \bar{f}_b^2 = \begin{bmatrix} 0 \\ -17.54 \end{bmatrix} \tag{4.233b}$$

$$\bar{\mathbf{P}}_4 = \bar{\mathbf{F}}_b^3 = \lambda^{3T} \bar{f}_b^3 = \begin{bmatrix} 6.58 \\ 8.77 \end{bmatrix} \tag{4.233c}$$

Note that the element and joint forces caused by the imperfection are symmetric with respect to the axis of symmetry of the perfect truss even though the configuration is asymmetric (Figure 4.25d). This is because the conditions of equilibrium were formulated for the perfect truss.

4.8 TEMPERATURE CHANGES

When a homogeneous body is unconstrained, a uniform change in temperature causes identical extensional strains in all directions; no shearing strains arise in the freely deforming body. The extensional strain can be approximated by the linear relation (Stippes et al., 1961).

$$\varepsilon_t = \alpha_t t \tag{4.234}$$

where ε_t is the thermal strain, α_t the coefficient of thermal expansion, and t the temperature change. A uniform change in temperature does not induce stresses in the unconstrained body.[3] This is reflected by the stress–strain relation in Figure 5.26. Specifically, if the body is free to expand, the strain is ε_t and the stress is zero. The equation of the line passing through this point and having slope E, the modulus of elasticity, is

$$\bar{\sigma} = E(\bar{\varepsilon} - \varepsilon_t) \tag{4.235}$$

---

[3] Nonuniform temperature distribution may cause stresses even if the body is externally unconstrained (Shames, 1975; Stippes et al., 1961).

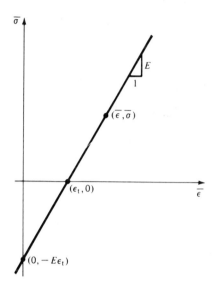

FIGURE 4.26 Stress–strain diagram.

where $\bar{\varepsilon}$ and $\bar{\sigma}$ are the actual strain and stress in the body. Equation (4.235) and Figure 4.26 indicate that if the body is completely constrained, that is, if $\bar{\varepsilon} = 0$, the resulting stress is $-E\varepsilon_t$. This suggests that thermal stresses may be controlled by placing releases, expansion joints, in the structure.

Temperature effects can be determined by the matrix displacement method (Figure 3.9) analogous to the effects of geometric imperfections (Section 4.7). Indeed, an isolated element subject to a temperature change becomes, in effect, an imperfect element. The fixed-end forces, required to transform the temperature distribution into equivalent joint loads (Section 3.6), can be computed by Eqs. (4.207). The components of the deformation vector (Figure 1.17b)

$$\mathbf{e} = \begin{bmatrix} e_1 \\ e_2 \\ e_3 \end{bmatrix} \tag{4.236}$$

can be expressed as [Eqs. (A.8), Appendix A]

$$e_1 = \int_0^L \varepsilon_0(x)\, dx$$

$$e_2 = \int_0^L (L - x)\phi_0(x)\, dx \tag{4.237}$$

$$e_3 = \int_0^L \Phi_0(x)\, dx$$

where ε_0 is the thermal strain at the centroidal axis of the element and ϕ_0 is the thermal curvature of the element.

Fixed-End Elements

Example 1. The element in Figure 4.27a is subjected to the temperature change

$$t(x, y) = t_0 + \frac{2}{h} y \, \Delta t \qquad (4.238)$$

which is uniform in the x direction and varies linearly in the y direction (Figure 4.27b). The quantities t_0, Δt are specific values of temperature, and h is the height of the element. The resulting fixed-end forces and the corresponding strain in the element are computed.

The thermal strain is (Figure 4.27c)

$$\varepsilon_t = \alpha_t \left(t_0 + \frac{2}{h} y \, \Delta t \right) \qquad (4.239)$$

or

$$\varepsilon_t = \varepsilon_0 - \phi_0 y \qquad (4.240)$$

where

$$\varepsilon_0 = \alpha_t t_0 = \text{constant}$$
$$\phi_0 = -\frac{2}{h} \alpha_t \, \Delta t = \text{constant} \qquad (4.241)$$

It follows from Eq. (1.46), where $\phi = d^2 v/dx^2$, and Eq. (4.240) that ϕ_0 is the thermal curvature. Since ϕ_0 is constant, the unconstrained element is bent into a circular arc of radius $1/\phi_0$.

From Eqs. (4.236), (4.237), and (4.241) we obtain the deformation vector

$$\mathbf{e} = \begin{bmatrix} \varepsilon_0 L \\ \frac{1}{2} \phi_0 L^2 \\ \phi_0 L \end{bmatrix} \qquad (4.242)$$

Equations (1.123), (1.153), (4.207), and (4.242) yield the fixed-end force vectors (Figure 4.27d)

$$\hat{\mathbf{f}}_b = \begin{bmatrix} N_t \\ 0 \\ M_t \end{bmatrix} = -\hat{\mathbf{f}}_a \qquad (4.243)$$

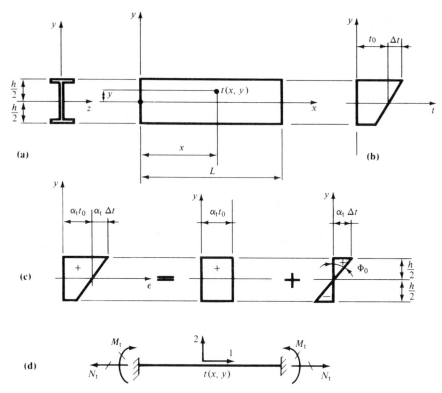

FIGURE 4.27 Beam with temperature change. **(a)** Beam; **(b)** temperature variation; **(c)** strain diagram; **(d)** fixed-end forces.

where

$$N_t = -EA\varepsilon_0, \quad M_t = -EI\phi_0 = EI\left(\frac{2}{h}\alpha_t\,\Delta t\right) \qquad (4.244)$$

The same result is obtained by the finite element method in Example 4 of Section 3.7.

Accordingly, the axial force and the bending moment of the end-constrained element are (see sign conventions in Figures 1.7d and 1.8b)

$$\hat{N}(x) = N_t, \quad \hat{M}(x) = M_t \qquad (4.245)$$

and the stress is [Eqs. (1.35) and (1.54c)]

$$\hat{\sigma}(x, y) = \frac{\hat{N}}{A} - \frac{\hat{M}y}{I} = -E(\varepsilon_0 - \phi_0 y) \qquad (4.246)$$

By Eq. (4.235)

$$\hat{\sigma} = E(\hat{\varepsilon} - \varepsilon_t) \qquad (4.247)$$

Equations (4.240), (4.246), and (4.247) yield

$$\hat{\varepsilon}(x, y) = \frac{\hat{\sigma}}{E} + \varepsilon_t = 0 \tag{4.248}$$

Thus, every point in the element is completely constrained through the end constraints. This is not true for every temperature distribution (see Example 2).

Example 2. The element in Figure 4.27a is subjected to the temperature change

$$t(x) = t_b \frac{x}{L} \tag{4.249}$$

which varies linearly in the x direction and is uniform in the y direction; t_b is the temperature increase at the b end. This results in the thermal strain

$$\varepsilon_t = \varepsilon_b \frac{x}{L}, \quad \varepsilon_b = \alpha_t t_b \tag{4.250}$$

the deformation vector

$$\mathbf{e} = \begin{bmatrix} \frac{1}{2}\varepsilon_b L \\ 0 \\ 0 \end{bmatrix} \tag{4.251}$$

and the fixed-end force vectors are

$$\hat{\mathbf{f}}_b = \begin{bmatrix} N_t \\ 0 \\ 0 \end{bmatrix} = -\hat{\mathbf{f}}_a \tag{4.252}$$

where

$$N_t = -\tfrac{1}{2}EA\varepsilon_b \tag{4.253}$$

The stress in the element is

$$\hat{\sigma} = \frac{N_t}{A} = -\tfrac{1}{2}E\varepsilon_b \tag{4.254}$$

From Eqs. (4.247), (4.250), and (4.254), we obtain the strain in the end-constrained element

$$\hat{\varepsilon} = \frac{\hat{\sigma}}{E} + \varepsilon_t = \varepsilon_b\left(-\frac{1}{2} + \frac{x}{L}\right) \tag{4.255}$$

Clearly, for this temperature distribution, the end constraints do not suppress the strain uniformly over the element.

Analysis of Structures

Example 3. Element 2 of the structure in Figure 4.28a undergoes the temperature increase defined in Eq. (4.249) of Example 2; the temperature of element 1 remains unchanged. The resulting element forces and stress and strain distributions are computed. Since the elements experience only axial deformations, the structure can be represented by truss elements [Figure 4.28d and Eq. (3.76)], and no coordinate transformations are required. The analysis is based on Figure 3.9 (steps 1–5).

1. *Unknowns*

$$q_1 \tag{4.256}$$

2. *Element Models*

$$\bar{\mathbf{f}}^i = \mathbf{k}\mathbf{d}^i + \hat{\mathbf{f}}^i, \quad i = 1, 2 \tag{4.257}$$

FIGURE 4.28 Continuous beams with temperature change. **(a)** $\bar{\mathscr{L}}$; **(b)** $\hat{\mathscr{L}}$; **(c)** $\mathscr{L}$; **(d)** elements; **(e)** fixed-end forces; **(f)** element forces.

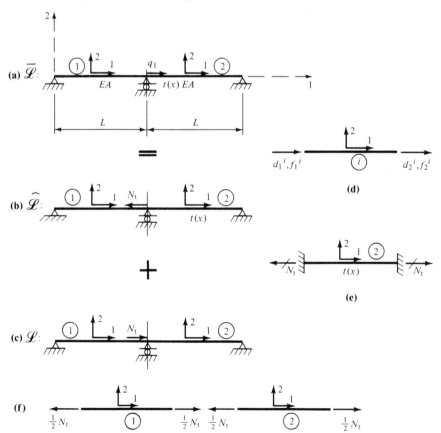

where

$$\mathbf{k} = \gamma \begin{bmatrix} 1 & -1 \\ -1 & 1 \end{bmatrix}, \quad \gamma = \frac{EA}{L} \tag{4.258}$$

$$\hat{\mathbf{f}}^1 = \mathbf{0}, \quad \hat{\mathbf{f}}^2 = \begin{bmatrix} -N_t \\ N_t \end{bmatrix}_0^1, \quad N_t = -\tfrac{1}{2}\varepsilon_b EA \tag{4.259}$$

$\hat{\mathbf{f}}^2$ follows from Figure 4.28e and Eq. (4.253). According to Figures 4.28a, d

$$\mathbf{M} = \begin{bmatrix} 0 & 1 \\ 1 & 0 \end{bmatrix} \tag{4.260}$$

3. System Model

System Stiffness Matrix

$$\mathbf{k}^1 = \gamma \begin{matrix} & 0 & 1 \\ & \begin{bmatrix} 1 & -1 \\ -1 & 1 \end{bmatrix} & \begin{matrix} 0 \\ 1 \end{matrix} \end{matrix} \overset{\mathbf{M}}{\to} K^{(1)} = \gamma$$

$$\tag{4.261}$$

$$\mathbf{k}^2 = \gamma \begin{matrix} & 1 & 0 \\ & \begin{bmatrix} 1 & -1 \\ -1 & 1 \end{bmatrix} & \begin{matrix} 1 \\ 0 \end{matrix} \end{matrix} \overset{\mathbf{M}}{\to} K^{(2)} = \gamma$$

$$K = \sum_{i=1}^{2} K^{(i)} = 2\gamma \tag{4.262}$$

Equivalent Joint Load Vector (Figures 4.28a–c)

$\mathscr{L}$:

$$\bar{\mathbf{Q}} = 0 \tag{4.263}$$

$\hat{\mathscr{L}}$:

$$\hat{F}^{(1)} = 0, \quad \hat{\mathbf{f}}^2 \overset{\mathbf{M}}{\to} \hat{F}^{(2)} = -N_t \tag{4.264}$$

$$\hat{Q} = \sum_{i=1}^{2} \hat{F}^{(i)} = -N_t$$

$\mathscr{L}$:

$$Q = \bar{Q} - \hat{Q} = N_t \tag{4.265}$$

System Model

$$2\gamma q_1 = N_t \tag{4.266}$$

4. Solution

$$q_1 = \frac{N_t}{2\gamma} = -\frac{1}{4}\varepsilon_b L \tag{4.267}$$

5. *Element Forces* (Figure 4.28f)
By Eqs. (4.257)–(4.260), and (4.267)

$$\mathbf{d}^1 = \begin{bmatrix} 0 \\ q_1 \end{bmatrix}\!\!\begin{matrix} 0 \\ 1 \end{matrix}, \quad \mathbf{d}^2 = \begin{bmatrix} q_1 \\ 0 \end{bmatrix}\!\!\begin{matrix} 1 \\ 0 \end{matrix} \tag{4.268}$$

and

$$\bar{\mathbf{f}}^1 = \bar{\mathbf{f}}^2 = \tfrac{1}{2} N_t \begin{bmatrix} -1 \\ 1 \end{bmatrix} \tag{4.269}$$

6. *Stresses and Strains*
The actual element stresses are (Figure 4.28f)

$$\bar{\sigma}^1 = \bar{\sigma}^2 = \frac{N_t}{2A} = -\frac{1}{4} E\varepsilon_b \tag{4.270}$$

It follows from Eq. (4.235) that

$$\bar{\varepsilon}^i = \frac{\bar{\sigma}^i}{E} + \varepsilon_t^i, \quad i = 1, 2 \tag{4.271}$$

where

$$\varepsilon_t^1 = 0 \tag{4.272a}$$

and by Eq. (4.250)

$$\varepsilon_t^2 = \varepsilon_b \frac{x}{L} \tag{4.272b}$$

Equations (4.270)–(4.272) yield the actual element strains

$$\bar{\varepsilon}^1 = -\frac{1}{4}\varepsilon_b, \quad \bar{\varepsilon}^2 = \varepsilon_b\!\left(-\frac{1}{4} + \frac{x}{L}\right) \tag{4.273}$$

Note that the expansion of element 2 caused by the temperature increase [Eq. (4.249)] is resisted by element 1. Consequently, both elements are stressed [Eqs. (4.270)].

Example 4. The elements of the frame in Figure 4.29a have identical properties. The centroidal axes of the elements are assumed to be inextensible (Section 4.3). Element 2 experiences the temperature change defined in Eq. (4.238) with $t_0 = 0$. The temperature in element 1 remains constant. Thus, the thermal strains in the elements are

$$\varepsilon_t^1 = 0, \quad \varepsilon_t^2 = -\phi_0 y \tag{4.274}$$

230

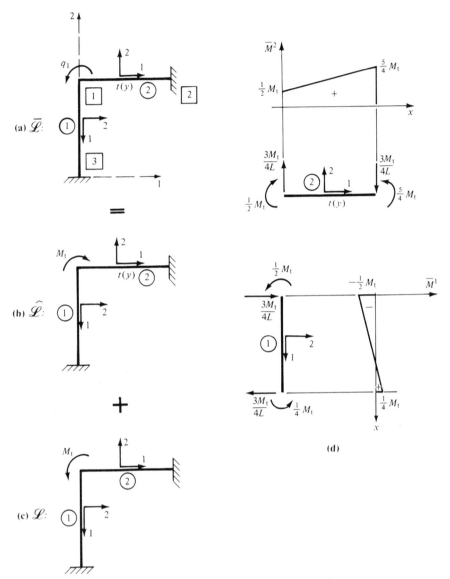

FIGURE 4.29 Frame with temperature change. **(a)** $\overline{\mathscr{L}}$; **(b)** $\widehat{\mathscr{L}}$; **(c)** $\mathscr{L}$; **(d)** element forces.

The resulting element forces and stress and strain distributions are determined. The analysis is based on the flow chart in Figure 3.9 (steps 1–5).

1. *Unknowns*

$$q_1 \tag{4.275}$$

2. *Element Models*

$$\bar{\mathbf{f}}^i = \mathbf{f}^i + \hat{\mathbf{f}}^i, \quad \mathbf{f}^i = \mathbf{k}\mathbf{d}^i, \quad i = 1, 2 \tag{4.276}$$

where $\mathbf{k}$ is defined in Eq. (3.37),

$$\hat{\mathbf{f}}^1 = \mathbf{0}, \quad \hat{\mathbf{f}}^2 = \begin{bmatrix} 0 & 0 \\ -M_t & 1 \\ 0 & 0 \\ M_t & 0 \end{bmatrix}, \quad M_t = -EI\phi_0 \tag{4.277}$$

$\hat{\mathbf{f}}^2$ follows from Figure 4.27d for $N_t = 0$. The member code is defined in Eq. (4.60).

3. *System Model*

System Stiffness Matrix
The system stiffness matrix is defined in Eq. (4.62).

Equivalent Joint Load Vector (Figures 4.29a–c)

$\bar{\mathscr{L}}$:

$$\bar{Q} = 0 \tag{4.278}$$

$\hat{\mathscr{L}}$:

$$\hat{F}^{(1)} = 0, \quad \hat{\mathbf{f}}^2 \overset{\mathbf{M}}{\to} \hat{F}^{(2)} = -M_t \tag{4.279}$$

$$\hat{Q} = \sum_{i=1}^{2} \hat{F}^{(i)} = -M_t \tag{4.280}$$

$\mathscr{L}$:

$$Q = \bar{Q} - \hat{Q} = M_t \tag{4.281}$$

System Model

$$\alpha 8L^2 q_1 = M_t \tag{4.282}$$

4. *Solution*

$$q_1 = \frac{M_t}{\alpha 8L^2} = -\frac{\phi_0 L}{8} \tag{4.283}$$

5. *Element Forces* (Figures 4.29d)

By Eqs. (3.37), (4.69), (4.276), (4.277), and (4.283)

$$\mathbf{d}^1 = \mathbf{d}^2 = \begin{bmatrix} 0 \\ q_1 \\ 0 \\ 0 \end{bmatrix} \tag{4.284}$$

$$\bar{\mathbf{f}}^1 = \mathbf{f}^1 = \mathbf{f}^2 = M_t \begin{bmatrix} \dfrac{3}{4L} \\[2mm] \dfrac{1}{2} \\[2mm] -\dfrac{3}{4L} \\[2mm] \dfrac{1}{4} \end{bmatrix}, \quad \bar{\mathbf{f}}^2 = M_t \begin{bmatrix} \dfrac{3}{4L} \\[2mm] -\dfrac{1}{2} \\[2mm] -\dfrac{3}{4L} \\[2mm] \dfrac{5}{4} \end{bmatrix} \tag{4.285}$$

6. *Stresses and Strains*

According to Figures 4.29d, the actual bending moments are

$$\overline{M}^1 = -M_t\left(\frac{1}{2} - \frac{3x}{4L}\right), \quad \overline{M}^2 = M_t\left(\frac{1}{2} + \frac{3x}{4L}\right) \tag{4.286}$$

The substitutions for M_t from Eqs. (4.277) and Eqs. (4.286) into

$$\bar{\sigma}^i = -\frac{\overline{M}^i y}{I} \tag{4.287}$$

yield the actual element stresses

$$\bar{\sigma}^1 = -E\phi_0 y\left(\frac{1}{2} - \frac{3x}{4L}\right), \quad \bar{\sigma}^2 = E\phi_0 y\left(\frac{1}{2} + \frac{3x}{4L}\right) \tag{4.288}$$

Equations (4.271), (4.274), and (4.288) yield the actual element strains

$$\bar{\varepsilon}^1 = \bar{\varepsilon}^2 = -\phi_0 y\left(\frac{1}{2} - \frac{3x}{4L}\right) \tag{4.289}$$

4.9 JOINT REFERENCE FRAMES

Occasionally, it is convenient to define the displacements and forces at a joint relative to a reference frame that is distinct from the global reference frame. For example, this situation arises when joint constraints do not

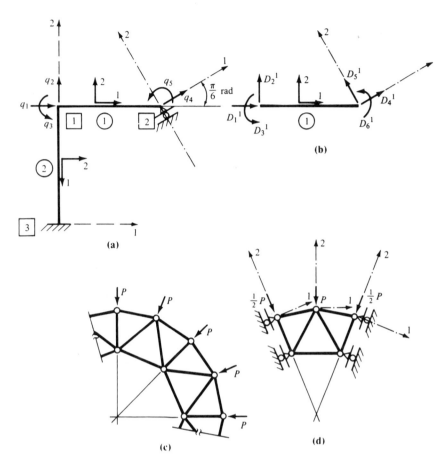

FIGURE 4.30 Joint reference frames. **(a)** Frame; **(b)** element 1; **(c)** truss ring; **(d)** radial section.

correspond to global axes as illustrated in Figure 4.30a. By aligning the coordinate axes of joint 2 with the constraint, we can express the corresponding conditions of compatibility directly as (Figure 4.30b)

$$D_4^1 = q_4, \quad D_5^1 = 0, \quad D_6^1 = q_5 \tag{4.290}$$

Similarly, we can exploit certain symmetry conditions by selecting suitable joint coordinate axes. For example, consider the axisymmetric truss ring in Figure 4.30c, whose joints lie on concentric circles. Under the uniform radial load, the joints deflect only in the radial directions. Thus, by selecting joint coordinate axes that are normal and tangential to the circle at each joint, we can eliminate all tangential displacements from the analysis. Moreover, the analysis can be confined to the radial section of the truss shown in

Figure 4.30d. In this substructure, which is still symmetric about the center line, the areas of the radial elements must be reduced to one half of their original values (Section 4.2).

To apply the matrix displacement method to structures with distinct joint reference frames, we need to specify coordinate transformations for each element end and formulate the element stiffness matrices in joint coordinates. Otherwise, the analysis proceeds as in Figures 3.2 and 3.9.

Element Stiffness Matrices in Joint Coordinates

The coordinate transformations (Section 2.6) at each element end can be expressed as

$$
\begin{aligned}
\mathbf{d}_a &= \lambda_a \mathbf{D}_a, \quad \mathbf{d}_b = \lambda_b \mathbf{D}_b \\
\mathbf{F}_a &= \lambda_a^T \mathbf{f}_a, \quad \mathbf{F}_b = \lambda_b^T \mathbf{f}_b
\end{aligned}
\tag{4.291}
$$

where λ_a and λ_b are the rotation matrices at the a and b end, respectively. We combine the coordinate transformations with the partitioned local element model

$$
\begin{aligned}
\mathbf{f}_a &= \mathbf{k}_{aa}\mathbf{d}_a + \mathbf{k}_{ab}\mathbf{d}_b \\
\mathbf{f}_b &= \mathbf{k}_{ba}\mathbf{d}_a + \mathbf{k}_{bb}\mathbf{d}_b
\end{aligned}
\tag{4.292}
$$

to obtain the element model in joint coordinates

$$
\begin{aligned}
\mathbf{F}_a &= \lambda_a^T \mathbf{k}_{aa} \lambda_a \mathbf{D}_a + \lambda_a^T \mathbf{k}_{ab} \lambda_b \mathbf{D}_b \\
\mathbf{F}_b &= \lambda_b^T \mathbf{k}_{ba} \lambda_a \mathbf{D}_a + \lambda_b^T \mathbf{k}_{bb} \lambda_b \mathbf{D}_b
\end{aligned}
\tag{4.293}
$$

or

$$
\mathbf{F} = \mathbf{K}\mathbf{D}
\tag{4.294}
$$

where

$$
\begin{aligned}
\mathbf{K}_{aa} &= \lambda_a^T \mathbf{k}_{aa} \lambda_a \\
\mathbf{K}_{ab} &= \lambda_a^T \mathbf{k}_{ab} \lambda_b = \mathbf{K}_{ba}^T \\
\mathbf{K}_{bb} &= \lambda_b^T \mathbf{k}_{bb} \lambda_b
\end{aligned}
\tag{4.295}
$$

To specialize Eqs. (4.295) for frame and truss elements, let

$$
\lambda_a = \lambda \quad \text{and} \quad \lambda_b = \lambda^*
\tag{4.296}
$$

Accordingly, the direction cosines of λ and λ^* are denoted c_i and c_i^*, respectively. Note that for $\lambda_a = \lambda_b = \lambda$, Eqs. (4.295) reduce to Eqs. (3.61).

Frame Element

By Eqs. (2.34), (3.59), (4.295), and (4.296), the element stiffness matrix in joint coordinates can be expressed as

$$
\mathbf{K} = \left[
\begin{array}{ccc:ccc}
g_1 & g_2 & g_4 & g_8 & g_9 & g_4 \\
 & g_3 & g_5 & g_{11} & g_{10} & g_5 \\
 & & g_6 & -g_4^* & -g_5^* & g_7 \\
\hdashline
 & & & g_1^* & g_2^* & -g_4^* \\
 & & & & g_3^* & -g_5^* \\
\text{sym.} & & & & & g_6^*
\end{array}
\right]
\tag{4.297}
$$

where $g_1\text{-}g_7$ are defined in Eqs. (3.64), g_i^* can be obtained from g_i by replacing c_1 and c_2 with c_1^* and c_2^*, respectively, and

$$
\begin{aligned}
g_8 &= -\alpha(\beta c_1 c_1^* + 12 c_2 c_2^*) \\
g_9 &= -\alpha(\beta c_1 c_2^* - 12 c_2 c_1^*) \\
g_{10} &= -\alpha(\beta c_2 c_2^* + 12 c_1 c_1^*) \\
g_{11} &= -\alpha(\beta c_2 c_1^* - 12 c_1 c_2^*)
\end{aligned}
\tag{4.298}
$$

Truss Element

From Eqs. (2.52), (3.76), (4.295), and (4.296), we obtain the element stiffness matrix in joint coordinates

$$
\mathbf{K} = \gamma \left[
\begin{array}{cccc}
c_1^2 & c_1 c_2 & -c_1 c_1^* & -c_1 c_2^* \\
 & c_2^2 & -c_2 c_1^* & -c_2 c_2^* \\
 & & c_1^{*2} & c_1^* c_2^* \\
\text{sym.} & & & c_2^{*2}
\end{array}
\right]
\tag{4.299}
$$

Matrix Displacement Analysis

Example. The elements of the truss in Figure 4.31a have identical, properties A, E, L. The reference frames of joints 1 and 2 coincide with the global reference frame, but joint 3 has a distinct reference frame whose orientation is determined by the constraint. The joint loads are $Q_1 = P$, $Q_2 = Q_3 = 0$. The analysis is based on Figure 3.2.

236

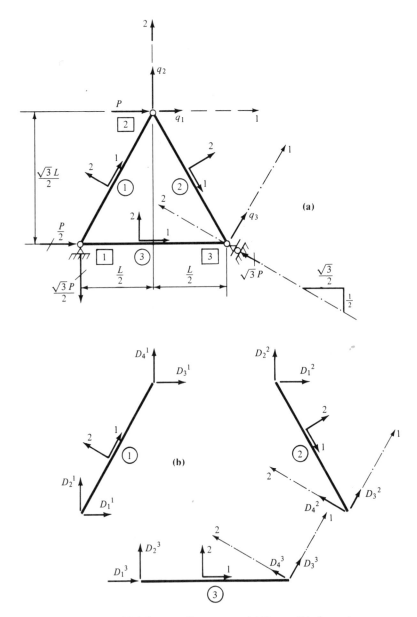

FIGURE 4.31 Truss with joint coordinate axes. **(a)** Truss; **(b)** elements.

1. *Unknowns*

$$q_k, \quad k = 1, 2, 3 \tag{4.300}$$

2. *Element Models*

$$\mathbf{f}^i = \mathbf{k}\mathbf{d}^i, \quad i = 1, 2, 3 \tag{4.301}$$

k is defined in Eq. (3.76).

3. *System Model*

The rotation matrices and the member code matrix are (Figures 4.31a, b)

$$\lambda^1 = \tfrac{1}{2}[1 \quad \sqrt{3}]$$
$$\lambda^2 = \tfrac{1}{2}[1 \quad -\sqrt{3}], \quad \lambda^{*2} = \tfrac{1}{2}[-1 \quad -\sqrt{3}] \tag{4.302}$$
$$\lambda^3 = [1 \quad 0], \quad \lambda^{*3} = \tfrac{1}{2}[1 \quad -\sqrt{3}]$$

$$\mathbf{M}^T = \begin{bmatrix} 0 & 0 & 1 & 2 \\ 1 & 2 & 3 & 0 \\ 0 & 0 & 3 & 0 \end{bmatrix} \tag{4.303}$$

System Stiffness Matrix

By Eqs. (2.52), (3.78), (4.299), (4.302), and (4.303)

$$\mathbf{K}^1 = \frac{\gamma}{4} \begin{matrix} & 0 & 0 & 1 & 2 \\ & \begin{bmatrix} & & & \\ & & 1 & \sqrt{3} \\ \text{sym.} & & & 3 \end{bmatrix} & \begin{matrix} 0 \\ 0 \\ 1 \\ 2 \end{matrix} \end{matrix} \tag{4.304a}$$

$$\mathbf{K}^2 = \frac{\gamma}{4} \begin{matrix} & 1 & 2 & 3 & 0 \\ & \begin{bmatrix} 1 & -\sqrt{3} & 1 & \\ & 3 & -\sqrt{3} & \\ & & 1 & \\ \text{sym.} & & & \end{bmatrix} & \begin{matrix} 1 \\ 2 \\ 3 \\ 0 \end{matrix} \end{matrix} \tag{4.304b}$$

$$\mathbf{K}^3 = \frac{\gamma}{4} \begin{matrix} & 0 & 0 & 3 & 0 \\ & \begin{bmatrix} & & & \\ & & 1 & \\ \text{sym.} & & & \end{bmatrix} & \begin{matrix} 0 \\ 0 \\ 3 \\ 0 \end{matrix} \end{matrix} \tag{4.304c}$$

Equations (4.304) yield the system stiffness matrix

$$\mathbf{K} = \frac{\gamma}{4} \begin{matrix} & 1 & 2 & 3 \\ & \begin{bmatrix} 2 & 0 & 1 \\ & 6 & -\sqrt{3} \\ \text{sym.} & & 2 \end{bmatrix} & \begin{matrix} 1 \\ 2 \\ 3 \end{matrix} \end{matrix} \tag{4.305}$$

System Model

$$\frac{\gamma}{4}\begin{bmatrix} 2 & 0 & 1 \\ 0 & 6 & -\sqrt{3} \\ 1 & -\sqrt{3} & 2 \end{bmatrix}\begin{bmatrix} q_1 \\ q_2 \\ q_3 \end{bmatrix} = \begin{bmatrix} P \\ 0 \\ 0 \end{bmatrix} \tag{4.306}$$

4. *Solution*

$$q_1 = \frac{3P}{\gamma}, \quad q_2 = -\frac{\sqrt{3}P}{3\gamma}, \quad q_3 = -\frac{2P}{\gamma} \tag{4.307}$$

5. *Element Forces*

By Eqs. (3.76), (4.291), (4.296), (4.301)–(4.303), and (4.307)

$$\mathbf{D}^2 = \begin{bmatrix} \mathbf{D}_a^2 \\ ---- \\ \mathbf{D}_b^2 \end{bmatrix} = \begin{bmatrix} q_1 \\ q_2 \\ ---- \\ q_3 \\ 0 \end{bmatrix}\begin{matrix} 1 \\ 2 \\ \\ 3 \\ 0 \end{matrix}$$

$$\mathbf{d}^2 = \begin{bmatrix} \lambda^2 D_a^2 \\ -------- \\ \lambda^{*2} D_b^2 \end{bmatrix} = \frac{P}{\gamma}\begin{bmatrix} 2 \\ 1 \end{bmatrix}, \quad \mathbf{f}^2 = \begin{bmatrix} P \\ ----- \\ -P \end{bmatrix} \tag{4.308}$$

Similarly, we obtain

$$\mathbf{f}^3 = \mathbf{f}^2 = -\mathbf{f}^1 \tag{4.309}$$

6. *Joint Forces* (Figure 4.31a)

$$\mathbf{P}_3 = \mathbf{F}_b^2 + \mathbf{F}_b^3 = \lambda^{*2T}f_b^2 + \lambda^{*3T}f_b^3 = \begin{bmatrix} 0 \\ \sqrt{3}P \end{bmatrix} \tag{4.310}$$

Similarly, we obtain

$$\mathbf{P}_1 = \frac{1}{2}\begin{bmatrix} P \\ -\sqrt{3}P \end{bmatrix}, \quad \mathbf{P}_2 = \begin{bmatrix} P \\ 0 \end{bmatrix} \tag{4.311}$$

4.10 INFLUENCE LINES

Influence lines are used to determine extreme loading conditions. This topic is covered in textbooks on elementary structural analysis (see references in Chapter 1). Our objective is to construct influence functions by the matrix displacement method.

An *influence function* is a relation between a unit load at any point of a structure and the response (usually an internal force) at a specific point of the structure. An *influence line* is the graph of an influence function. For example, in Figure 4.32a, $M_b(x)$ and $V_b(x)$ represent, respectively, the bending moment and shear force at b induced by a unit load at x. The corresponding influence lines are shown in Figures 4.32b and c.

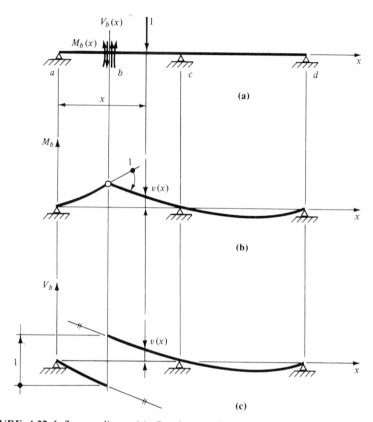

FIGURE 4.32 Influence lines. (a) Continuous beams; (b) influence line for M_b; (c) influence line for V_b.

Influence lines for any response parameter R can be constructed by the *Müller–Breslau* principle with the aid of an auxiliary structure. The *auxiliary structure* is obtained by introducing a release corresponding to R in the structure. According to the Müller–Breslau principle, the influence line for R is the configuration of the *auxiliary structure* obtained by introducing a unit displacement at the release. The Müller–Breslau principle is illustrated in Figures 4.32a–c. The results can be verified by Betti's law. For example, the application of Betti's law to Figures 4.32a and b yields

$$M_b(x) \cdot 1 - 1 \cdot v(x) = 0$$

or

$$M_b(x) = v(x) \qquad (4.312)$$

Equation (4.312) states that the bending moment at b induced by a unit load at x is equal to the deflection at x of the auxiliary structure.

Thus, by the Müller–Breslau principle, the construction of the influence function for R is reduced to the analysis of the auxiliary structure with a unit displacement corresponding to R. Specifically, we can regard the unit displacement at the release as an element action[4] and transform it into equivalent joint loads (Section 3.6). Accordingly, the analysis of the auxiliary structure is based on Figure 3.9 (steps 1–4), and the element displacement functions are expressed as

$$\bar{v}^i(x) = \hat{v}^i(x) + v^i(x) \tag{4.313}$$

Frames and Continuous Beams

The fixed-end forces for beam elements with unit displacements at moment and shear releases and the corresponding displacement functions $\hat{v}(x)$ are defined in Appendix A. The displacement function $v(x)$ can be expressed in terms of the element-end displacements by Eqs. (1.182)–(1.184), which satisfy the homogeneous differential equation, Eq. (1.75).

Example 1. The influence function for the bending moment at c in Figure 4.33a is determined. By the Müller–Breslau principle

$$M_c(x) = \bar{v}(x) \tag{4.314}$$

where $\bar{v}(x)$ is the displacement function of the auxiliary structure in Figure 4.33b. The analysis of the auxiliary structure is based on Figure 3.9 (steps 1–4).

1. *Unknown*

$$q_1 \tag{4.315}$$

2. *Element Model*

$$\bar{\mathbf{f}} = \mathbf{kd} + \hat{\mathbf{f}} \tag{4.316}$$

where $\mathbf{k}$ is defined in Eq. (3.37),

$$\hat{\mathbf{f}} = \alpha L^2 \begin{bmatrix} 0 & 0 \\ -1 & 0 \\ 0 & 0 \\ 1 & 1 \end{bmatrix} \tag{4.317}$$

and

$$\mathbf{M}^T = [0 \quad 0 \quad 0 \quad 1] \tag{4.318}$$

Equation (4.317) is obtained by specializing Eq. (A.56) for $a = \frac{1}{2}$.

---

[4] Alternatively, the release can be regarded as a joint release (Tezcan, 1966).

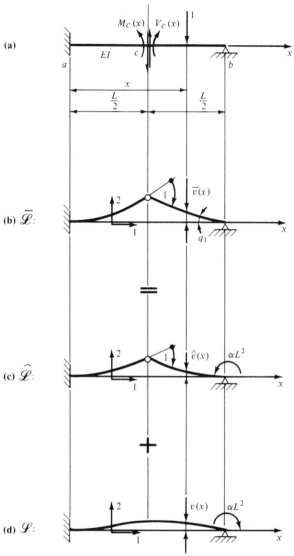

FIGURE 4.33 Construction of influence line for M_c. **(a)** Beam; **(b)** $\overline{\mathscr{L}}$; **(c)** $\widehat{\mathscr{L}}$; **(d)** $\mathscr{L}$.

3. *System Model*

System Stiffness Matrix

$$
\mathbf{k} = \alpha \begin{matrix} \begin{matrix} 0 & \;\; 0 & \;\; 0 & \;\; 1 \end{matrix} \\ \begin{bmatrix} & & & \\ & & & \\ & & & \\ \text{sym.} & & 4L^2 \end{bmatrix} \begin{matrix} 0 \\ 0 \\ 0 \\ 1 \end{matrix} \end{matrix} \overset{\mathbf{M}}{\to} K = 4\alpha L^2 \tag{4.319}
$$

Equivalent Joint Load Vector (Figures 4.33b–d)

$\overline{\mathscr{L}}$:

$$\overline{Q} = 0 \tag{4.320}$$

$\hat{\mathscr{L}}$:

$$\hat{\mathbf{f}} \overset{M}{\to} \hat{Q} = \alpha L^2 \tag{4.321}$$

$\mathscr{L}$:

$$Q = \overline{Q} - \hat{Q} = -\alpha L^2 \tag{4.322}$$

System Model

$$4\alpha L^2 q_1 = -\alpha L^2 \tag{4.323}$$

4. *Solution*

$$q_1 = -\tfrac{1}{4} \tag{4.324}$$

5. *Displacment Function*

$$\overline{v}(x) = \hat{v}(x) + v(x) \tag{4.325}$$

where by Eqs. (A.57) for $a = \tfrac{1}{2}$

$$\hat{v} = \begin{cases} \dfrac{L}{2}\,\xi^2, & 0 \le \xi \le \dfrac{1}{2} \\[2mm] \dfrac{L}{2}\,(\xi - 1)^2, & \dfrac{1}{2} \le \xi \le 1 \end{cases} \tag{4.326}$$

and by Eqs. (1.182)–(1.184) for $d_1 = d_2 = d_3 = 0$, $d_4 = -\tfrac{1}{4}$ [Eqs. (4.318) and (4.324)]

$$v = \frac{L}{4}(\xi^2 - \xi^3), \qquad 0 \le \xi \le 1 \tag{4.327}$$

For example, Eqs. (4.314), (4.325)–(4.327) yield

$$M_c\!\left(\frac{L}{2}\right) = \overline{v}\!\left(\frac{L}{2}\right) = \frac{5L}{32} \tag{4.328}$$

Example 2. The influence function for the shear force to the right of joint 2 in Figure 4.34a is determined. By the Müller–Breslau principle

$$V_2^R = \overline{v}(x) \tag{4.329}$$

where $\overline{v}(x)$ is the displacement function of the auxiliary structure in Figure 4.34b, which is analyzed by the matrix displacement method (Figure 3.9, steps 1–4).

1. *Unknowns*

$$q_1, q_2 \tag{4.330}$$

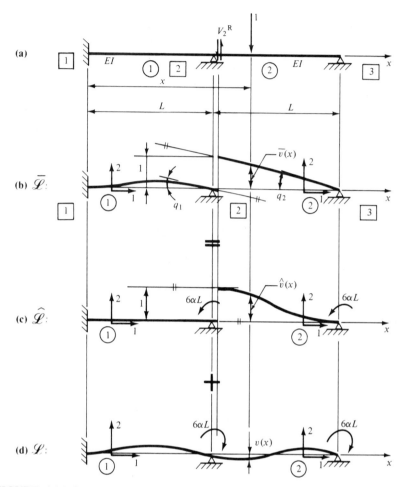

FIGURE 4.34 Construction of influence line for V_2^R. (a) Continuous beams; (b) $\overline{\mathscr{L}}$; (c) $\hat{\mathscr{L}}$; (d) $\mathscr{L}$.

2. *Element Models*

$$\overline{\mathbf{f}}^i = \mathbf{k}\mathbf{d}^i + \hat{\mathbf{f}}^i \qquad (4.331)$$

where **k** is defined in Eq. (3.37), by Eqs. (A.58)

$$\hat{\mathbf{f}}^1 = \mathbf{0}, \quad \hat{\mathbf{f}}^2 = \alpha \begin{bmatrix} 12 \\ 6L \\ -12 \\ 6L \end{bmatrix} \begin{matrix} 0 \\ 1 \\ 0 \\ 2 \end{matrix} \qquad (4.332)$$

and

$$\mathbf{M}^T = \begin{bmatrix} 0 & 0 & 0 & 1 \\ 0 & 1 & 0 & 2 \end{bmatrix} \qquad (4.333)$$

3. System Model

System Stiffness Matrix
The system stiffness matrix is defined in Eq. (3.15).

Equivalent Joint Load Vector (Figures 4.34b–d)
$\bar{\mathscr{L}}$:

$$\bar{Q} = 0 \tag{4.334}$$

$\hat{\mathscr{L}}$:

$$\hat{Q} = \hat{F}^{(2)} = 6\alpha L \begin{bmatrix} 1 \\ 1 \end{bmatrix} \tag{4.335}$$

$\mathscr{L}$:

$$Q = \bar{Q} - \hat{Q} = -6\alpha L \begin{bmatrix} 1 \\ 1 \end{bmatrix} \tag{4.336}$$

System Model

$$\alpha \begin{bmatrix} 8L^2 & 2L^2 \\ 2L^2 & 4L^2 \end{bmatrix} \begin{bmatrix} q_1 \\ q_2 \end{bmatrix} = -6\alpha L \begin{bmatrix} 1 \\ 1 \end{bmatrix} \tag{4.337}$$

4. Solution

$$q_1 = -\frac{3}{7L}, \quad q_2 = -\frac{9}{7L} \tag{4.338}$$

5. Displacement Functions

By Eq. (4.333)

$$\mathbf{d}^1 = \begin{bmatrix} 0 & 0 \\ 0 & 0 \\ 0 & 0 \\ q_1 & 1 \end{bmatrix}, \quad \mathbf{d}^2 = \begin{bmatrix} 0 & 0 \\ q_1 & 1 \\ 0 & 0 \\ q_2 & 2 \end{bmatrix} \tag{4.339}$$

Thus,

$$\bar{v}^i(x) = \hat{v}^i(x) + v^i(x), \quad i = 1, 2 \tag{4.340}$$

where by Eqs. (A.59)

$$\hat{v}^1 = 0$$
$$\hat{v}^2 = 2\xi^3 - 3\xi^2 + 1, \quad 0 < \xi \le 1 \tag{4.341}$$

and by Eqs. (1.182), (1.183), (4.338), and (4.339)

$$v^1 = -\tfrac{3}{7}(\xi^3 - \xi^2)$$
$$v^2 = -\tfrac{3}{7}(4\xi^3 - 5\xi^2 + \xi) \tag{4.342}$$

$$0 \le \xi \le 1, \quad \xi = \frac{x}{L} \tag{4.343}$$

Trusses

The construction of influence lines for trusses is discussed with reference to Figure 4.35. Suppose a unit load moves along the bottom chord (Figure 4.35a), and the influence line for the force in element j is desired. By the Müller–Breslau principle, the influence line is the configuration of the bottom chord of the truss in Figure 4.35b. This can be verified by the application of Betti's law to Figures 4.35a and b:

$$N^j(x) \cdot 1 + 1 \cdot |\bar{v}(x)| = 0$$

FIGURE 4.35 Construction of influence line for N^j. **(a)** Truss; **(b)** $\overline{\mathscr{L}}$; **(c)** $\hat{\mathscr{L}}$; **(d)** $\mathscr{L}$.

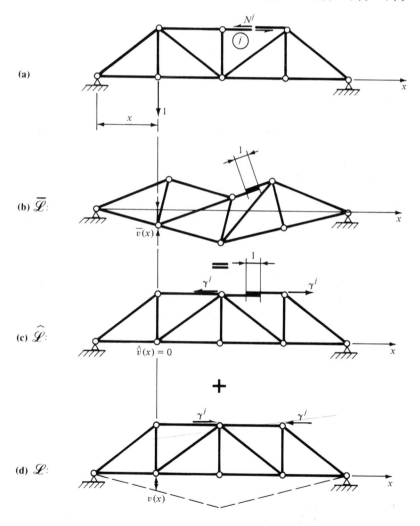

or

$$N^j(x) = -|\bar{v}(x)| = \bar{v}(x) < 0 \qquad (4.344)$$

To determine the configuration of the bottom chord in Figure 4.35b by the matrix displacement method, we need only analyze the truss subjected to the equivalent joint loads in Figure 4.35d because $\hat{v}^i(x) = 0$, and, hence,

$$\bar{v}^i(x) = v^i(x) \qquad (4.345)$$

Moreover, since the ordinates of the influence lines are of interest only at the joints, only the joint deflections are required.

The fixed-end forces of the element subjected to the unit axial deflection at the release (Figure 4.35c) are equal in magnitude to the extensional stiffness of the element, γ.

PROBLEMS

4.1 Apply the definitions of symmetry and antisymmetry and the conditions of equilibrium to the joint in Figure 4.6b to verify Eqs. (4.4) and (4.5).

4.2 The structures are symmetric (Section 4.2). Decompose the loads into symmetric and antisymmetric components and sketch the substructures for symmetric and antisymmetric response. All elements have the properties E, A, I.

4.3 The elements of the symmetric truss (Section 4.2) have the identical extensional stiffness γ.
   a. Decompose the load into symmetric and antisymmetric components.
   b. Analyze the substructure with the symmetric load.
   c. Analyze the substructure with the antisymmetric load.
   d. Use Eqs. (4.20)–(4.23) to determine from (b) and (c) the symmetric and antisymmetric responses of the truss; superpose them to obtain the complete response of the truss.

4.4 Consider the frame in Figure 4.10b:
   a. Use the member code matrix to generate the system stiffness matrix.
   b. Use the definition of a system stiffness coefficient, Eq. (3.35), to verify the coefficients of the first column of the system stiffness matrix in (a). (Note that Figure 4.10d represents the corresponding configuration.)

4.5 Analogous to Eqs. (2.78), we can establish contragredient transformations for the elements of the frame in Figure 4.10b of the form

$$\mathbf{d}^i = \mathbf{A}^i\mathbf{q}$$

$$\mathbf{F}^{(i)} = \mathbf{A}^{iT}\mathbf{f}^i$$

   a. Combine the transformations with the beam model

$$\mathbf{f}^i = \mathbf{k}^i\mathbf{d}^i$$

   to obtain the generalized element stiffness matrix

$$\mathbf{K}^{(i)} = \mathbf{A}^{iT}\mathbf{k}^i\mathbf{A}^i$$

   b. By Eqs. (4.35), define $\mathbf{A}^1$ and use the formula in (a) to determine the generalized stiffness matrix for element 1, $\mathbf{K}^{(1)}$; verify $\mathbf{K}^{(1)}$ by the member code approach.

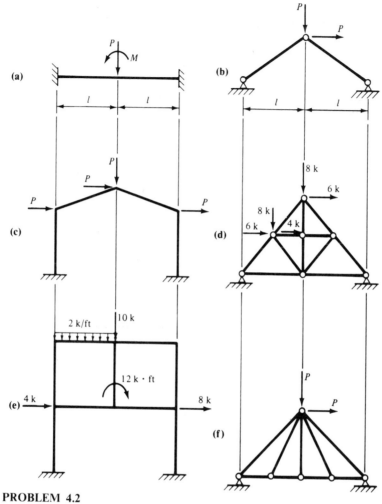

PROBLEM 4.2

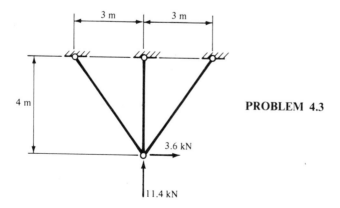

PROBLEM 4.3

4.6   Assume that the element centroidal axes of the frame in Problem 2.6 are in-extensible (Section 4.3). Analyze the frame by the matrix displacement method.

4.7   Assume that the element centroidal axes of the frame in Problem 2.7 are inextensible (Section 4.3). Take advantage of symmetry to analyze the frame (Section 4.2).

4.8   Assume that the element centroidal axes are inextensible (Section 4.3). Utilize structural symmetry to analyze the frame (Section 4.2).

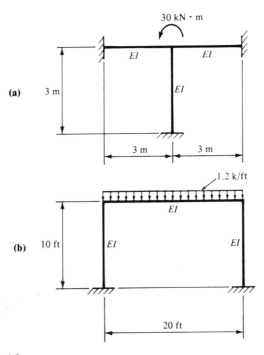

PROBLEM 4.8

4.9   Analyze the structure with the shear release at joint 2 (Section 4.4):
   a. Disregard structural symmetry.
   b. Decompose the load into symmetric and antisymmetric components and exploit symmetry (Section 4.2).

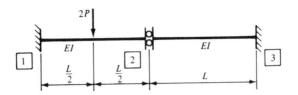

PROBLEM 4.9

4.10 Utilize symmetry (Section 4.2) to analyze the structure in Figure 4.16a.

4.11 The structure consists of an infinitesimal rotational spring whose stiffness is defined in Eq. (4.97), a beam element, and an extensional spring that can be represented as a truss element with extensional stiffness $\gamma = \alpha\beta$ (Section 4.5).
   a. Construct the system stiffness matrix.
   b. Verify the coefficients of the first and second columns of the system stiffness matrix by Eq. (3.35).

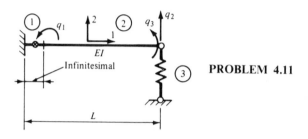

PROBLEM 4.11

4.12 Specialize Eqs. (4.127)–(4.129)
   a. for $s_a = \infty,\, s_b = 0$;
   b. For $s_a = s_b = \infty$.

4.13 Analyze the structure in Figure 4.21a:
   a. Place the hinge at joint 2 (Section 4.4).
   b. Place the hinge near the b end of element 1.

4.14 The truss forms an equilateral triangle with the internal node at the centroid. Regard the truss as a complex element and formulate the condensed model by eliminating the internal degrees of freedom (Section 4.6).

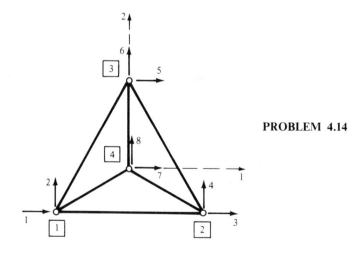

PROBLEM 4.14

4.15 Modify the frame in Figure 3.7a by fixing joint 3. Suppose that joint 1 experiences the counterclockwise rotation θ (Section 4.7).

a. Determine the element-end forces.
b. For $\theta = 0.05$ rad and the section modulus $S = 12.5$ in^3, compute the maximum stress.

4.16 Consider Example 3 of Section 4.7. Assume that there are no element imperfections but joint 4 is displaced in the direction of the global 2 axis by $\frac{5}{8}$ in. Compute the resulting element and joint forces.

4.17 The elements of the truss have the identical extensional stiffness $\gamma = 40$ k/in. Element 5 is $\frac{1}{2}$ inch too long. Compute the resulting element and joint forces (Section 4.7).

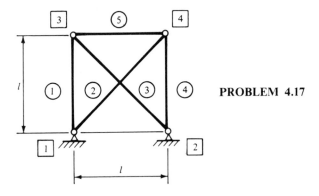

PROBLEM 4.17

4.18 The element in Figure 4.27a is subjected to the temperature change

$$t(x, y) = -\frac{2}{h} \Delta t \, y \left(\frac{x}{L}\right)^2$$

Compute the temperature-induced fixed-end forces and the corresponding stress and strain distributions.

4.19 Element 1 of the frame experiences the uniform temperature increase t_0 (Section 4.8). Analyze the frame: Sketch the resolution $\mathcal{P} = \hat{\mathcal{P}} + \mathcal{L}$; determine the element-end forces, the joint forces, and the stress and strain distributions of element 1. The elements have identical properties.

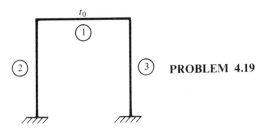

PROBLEM 4.19

4.20   Element 1 of the continuous beams undergoes the temperature increase defined in Eq. (4.249). Compute the resulting element forces and the stress and strain distributions. As in Example 3 of Section 4.8, the structure can be represented by truss elements.

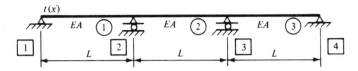

PROBLEM 4.20

4.21   Use the matrix displacement method (Figure 3.2) to analyze the truss (Section 4.9); show the joint forces on the truss.

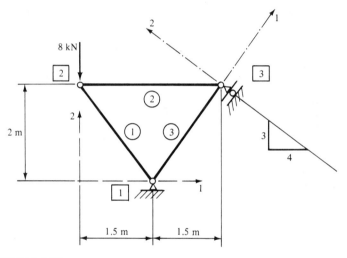

PROBLEM 4.21

4.22   Construct the system stiffness matrix for the frame in Figure 4.30a. The elements have identical properties: A, I, E, L.

4.23   Use the matrix displacement method to determine the influence function for the shear force at c in Figure 4.33a.

4.24   Consider the continuous beams in Figure 4.34a. Use the matrix displacement method to determine the influence functions
   a. for the reactive force at joint 1;
   b. for the reactive force at joint 2;
   c. for the bending moment at the midspan of element 1;
   d. for the shear force at the midspan of element 1;
   e. for the bending moment at joint 2.

REFERENCES

American Heritage Dictionary of the English Language. 1971. American Heritage and Houghton, Boston.

Glockner, P. G. 1973. "Symmetry in Structural Mechanics," *Journal of the Structural Division, ASCE,* 99, 71–89.

Hoff, N. J. 1956. *The Analysis of Structures.* Wiley, New York.

Livesley, R. K. 1964. *Matrix Methods of Structural Analysis.* Pergamon Press, Oxford.

Livesley, R. K. 1975. *Matrix Methods of Structural Analysis,* 2nd ed. Pergamon Press, Oxford.

McGuire, W., and R. H. Gallagher. 1979. *Matrix Structural Analysis.* Wiley, New York.

Shames, I. H. 1975. *Introduction to Solid Mechanics.* Prentice-Hall, Englewood Cliffs, NJ.

Stippes, M., G. Wempner, M. Stern, and R. Beckett, 1961. *An Introduction to the Mechanics of Deformable Bodies.* Merrill, Columbus, OH.

Tezcan, S. S. 1966. "Computer Analysis of Plane and Space Structures," *Journal of the Structural Division, ASCE,* 92, 143–173.

Thomas, G. B., Jr. 1956. *Calculus and Analytic Geometry,* 2nd ed. Addison-Wesley, Reading, MA.

Tong, P., and J. N. Rossettos. 1977. *Finite-Element Method.* MIT Press, Cambridge, MA.

Zienkiewicz, O. C. 1977. *The Finite Element Method,* 3d ed. McGraw-Hill, New York.

Additional Reference

Glockner, P. G., and M. C. Singh, eds. 1974. *Symmetry, Similarity and Group Theoretic Methods in Mechanics,* Proceedings of Symposium held at the University of Calgary, August 19–21.

FIVE

MATRIX DISPLACEMENT METHOD:
SPACE STRUCTURES

5.1 INTRODUCTION

The matrix displacement method is extended to skeletal space structures represented as space trusses (Section 5.4), space frames (Section 5.5), and grids (Section 5.6). The formulation is confined to prismatic elements with bisymmetric cross sections.

The representation of a structure by a three-dimensional model may be necessary to test the adequacy of a two-dimensional model, or because the form of the structure—for example, a reticulated shell (Makowski, 1962, 1967; Sherman et al., 1976)—precludes a two-dimensional model.

5.2 GENERALIZED DISPLACEMENTS

Frame Element

Consider the prismatic element in Figure 5.1a, whose cross section (Figure 5.1b) has biaxial symmetry. In the absence of shearing deformation,[1] the state of the element is characterized by axial deformation, flexural deformations about the two principal axes, and torsional deformation (Table 5.1). Moreover, these deformations are independent (that is, they are uncoupled) if the element behaves linearly (Section 1.5) and if the cross sections are free to warp. The independence of axial and flexural deformations is reflected, for example, in Eq. (1.99) by the vanishing of stiffness coefficients that link

---

[1] For slender elements the effect of shearing deformation is negligible (Cowper, 1966; Timoshenko and Gere, 1972). Element stiffness matrices that include shearing deformation are presented by Weaver and Gere (1980). Nickel and Secor (1972), and Yoo and Fehrenbach (1981).

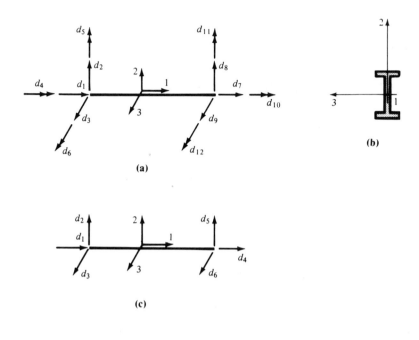

(a)

(b)

(c)

(d)

FIGURE 5.1 Space elements. **(a)** Frame element; **(b)** cross section; **(c)** truss element; **(d)** axial deformation element.

TABLE 5.1 Space Frame Element

Deformations	Generalized displacements	Degrees of freedom
Axial	d_1, d_7	2
Torsional	d_4, d_{10}	2
Bending about the 3 axis	d_2, d_6, d_8, d_{12}	4
Bending about the 2 axis	d_3, d_5, d_9, d_{11}	4
Combined	d_1-d_{12}	12

the two deformations. To explain the significance of warping, a brief discussion of torsion of thin-walled open sections is presented.

In the bisymmetric cross section of Figure 5.1b, the centroid coincides with the shear center and the center of twist (Boresi et al., 1978). The shear center is the point in the plane of the cross section through which a transverse force must pass to induce bending without twisting. The center of twist is the point through which a torque must pass to induce twisting without bending. The coincidence of the shear center and the center of twist for linearly elastic elements follows from Betti's law. For noncircular cross sections, twisting causes plane sections to deform—to warp—out of their planes. As a consequence, longitudinal fibers are strained and normal stresses arise. For the normal strains and stresses to be negligible, and torsional deformations to remain uncoupled, the cross sections must be free to warp (Boresi et al., 1978).

If warping is not restrained, the torsional behavior of thin-walled open sections can be represented by Eqs. (1.90) if J is approximated by the relation (Boresi et al., 1978)

$$J = \tfrac{1}{3} \sum bt^3 \tag{5.1}$$

where b and t denote the length and thickness of each rectangular segment of the cross section. Accordingly, J is no longer the polar moment of inertia of the cross section.

Truss Element

In the absence of flexural and torsional deformations, the 12 degree-of-freedom space frame element (Figure 5.1a) reduces to the six degree-of-freedom space truss element (Figure 5.1c). Moreover, as in the plane truss element (Figure 2.12a), the axial strain, and hence the axial force, is a function only of the axial deflections. Thus, the local space truss element can be represented as shown in Figure 5.1d. The corresponding element model is defined in Eqs. (3.76). This simplification is a consequence of the assumption in the linear theory that the effect of rotation on the axial strain is negligible; this is reflected in Eq. (1.31).

Assemblies of Elements

Equation (2.1) can be extended to define the number of degrees of freedom of space structures composed of the elements in Figure 5.1. Moreover, if we include the possibility of internal releases (Section 4.4), it assumes the form

$$n = l \cdot NJ - NC + NR \tag{5.2}$$

$$l = \begin{cases} 3 & \text{for space trusses and grids (Section 5.6)} \\ 6 & \text{for space frames} \end{cases}$$

where l represents the degrees of freedom of a free joint, NJ the number of joints, NC the number of joint constraints, and NR the number of internal releases.

Static Determinacy

Although static determinacy is of no consequence in the matrix displacement analysis, it is frequently of interest to the analyst, and it can easily be computed once n is evaluated by Eq. (2.1) or (5.2). Specifically, for a stable structure, that is, a structure sufficiently constrained to prevent rigid-body displacements, the degree of static indeterminacy is defined by the relation

$$i = s \cdot \text{NE} - n \qquad (5.3)$$

where

$$s = \begin{cases} 1 & \text{for trusses (plane or space)} \\ 3 & \text{for plane frames and grids} \\ 6 & \text{for space frames} \end{cases}$$

In Eq. (5.3), s denotes the number of statically independent forces per element (Section 1.8) and NE is the number of elements. Moreover, it follows from Section 2.7 that the number of degrees of freedom, n, is equal to the number of equations of equilibrium. Accordingly, the structure is statically determinate if $i = 0$ and statically indeterminate if $i > 0$.

For example, for the space truss shown in Figure 5.3 (page 260), Eqs. (5.2) and (5.3) yield $n = 3 \cdot 5 - 12 = 3$ and $i = 1 \cdot 4 - 3 = 1$, respectively. However, if the space truss in Figure 5.3 is replaced by a space frame, Eqs. (5.2) and (5.3) yield $n = 6 \cdot 5 - 12 = 18$ and $i = 6 \cdot 4 - 18 = 6$, respectively.

5.3 NOTATION AND CONVENTION

The symbols and conventions of Section 2.3 are extended to space structures subject to the following changes: The global 2-axis is assumed to be vertical; the orientation of the local 3 axis is independent of the global 3 axis.

5.4 SPACE TRUSSES

To apply the matrix displacement method (Figure 3.2) to space trusses, we need to specialize the coordinate transformations of Appendix C for the local and global space truss elements (Figures 5.2a and b) and formulate the global element stiffness matrix.

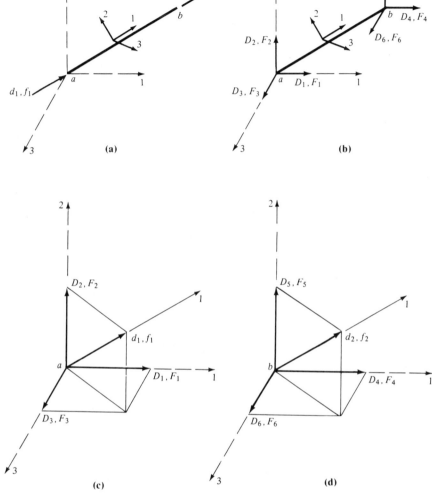

FIGURE 5.2 Space truss elements. **(a)** Local; **(b)** global; **(c)** a end; **(d)** b end.

Coordinate Transformations

By Eqs. (C.13), (C.14), and (C.16), displacement and force transformations at the a end can be expressed as

$$
\begin{bmatrix} d_1 \\ d_2 \\ d_3 \end{bmatrix} = \begin{bmatrix} \mathbf{i}_1 \\ \mathbf{i}_2 \\ \mathbf{i}_3 \end{bmatrix} \mathbf{D}_a, \quad \mathbf{F}_a = [\mathbf{i}_1^T \quad \mathbf{i}_2^T \quad \mathbf{i}_3^T] \begin{bmatrix} f_1 \\ f_2 \\ f_3 \end{bmatrix} \tag{5.4}
$$

where

$$\mathbf{D}_a = \begin{bmatrix} D_1 \\ D_2 \\ D_3 \end{bmatrix}, \quad \mathbf{F}_a = \begin{bmatrix} F_1 \\ F_2 \\ F_3 \end{bmatrix} \tag{5.5}$$

Since at the a end (Figure 5.2c) d_1 is the only local displacement of interest and f_1 is the only nonzero local force, Eqs. (5.4) can be reduced to

$$d_1 = \mathbf{i}_1 \mathbf{D}_a, \quad \mathbf{F}_a = \mathbf{i}_1^T f_1 \tag{5.6}$$

The substitutions

$$d_a = d_1, \quad f_a = f_1, \quad \boldsymbol{\lambda} = \mathbf{i}_1 \tag{5.7}$$

transform Eqs. (5.6) to the standard form

$$d_a = \boldsymbol{\lambda} \mathbf{D}_a, \quad \mathbf{F}_a = \boldsymbol{\lambda}^T f_a \tag{5.8}$$

Similarly, the transformations at the b end (Figure 5.2d) can be expressed as

$$d_b = \boldsymbol{\lambda} \mathbf{D}_b, \quad \mathbf{F}_b = \boldsymbol{\lambda}^T f_b \tag{5.9}$$

where

$$d_b = d_2, \quad \mathbf{D}_b = \begin{bmatrix} D_4 \\ D_5 \\ D_6 \end{bmatrix}, \quad f_b = f_2, \quad \mathbf{F}_b = \begin{bmatrix} F_4 \\ F_5 \\ F_6 \end{bmatrix} \tag{5.10}$$

Equations (5.8) and (5.9) yield the transformations for the space truss element (Figures 5.2a, b)

$$\mathbf{d} = \boldsymbol{\Lambda} \mathbf{D}, \quad \mathbf{F} = \boldsymbol{\Lambda}^T \mathbf{f} \tag{5.11}$$

where

$$\boldsymbol{\Lambda} = \begin{bmatrix} \boldsymbol{\lambda} & \mathbf{0} \\ \mathbf{0} & \boldsymbol{\lambda} \end{bmatrix} \tag{5.12}$$

$$\boldsymbol{\lambda} = \frac{\Delta \mathbf{X}}{|\Delta \mathbf{X}|} = [c_1 \quad c_2 \quad c_3] \tag{5.13}$$

$$\Delta \mathbf{X} = \mathbf{X}_k - \mathbf{X}_j$$

In Eqs. (5.13), c_i is the direction cosine of the local 1 axis relative to the global i axis, $\mathbf{X}_j$ and $\mathbf{X}_k$ are the coordinate vectors of the joints at the a end and b end of the element, respectively, and the difference vector $\Delta \mathbf{X}$ coincides with the element axis (Figure C.3b).

Element Models

The local element model is [Eqs. (3.76)]

$$\mathbf{f} = \mathbf{kd} \tag{5.14a}$$

where

$$\mathbf{k} = \gamma \begin{bmatrix} 1 & -1 \\ -1 & 1 \end{bmatrix}, \quad \gamma = \frac{EA}{L} \qquad (5.14b)$$

By Eqs. (3.60), (3.61), (3.76), and (5.13) we obtain the global element stiffness matrix

$$\mathbf{K} = \gamma \begin{bmatrix} c_1^2 & c_1c_2 & c_1c_3 & -c_1^2 & -c_1c_2 & -c_1c_3 \\ & c_2^2 & c_2c_3 & -c_1c_2 & -c_2^2 & -c_2c_3 \\ & & c_3^2 & -c_1c_3 & -c_2c_3 & -c_3^2 \\ \hline & & & c_1^2 & c_1c_2 & c_1c_3 \\ & & & & c_2^2 & c_2c_3 \\ \text{sym.} & & & & & c_3^2 \end{bmatrix} \qquad (5.15)$$

Matrix Displacement Analysis

Example. The truss in Figure 5.3 is subjected to the joint forces $Q_1 = Q_2 = 0$, $Q_3 = P$. The elements have identical extensional stiffnesses. The analysis is based on Figure 3.2.

1. *Unknowns*

$$q_k, \quad k = 1, 2, 3 \qquad (5.16)$$

2. *Element Models*

$$\mathbf{f}^i = \mathbf{k}\mathbf{d}^i, \quad i = 1, \dots, 4 \qquad (5.17)$$

where $\mathbf{k}$ is defined in Eqs. (5.14b).

3. *System Model*
The rotation matrices are

$$\begin{aligned} \boldsymbol{\lambda}^1 = \tfrac{1}{2}[1 \quad \sqrt{2} \quad 1], \quad \boldsymbol{\lambda}^2 = \tfrac{1}{2}[1 \quad \sqrt{2} \quad -1] \\ \boldsymbol{\lambda}^3 = \tfrac{1}{2}[-1 \quad \sqrt{2} \quad -1], \quad \boldsymbol{\lambda}^4 = \tfrac{1}{2}[-1 \quad \sqrt{2} \quad 1] \end{aligned} \qquad (5.18)$$

Equations (5.18) follow from Eqs. (5.13), and Figure 5.3. For example, for element 1,

$$\Delta \mathbf{X} = \mathbf{X}_5 - \mathbf{X}_1 = l[1 \quad \sqrt{2} \quad 1], \quad |\Delta \mathbf{X}| = 2l$$

Thus,   (5.19)

$$\boldsymbol{\lambda}^1 = \frac{\Delta \mathbf{X}}{|\Delta \mathbf{X}|} = \tfrac{1}{2}[1 \quad \sqrt{2} \quad 1] = [c_1 \quad c_2 \quad c_3]$$

The member codes are

$$\mathbf{M}^{iT} = [0 \quad 0 \quad 0 \quad 1 \quad 2 \quad 3]; \quad i = 1, \dots, 4 \qquad (5.20)$$

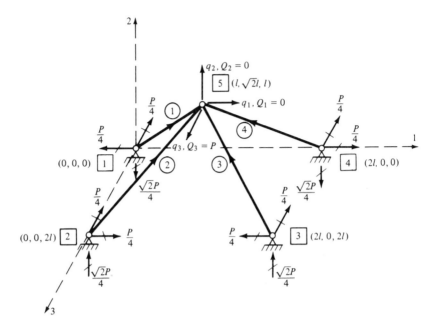

FIGURE 5.3 Space truss.

System Stiffness Matrix

By Eqs. (5.15), (5.18), and (5.20) we obtain for element 1

$$\mathbf{K}^1 = \frac{\gamma}{4} \begin{array}{cccccc} 0 & 0 & 0 & 1 & 2 & 3 \\ \left[\begin{array}{cccccc} & & & & & \\ & & & & & \\ & & & 1 & \sqrt{2} & 1 \\ & & & & 2 & \sqrt{2} \\ \text{sym.} & & & & & 1 \end{array} \right] \begin{array}{l} 0 \\ 0 \\ 0 \\ 1 \\ 2 \\ 3 \end{array} \end{array} \xrightarrow{\mathbf{M}} \mathbf{K}^{(1)} = \frac{\gamma}{4} \left[\begin{array}{ccc} 1 & \sqrt{2} & 1 \\ & 2 & \sqrt{2} \\ \text{sym.} & & 1 \end{array} \right]$$

$$(5.21)$$

Similarly we obtain the generalized stiffness matrices for elements 2–4:

$$\mathbf{K}^{(2)} = \frac{\gamma}{4} \left[\begin{array}{ccc} 1 & \sqrt{2} & -1 \\ & 2 & -\sqrt{2} \\ \text{sym.} & & 1 \end{array} \right], \quad \mathbf{K}^{(3)} = \frac{\gamma}{4} \left[\begin{array}{ccc} 1 & -\sqrt{2} & 1 \\ & 2 & -\sqrt{2} \\ \text{sym.} & & 1 \end{array} \right]$$

$$\mathbf{K}^{(4)} = \frac{\gamma}{4} \left[\begin{array}{ccc} 1 & -\sqrt{2} & -1 \\ & 2 & \sqrt{2} \\ \text{sym.} & & 1 \end{array} \right]$$

$$(5.22)$$

Thus,

$$\mathbf{K} = \sum_{i=1}^{4} \mathbf{K}^{(i)} = \gamma \begin{bmatrix} 1 & 0 & 0 \\ & 2 & 0 \\ \text{sym.} & & 1 \end{bmatrix} \tag{5.23}$$

System Model

$$\gamma \begin{bmatrix} 1 & 0 & 0 \\ 0 & 2 & 0 \\ 0 & 0 & 1 \end{bmatrix} \begin{bmatrix} q_1 \\ q_2 \\ q_3 \end{bmatrix} = \begin{bmatrix} 0 \\ 0 \\ P \end{bmatrix} \tag{5.24}$$

4. *Solution*

$$\mathbf{q} = \frac{P}{\gamma} \begin{bmatrix} 0 \\ 0 \\ 1 \end{bmatrix} \tag{5.25}$$

5. *Element Forces*
It follows from Eq. (5.20) that

$$\mathbf{D}_a^i = \mathbf{0}, \quad \mathbf{D}_b^i = \mathbf{q} \tag{5.26}$$

Accordingly, for element 1

$$d_a^1 = 0, \quad d_b^1 = \boldsymbol{\lambda}^1 \mathbf{D}_b^1 = \frac{P}{2\gamma} \tag{5.27}$$

and

$$\mathbf{f}^1 = \gamma \begin{bmatrix} 1 & -1 \\ -1 & 1 \end{bmatrix} \begin{bmatrix} 0 \\ 1 \end{bmatrix} \frac{P}{2\gamma} = \frac{P}{2} \begin{bmatrix} -1 \\ 1 \end{bmatrix} \tag{5.28}$$

Similarly we obtain

$$\mathbf{f}^4 = \mathbf{f}^1, \quad \mathbf{f}^2 = \mathbf{f}^3 = -\mathbf{f}^1 \tag{5.29}$$

6. *Joint Forces*

$$\mathbf{P}_1 = \mathbf{F}_a^1 = \boldsymbol{\lambda}^{1T} f_a^1 = \frac{P}{4} \begin{bmatrix} -1 \\ -\sqrt{2} \\ -1 \end{bmatrix} \tag{5.30}$$

Similarly we obtain

$$\mathbf{P}_2 = \frac{P}{4} \begin{bmatrix} 1 \\ \sqrt{2} \\ -1 \end{bmatrix}, \quad \mathbf{P}_3 = \frac{P}{4} \begin{bmatrix} -1 \\ \sqrt{2} \\ -1 \end{bmatrix}, \quad \mathbf{P}_4 = \frac{P}{4} \begin{bmatrix} 1 \\ -\sqrt{2} \\ -1 \end{bmatrix} \tag{5.31}$$

The joint forces are shown in Figure 5.3. Equilibrium for the truss can be verified by summing the forces in the directions of the global axes and the moments about the global axes.

5.5 SPACE FRAMES

To extend the matrix displacement method to space frames without and with element actions (Figures 3.2 and 3.9) we need to apply the coordinate transformations of Appendix C to space frame elements (Figure 5.4) and formulate the local and global element stiffness matrices.

The formulation is confined to elements with coincident shear center and centroidal axes (Figure 5.1b). Elements with distinct shear center and centroidal axes are considered in the book by Weaver and Gere (1980).

Coordinate Transformations

As shown in Figure 5.4, the local and global element displacements and forces are numbered, analogous to Section 2.3, in the sequence of the co-ordinate axes from the a end to the b end. At each end, the deflections (forces) are numbered ahead of the rotations (moments). Correspondingly, at each joint, the deflections (forces) are numbered ahead of the rotations (moments).

It follows from Eqs. (C.12) and (C.13) that the local and global displacements are linked by the transformations

$$
\begin{bmatrix} d_1 \\ d_2 \\ d_3 \\ \hline d_4 \\ d_5 \\ d_6 \end{bmatrix}
=
\begin{bmatrix} \lambda & \vdots & 0 \\ \hline 0 & \vdots & \lambda \end{bmatrix}
\begin{bmatrix} D_1 \\ D_2 \\ D_3 \\ \hline D_4 \\ D_5 \\ D_6 \end{bmatrix}
\qquad
\begin{bmatrix} d_7 \\ d_8 \\ d_9 \\ \hline d_{10} \\ d_{11} \\ d_{12} \end{bmatrix}
=
\begin{bmatrix} \lambda & \vdots & 0 \\ \hline 0 & \vdots & \lambda \end{bmatrix}
\begin{bmatrix} D_7 \\ D_8 \\ D_9 \\ \hline D_{10} \\ D_{11} \\ D_{12} \end{bmatrix}
\tag{5.32}
$$

or

$$
\mathbf{d}_a = \mathbf{\Lambda D}_a, \quad \mathbf{d}_b = \mathbf{\Lambda D}_b
\tag{5.33}
$$

where

$$
\mathbf{\Lambda} = \begin{bmatrix} \lambda & 0 \\ 0 & \lambda \end{bmatrix}
\tag{5.34}
$$

Similarly, by Eqs. (C.12) and (C.16) we obtain the force transformations

$$
\mathbf{F}_a = \mathbf{\Lambda}^T \mathbf{f}_a, \quad \mathbf{F}_b = \mathbf{\Lambda}^T \mathbf{f}_b
\tag{5.35}
$$

Equations (5.33) and (5.35) yield the combined transformations

$$
\mathbf{d} = \mathbf{RD}, \quad \mathbf{F} = \mathbf{R}^T \mathbf{f}
\tag{5.36}
$$

where

$$
\mathbf{R} = \begin{bmatrix} \mathbf{\Lambda} & 0 \\ 0 & \mathbf{\Lambda} \end{bmatrix}
=
\begin{bmatrix} \lambda & 0 & 0 & 0 \\ 0 & \lambda & 0 & 0 \\ 0 & 0 & \lambda & 0 \\ 0 & 0 & 0 & \lambda \end{bmatrix}
\tag{5.37}
$$

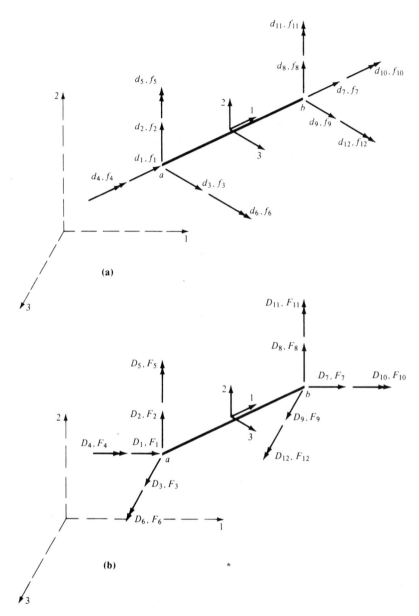

FIGURE 5.4 Space frame elements. **(a)** Local; **(b)** global.

Local Element Model

The stiffness matrix of the local element model

$$\bar{\mathbf{f}} = \mathbf{kd} + \hat{\mathbf{f}} \tag{5.38}$$

can be constructed by superposing the component stiffness matrices representing the four independent deformations described in Table 5.1. For this purpose it is convenient to use the code number technique illustrated for the plane frame element in Eqs. (1.120) and (1.121). The component stiffness matrices with the code numbers are (see Figures 5.4 and 5.5 and Table 5.1):

For axial deformation (Figure 5.5a)

$$\mathbf{k} = \gamma \begin{bmatrix} 1 & 7 \\ 1 & -1 \\ \text{sym.} & 1 \end{bmatrix} \begin{matrix} 1 \\ 7 \end{matrix}, \quad \gamma = \frac{EA}{L} \tag{5.39a}$$

For torsional deformation (Figure 5.5b)

$$\mathbf{k} = \delta \begin{bmatrix} 4 & 10 \\ 1 & -1 \\ \text{sym.} & 1 \end{bmatrix} \begin{matrix} 4 \\ 10 \end{matrix}, \quad \delta = \frac{GJ}{L} \tag{5.39b}$$

FIGURE 5.5 Deformations of space frame element. **(a)** Axial; **(b)** torsional; **(c)** bending about 3 axis; **(d)** bending about 2 axis.

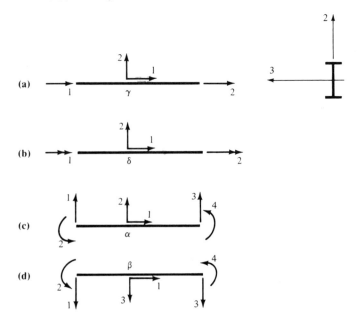

For bending about the 3 axis (Figure 5.5c)

$$\mathbf{k} = \alpha \begin{array}{c} \\ \begin{bmatrix} \begin{array}{cccc} 2 & 6 & 8 & 12 \end{array} \\ \begin{bmatrix} 12 & 6L & -12 & 6L \\ & 4L^2 & -6L & 2L^2 \\ & & 12 & -6L \\ \text{sym.} & & & 4L^2 \end{bmatrix} \begin{array}{c} 2 \\ 6 \\ 8 \\ 12 \end{array} \end{bmatrix} \end{array}, \quad \alpha = \frac{EI_3}{L^3} \qquad (5.39c)$$

For bending about the 2 axis (Figure 5.5d)

$$\mathbf{k} = \beta \begin{array}{c} \\ \begin{bmatrix} \begin{array}{cccc} 3 & 5 & 9 & 11 \end{array} \\ \begin{bmatrix} 12 & -6L & -12 & -6L \\ & 4L^2 & 6L & 2L^2 \\ & & 12 & 6L \\ \text{sym.} & & & 4L^2 \end{bmatrix} \begin{array}{c} 3 \\ 5 \\ 9 \\ 11 \end{array} \end{bmatrix} \end{array}, \quad \beta = \frac{EI_2}{L^3} \qquad (5.39d)$$

Equations (5.39a) and (5.39c) follow from Eqs. (1.96b) and (1.98b). Equation (5.39b) is obtained from Eqs. (1.90), and for thin-walled open sections J is defined in Eq. (5.1). Equation (5.39d) can be determined by Eq. (1.105). The superposition of Eqs. (5.39) in accordance with the code numbers yields the local stiffness matrix of the space frame element (Figure 5.4a)

$$\mathbf{k} = \begin{array}{c} \begin{array}{cccccccccccc} 1 & 2 & 3 & 4 & 5 & 6 & 7 & 8 & 9 & 10 & 11 & 12 \end{array} \\ \left[\begin{array}{cccccc|cccccc} \gamma & 0 & 0 & 0 & 0 & 0 & -\gamma & 0 & 0 & 0 & 0 & 0 \\ & 12\alpha & 0 & 0 & 0 & 6L\alpha & 0 & -12\alpha & 0 & 0 & 0 & 6L\alpha \\ & & 12\beta & 0 & -6L\beta & 0 & 0 & 0 & -12\beta & 0 & -6L\beta & 0 \\ & & & \delta & 0 & 0 & 0 & 0 & 0 & -\delta & 0 & 0 \\ & & & & 4L^2\beta & 0 & 0 & 0 & 6L\beta & 0 & 2L^2\beta & 0 \\ & & & & & 4L^2\alpha & 0 & -6L\alpha & 0 & 0 & 0 & 2L^2\alpha \\ \hline & & & & & & \gamma & 0 & 0 & 0 & 0 & 0 \\ & & & & & & & 12\alpha & 0 & 0 & 0 & -6L\alpha \\ & & & & & & & & 12\beta & 0 & 6L\beta & 0 \\ & & & & & & & & & \delta & 0 & 0 \\ & & & & & & & & & & 4L^2\beta & 0 \\ \text{sym.} & & & & & & & & & & & 4L^2\alpha \end{array} \right] \begin{array}{c} 1 \\ 2 \\ 3 \\ 4 \\ 5 \\ 6 \\ 7 \\ 8 \\ 9 \\ 10 \\ 11 \\ 12 \end{array} \end{array}$$

$$(5.40)$$

where

$$\alpha = \frac{EI_3}{L^3}, \quad \beta = \frac{EI_2}{L^3}, \quad \gamma = \frac{EA}{L}, \quad \delta = \frac{GJ}{L} \qquad (5.41)$$

Similarly, the fixed-end force vector $\hat{\mathbf{f}}$ in Eq. (5.38) can be obtained by superposing the fixed-end forces corresponding to the four independent deformations.

Global Element Model

From Eqs. (5.36) and (5.38) we obtain the global element model

$$\overline{\mathbf{F}} = \mathbf{KD} + \hat{\mathbf{F}} \qquad (5.42)$$

where

$$\mathbf{K} = \mathbf{R}^T \mathbf{kR} \quad \text{and} \quad \hat{\mathbf{F}} = \mathbf{R}^T \hat{\mathbf{f}} \qquad (5.43)$$

By substituting for $\mathbf{R}$ and $\mathbf{k}$ from Eqs. (5.37) and (5.40), and for λ from Eq. (C.12), we can express the global element stiffness matrix, analogous to Eq. (3.63), in terms of distinct coefficient functions g_i, $i = 1, 2, \ldots, 27$. It is convenient to locate these functions in the stiffness matrix indirectly through an index matrix, which stores the subscripts and negative signs of these functions. The index matrix is

$$
\mathbf{INDEX} =
\begin{array}{c}
\begin{array}{cccccccccccc}
1 & 2 & 3 & 4 & 5 & 6 & 7 & 8 & 9 & 10 & 11 & 12
\end{array} \\
\left[
\begin{array}{cccccc:cccccc}
1 & 2 & 3 & 13 & 14 & 15 & -1 & -2 & -3 & 13 & 14 & 15 \\
 & 4 & 5 & 16 & 17 & 18 & -2 & -4 & -5 & 16 & 17 & 18 \\
 & & 6 & 19 & 20 & 21 & -3 & -5 & -6 & 19 & 20 & 21 \\
 & & & 7 & 8 & 9 & -13 & -16 & -19 & 22 & 23 & 24 \\
 & & & & 10 & 11 & -14 & -17 & -20 & 23 & 25 & 26 \\
 & & & & & 12 & -15 & -18 & -21 & 24 & 26 & 27 \\
\hdashline
 & & & & & & 1 & 2 & 3 & -13 & -14 & -15 \\
 & & & & & & & 4 & 5 & -16 & -17 & -18 \\
 & & & & & & & & 6 & -19 & -20 & -21 \\
 & & & & & & & & & 7 & 8 & 9 \\
 & & & & & & & & & & 10 & 11 \\
\text{sym.} & & & & & & & & & & & 12
\end{array}
\right]
\begin{array}{c}
1 \\ 2 \\ 3 \\ 4 \\ 5 \\ 6 \\ 7 \\ 8 \\ 9 \\ 10 \\ 11 \\ 12
\end{array}
\end{array}
$$

$$(5.44)$$

and the distinct coefficient functions are

$$g_1 = \gamma l_{11}^2 + 12\alpha l_{21}^2 + 12\beta l_{31}^2$$

$$g_2 = \gamma l_{11} l_{12} + 12\alpha l_{21} l_{22} + 12\beta l_{31} l_{32}$$

$$g_3 = \gamma l_{11} l_{13} + 12\alpha l_{21} l_{23} + 12\beta l_{31} l_{33}$$

$$g_4 = \gamma l_{12}^2 + 12\alpha l_{22}^2 + 12\beta l_{32}^2$$

$$g_5 = \gamma l_{12} l_{13} + 12\alpha l_{22} l_{23} + 12\beta l_{32} l_{33}$$

$$g_6 = \gamma l_{13}^2 + 12\alpha l_{23}^2 + 12\beta l_{33}^2$$

$$g_7 = \delta l_{11}^2 + 4L^2 \alpha l_{31}^2 + 4L^2 \beta l_{21}^2$$

$$g_8 = \delta l_{11} l_{12} + 4L^2 \alpha l_{31} l_{32} + 4L^2 \beta l_{21} l_{22}$$

$$g_9 = \delta l_{11} l_{13} + 4L^2 \alpha l_{31} l_{33} + 4L^2 \beta l_{21} l_{23}$$

$$g_{10} = \delta l_{12}^2 + 4L^2 \alpha l_{32}^2 + 4L^2 \beta l_{22}^2$$

$$g_{11} = \delta l_{12} l_{13} + 4L^2 \alpha l_{32} l_{33} + 4L^2 \beta l_{22} l_{23}$$

$$g_{12} = \delta l_{13}^2 + 4L^2 \alpha l_{33}^2 + 4L^2 \beta l_{23}^2$$

$$g_{13} = 6L\alpha l_{21} l_{31} - 6L\beta l_{31} l_{21}$$

$$g_{14} = 6L\alpha l_{21} l_{32} - 6L\beta l_{31} l_{22} \qquad (5.45)$$

$$g_{15} = 6L\alpha l_{21} l_{33} - 6L\beta l_{31} l_{23}$$

$$g_{16} = 6L\alpha l_{22} l_{31} - 6L\beta l_{32} l_{21}$$

$$g_{17} = 6L\alpha l_{22} l_{32} - 6L\beta l_{32} l_{22}$$

$$g_{18} = 6L\alpha l_{22} l_{33} - 6L\beta l_{32} l_{23}$$

$$g_{19} = 6L\alpha l_{23} l_{31} - 6L\beta l_{33} l_{21}$$

$$g_{20} = 6L\alpha l_{23} l_{32} - 6L\beta l_{33} l_{22}$$

$$g_{21} = 6L\alpha l_{23} l_{33} - 6L\beta l_{33} l_{23}$$

$$g_{22} = -\delta l_{11}^2 + 2L^2 \alpha l_{31}^2 + 2L^2 \beta l_{21}^2$$

$$g_{23} = -\delta l_{11} l_{12} + 2L^2 \alpha l_{31} l_{32} + 2L^2 \beta l_{21} l_{22}$$

$$g_{24} = -\delta l_{11} l_{13} + 2L^2 \alpha l_{31} l_{33} + 2L^2 \beta l_{21} l_{23}$$

$$g_{25} = -\delta l_{12}^2 + 2L^2 \alpha l_{32}^2 + 2L^2 \beta l_{22}^2$$

$$g_{26} = -\delta l_{12} l_{13} + 2L^2 \alpha l_{32} l_{33} + 2L^2 \beta l_{22} l_{23}$$

$$g_{27} = -\delta l_{13}^2 + 2L^2 \alpha l_{33}^2 + 2L^2 \beta l_{23}^2$$

Through the index matrix, the global element stiffness coefficients are defined as

$$K_{jk} = \pm g_i \quad \text{if} \quad \text{INDEX}_{jk} = \pm i \qquad (5.46)$$

For example, by Eqs. (5.44) and (5.46) we obtain

$$K_{23} = g_5 \quad \text{and} \quad K_{27} = -g_2 \qquad (5.47)$$

Matrix Displacement Analysis

Example. The plane frame in Figure 5.6a is subjected to the concentrated load P acting normal to the plane of the frame. The elements have identical

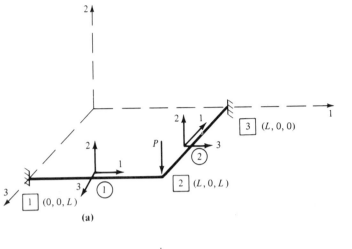

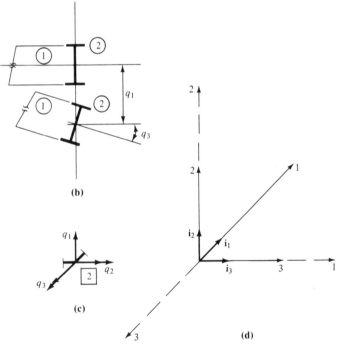

FIGURE 5.6 Grid represented as space frame. **(a)** Grid; **(b)** joint displacements; **(c)** degrees of freedom; **(d)** verification of λ^2.

properties. The load causes joint 2 to deflect normal to the plane of the frame and to rotate about the global 1 and 3 axes. Since the elements are rigidly interconnected, the deflection of element 1 produces twisting in element 2 (Figure 5.6b) and vice versa. Thus, the elements experience bending about the local 3 axes and twisting. Such structures are called grids (Section 5.6). It follows that a free joint of a grid has three degrees of freedom (Figure 5.6c).

In this example, the structure is regarded as a space frame, and the analysis is confined to the construction of the system stiffness matrix (Figure 3.2).

1. *Unknowns*

$$q_k, \quad k = 1, 2, 3 \tag{5.48}$$

2. *Element Models*

$$\mathbf{f}^i = \mathbf{k}\mathbf{d}^i, \quad i = 1, 2 \tag{5.49}$$

where $\mathbf{k}$ is defined in Eq. (5.40).

3. *System Model*
The rotation matrix of element 2 is

$$\lambda^2 = \begin{bmatrix} 0 & 0 & -1 \\ 0 & 1 & 0 \\ 1 & 0 & 0 \end{bmatrix} \tag{5.50}$$

Equation (5.50) follows from Eqs. (C.23)–(C.25), and (C.30). Specifically,

$$\Delta\mathbf{X} = \mathbf{X}_3 - \mathbf{X}_2 = [0 \quad 0 \quad -L], \quad |\Delta\mathbf{X}| = L$$

Thus,

$$\mathbf{i}_1 = [0 \quad 0 \quad -1] = [c_1 \quad c_2 \quad c_3] \tag{5.51}$$

Equations (5.51) and (C.30) yield Eq. (5.50).

Alternatively, λ^2 can be obtained on the basis of Eq. (C.14) by projecting unit lengths from the local axes onto the global axes (Figure 5.6d).

According to Figures 5.4 and 5.6, the member code matrix is

$$\mathbf{M}^T = \begin{bmatrix} 0 & 0 & 0 & 0 & 0 & 0 & 0 & 1 & 0 & 2 & 0 & 3 \\ 0 & 1 & 0 & 2 & 0 & 3 & 0 & 0 & 0 & 0 & 0 & 0 \end{bmatrix} \tag{5.52}$$

System Stiffness Matrix

Since the a end of element 1 and the b end of element 2 are constrained, the contributions to the system stiffness matrix come from the element submatrices $\mathbf{K}_{bb}^1$ and $\mathbf{K}_{aa}^2$. In particular,

$$\mathbf{K}_{bb}^1 = \mathbf{k}_{bb}^1 = \begin{bmatrix} & 0 & 1 & 0 & 2 & 0 & 3 & \\ & & & & & & & 0 \\ & & 12\alpha & & 0 & & -6L\alpha & 1 \\ & & & & & & & 0 \\ & & & & \delta & & 0 & 2 \\ & & & & & & & 0 \\ & \text{sym.} & & & & & 4L^2\alpha & 3 \end{bmatrix} \xrightarrow{M} \mathbf{K}^{(1)}$$

$$= \begin{bmatrix} 1 & 2 & 3 & \\ 12\alpha & 0 & -6L\alpha & 1 \\ & \delta & 0 & 2 \\ \text{sym.} & & 4L^2\alpha & 3 \end{bmatrix}$$

and

$$\mathbf{K}_{aa}^2 = \begin{bmatrix} & 0 & 1 & 0 & 2 & 0 & 3 & \\ & & & & & & & 0 \\ & & g_4 & & g_{16} & & g_{18} & 1 \\ & & & & & & & 0 \\ & & & & g_7 & & g_9 & 2 \\ & & & & & & & 0 \\ & \text{sym.} & & & & & g_{12} & 3 \end{bmatrix} \xrightarrow{M} \mathbf{K}^{(2)} = \begin{bmatrix} 1 & 2 & 3 & \\ 12\alpha & 6L\alpha & 0 & 1 \\ & 4L^2\alpha & 0 & 2 \\ \text{sym.} & & \delta & 3 \end{bmatrix} \quad (5.53)$$

because by Eqs. (5.44)–(5.46), (5.50), and (C.12), $g_4 = 12\alpha, g_7 = 4L^2\alpha, g_9 = 0$, $g_{12} = \delta, g_{16} = 6L\alpha$, and $g_{18} = 0$. Thus,

$$\mathbf{K} = \sum_{i=1}^{2} \mathbf{K}^{(i)} = \begin{bmatrix} 24\alpha & 6L\alpha & -6L\alpha \\ & \delta + 4L^2\alpha & 0 \\ \text{sym.} & & \delta + 4L^2\alpha \end{bmatrix} \quad (5.54)$$

5.6 GRIDS

A grid is a plane framework of elements loaded normal to its plane (Figure 5.7a). Bridge decks and floor systems are examples of grids. If the deformations of the elements are uncoupled (Section 5.2), a joint does not deflect in the plane of the grid nor rotate about an axis normal to the plane of the grid (see the Example in Section 5.5). Accordingly, a grid can be represented by joints with three degrees of freedom and elements with six degrees of freedom.

The matrix displacement method (Figure 3.2 or 3.9) is specialized for grids

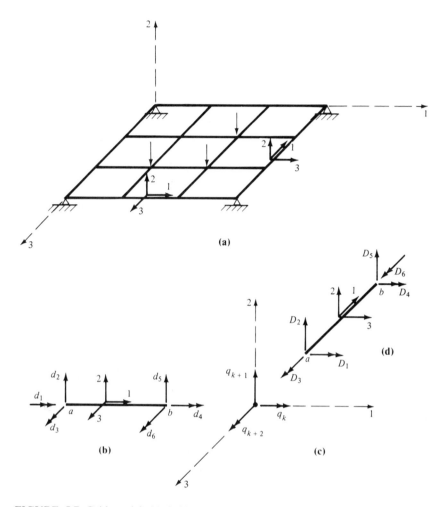

FIGURE 5.7 Grid model. **(a)** Grid; **(b)** local element; **(c)** joint; **(d)** global element.

with orthogonal elements arranged as shown in Figure 5.7a: The local and global reference frames are identical for elements parallel to the global 1 axis and distinct for elements parallel to the global 3 axis. To use the standard λ matrix, the displacements and forces are numbered in the sequence of the coordinate axes (Figures 5.7b, c, and d).

Element Models

The elements experience torsional deformations and bending about the local 3 axis (Figure 5.7b). Hence, the local element stiffness matrix can be constructed from the component stiffness matrices in Eqs. (5.39b) and

(5.39c). The superposition of Eqs. (5.39b) and (5.39c), with the code numbers defined in Eqs. (1.120), yields the local element stiffness matrix

$$\mathbf{k} = \begin{bmatrix} \delta & 0 & 0 & -\delta & 0 & 0 \\ & 12\alpha & 6L\alpha & 0 & -12\alpha & 6L\alpha \\ & & 4L^2\alpha & 0 & -6L\alpha & 2L^2 \\ \hline & & & \delta & 0 & 0 \\ & & & & 12\alpha & -6L\alpha \\ \text{sym.} & & & & & 4L^2\alpha \end{bmatrix} \tag{5.55}$$

where

$$\alpha = \frac{EI_3}{L^3}, \quad \delta = \frac{GJ}{L}, \quad G = \frac{E}{2(1+v)} \tag{5.56}$$

The global stiffness matrix for elements parallel to the global 3 axis (Figure 5.7d) has the form

$$\mathbf{K} = \mathbf{\Lambda}^T \mathbf{k} \mathbf{\Lambda}, \quad \mathbf{\Lambda} = \begin{bmatrix} \lambda & 0 \\ 0 & \lambda \end{bmatrix} \tag{5.57}$$

where by Eq. (5.50)

$$\lambda = \begin{bmatrix} 0 & 0 & -1 \\ 0 & 1 & 0 \\ 1 & 0 & 0 \end{bmatrix} \tag{5.58}$$

Equations (5.55)–(5.58) yield

$$\mathbf{K} = \begin{bmatrix} 4L^2\alpha & 6L\alpha & 0 & 2L^2\alpha & -6L\alpha & 0 \\ & 12\alpha & 0 & 6L\alpha & -12\alpha & 0 \\ & & \delta & 0 & 0 & -\delta \\ & & & 4L^2\alpha & -6L\alpha & 0 \\ & & & & 12\alpha & 0 \\ \text{sym.} & & & & & \delta \end{bmatrix} \tag{5.59}$$

Matrix Displacement Analysis

Example. The frame in Figure 5.8a is analyzed as a grid on the basis of Figure 3.2. The elements have identical properties, and the joint loads are $Q_1 = Q_3 = 0$, $Q_2 = -P$.[2] The effect of the torsional stiffness on the response is examined for W8 × 15 beams with $J/I_3 = 0.0029$ (American

---

[2] Clearly, nonzero values for Q_1 and Q_3 are permissible. They arise, for example, when element loads are transformed into equivalent joint loads (Figure 3.9).

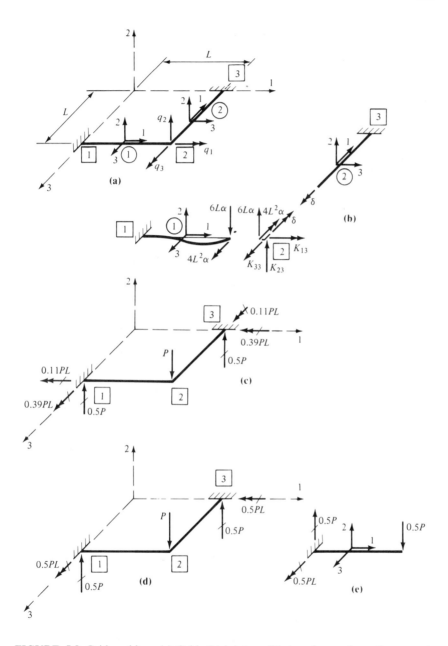

FIGURE 5.8 Grid problem. **(a)** Grid; **(b)** joint equilibrium for configuration $q_3 = 1$, $q_1 = q_2 = 0$; **(c)** joint forces, grid composed of pipes; **(d)** joint forces, grid composed of W8 × 15, beams; **(e)** cantilever.

Institute of Steel Construction, 1980) and circular pipes with $J/I_3 = 2$. We assume that $v = 0.29$.

1. *Unknowns*

$$q_k, \quad k = 1, 2, 3 \tag{5.60}$$

2. *Element Models*

$$\mathbf{f}^i = \mathbf{k}\mathbf{d}^i, \quad i = 1, 2 \tag{5.61}$$

where $\mathbf{k}$ is defined in Eq. (5.55).

3. *System Model*
According to Figures 5.7b, 5.7d, and 5.8a

$$\mathbf{M}^T = \begin{bmatrix} 0 & 0 & 0 & \vdots & 1 & 2 & 3 \\ 1 & 2 & 3 & \vdots & 0 & 0 & 0 \end{bmatrix} \tag{5.62}$$

It follows from Eqs. (5.55), (5.59), and (5.62) that

$$\mathbf{K}^{(1)} = \mathbf{k}_{bb}^1 = \begin{bmatrix} \delta & 0 & 0 \\ & 12\alpha & -6L\alpha \\ \text{sym.} & & 4L^2\alpha \end{bmatrix}$$

$$\mathbf{K}^{(2)} = \mathbf{K}_{aa}^2 = \begin{bmatrix} 4L^2\alpha & 6L\alpha & 0 \\ & 12\alpha & 0 \\ \text{sym.} & & \delta \end{bmatrix} \tag{5.63}$$

Thus,

$$\mathbf{K} = \sum_{i=1}^{2} \mathbf{K}^{(i)} = \begin{bmatrix} \delta + 4L^2\alpha & 6L\alpha & 0 \\ & 24\alpha & -6L\alpha \\ \text{sym.} & & \delta + 4L^2\alpha \end{bmatrix} \tag{5.64}$$

and

$$\begin{bmatrix} \delta + 4L^2\alpha & 6L\alpha & 0 \\ 6L\alpha & 24\alpha & -6L\alpha \\ 0 & -6L\alpha & \delta + 4L^2\alpha \end{bmatrix} \begin{bmatrix} q_1 \\ q_2 \\ q_3 \end{bmatrix} = \begin{bmatrix} 0 \\ -P \\ 0 \end{bmatrix} \tag{5.65}$$

The system stiffness coefficients can be verified by Eq. (3.35). For example, the conditions of equilibrium of the joint in Figure 5.8b yield the coefficients of the third column of $\mathbf{K}$: $K_{13} = 0$, $K_{23} = -6L\alpha$, $K_{33} = \delta + 4L^2\alpha$.

4. *Solution*

$$q_1 = \frac{PL}{4(\delta + L^2\alpha)} = -q_3, \quad q_2 = -\frac{P(\delta + 4L^2\alpha)}{24\alpha(\delta + L^2\alpha)} \tag{5.66}$$

5. Element Forces
It follows from Figure 5.8a and Eqs. (5.55), (5.56), (5.58), (5.61), (5.62), and (5.66) that

$$\mathbf{d}_a^1 = \mathbf{d}_b^2 = 0, \quad \mathbf{d}_b^1 = \mathbf{q}, \quad \mathbf{d}_a^2 = \lambda^2 \mathbf{q} = \begin{bmatrix} q_1 \\ q_2 \\ q_1 \end{bmatrix} \tag{5.67}$$

$$\mathbf{f}^1 = \begin{bmatrix} -\dfrac{PL}{4}\left(\dfrac{c}{1+c}\right) \\[2mm] \dfrac{P}{2} \\[2mm] \dfrac{PL}{4}\left(\dfrac{2+c}{1+c}\right) \\[2mm] \dfrac{PL}{4}\left(\dfrac{c}{1+c}\right) \\[2mm] -\dfrac{P}{2} \\[2mm] \dfrac{PL}{4}\left(\dfrac{c}{1+c}\right) \end{bmatrix}, \quad \mathbf{f}^2 = \begin{bmatrix} \dfrac{PL}{4}\left(\dfrac{c}{1+c}\right) \\[2mm] -\dfrac{P}{2} \\[2mm] -\dfrac{PL}{4}\left(\dfrac{c}{1+c}\right) \\[2mm] -\dfrac{PL}{4}\left(\dfrac{c}{1+c}\right) \\[2mm] \dfrac{P}{2} \\[2mm] -\dfrac{PL}{4}\left(\dfrac{2+c}{1+c}\right) \end{bmatrix} \tag{5.68}$$

where

$$c = \frac{\delta}{L^2 \alpha} = \frac{J}{2(1+v)I_3} \tag{5.69}$$

(c is a relative measure of torsional stiffness).

6. Joint Forces

$$\mathbf{P}_1 = \mathbf{f}_a^1, \quad \mathbf{P}_2 = \mathbf{f}_b^1 + \lambda^{2T}\mathbf{f}_a^2 = \begin{bmatrix} 0 \\ -P \\ 0 \end{bmatrix}$$

$$\mathbf{P}_3 = \lambda^{2T}\mathbf{f}_b^2 = \begin{bmatrix} -\dfrac{PL}{4}\left(\dfrac{2+c}{1+c}\right) \\[2mm] \dfrac{P}{2} \\[2mm] \dfrac{PL}{4}\left(\dfrac{c}{1+c}\right) \end{bmatrix} \tag{5.70}$$

When we specialize Eqs. (5.70), we obtain for the frame composed of circular pipes (Figure 5.8c)

$$\mathbf{P}_1 = \begin{bmatrix} -0.11PL \\ 0.50P \\ 0.39PL \end{bmatrix}, \quad \mathbf{P}_3 = \begin{bmatrix} -0.39PL \\ 0.50P \\ 0.11PL \end{bmatrix} \tag{5.71}$$

and for the frame composed of W8 × 15 beams (Figure 5.8d)

$$\mathbf{P}_1 = \begin{bmatrix} 0.00 \\ 0.50P \\ 0.50PL \end{bmatrix}, \quad \mathbf{P}_3 = \begin{bmatrix} -0.50PL \\ 0.50P \\ 0.00 \end{bmatrix} \tag{5.72}$$

It is apparent that the torsional stiffness of the W8 × 15 beam has a negligible effect on the load-carrying mechanism of the frame, whose members act essentially as independent cantilevers (Figure 5.8e). Observe that the grids in Figures 5.8c and d are in equilibrium (sum forces in the direction of the global 2 axis and moments about the global 1 and 3 axes).

PROBLEMS

5.1 The elements of the space truss have identical properties. The load at joint 4 acts in the negative direction of the global 2 axis. Analyze the truss by the matrix displacement method. Show the joint forces and verify equilibrium of the truss.

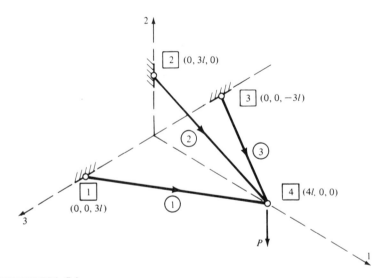

PROBLEM 5.1

5.2 The elements of the space truss have the same extensional stiffness γ and $\mathbf{q}^T = (25/\gamma)[1 \quad 1 \quad 0]$.
a. Compute $\mathbf{f}^1$, $\mathbf{P}_2$, and draw the free-body diagrams of element 1 and joint 2.

b. Compute $\mathbf{P}_1$.

c. Regard the space truss as a space frame with vertical principal element planes: Use Eq. (C.30) to compute $\boldsymbol{\lambda}^1$ and draw the unit local vectors $\mathbf{i}_1, \mathbf{i}_2, \mathbf{i}_3$ of element 1.

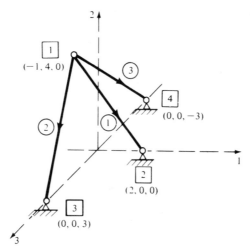

PROBLEM 5.2

5.3 The elements of the space frame have identical properties.

    a. Construct the system stiffness matrix; note that $\mathbf{K}^{(1)} = \mathbf{k}_{bb}^1$, $\mathbf{K}^{(2)} = \mathbf{K}_{aa}^2$, $\mathbf{K}^{(3)} = \mathbf{K}_{aa}^3$.

    b. Use Eq. (3.35) to verify the coefficients of the fourth column of the system stiffness matrix (see Figure 5.8b).

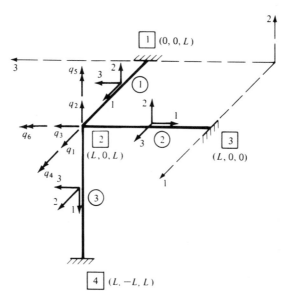

PROBLEM 5.3

5.4 The elements of the grid have identical properties. Each element is subjected to a uniformly distributed load of intensity p.

   a. Formulate the system stiffness matrix and verify the coefficients of the first column of $\mathbf{K}$ by Eq. (3.35) (see Figure 5.8b).

   b. Express $\bar{\mathbf{f}}^1$ in terms of p, L, and c [Eq. (5.69)] and draw the free-body diagram of element 1.

   c. Obtain $\bar{\mathbf{f}}^2$ by changing the signs of the first and fourth components of $\bar{\mathbf{f}}^1$ and compute $\bar{\mathbf{P}}_j$, $j = 1, 2, 3$.

   d. Show the joint forces and verify equilibrium for the grid.

   e. Show the joint forces for $c = 0$.

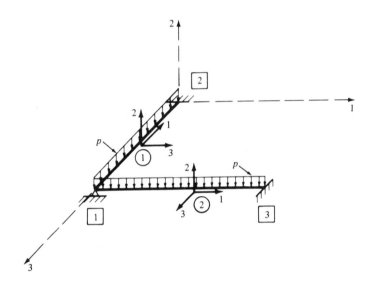

PROBLEM 5.4

REFERENCES

American Institute of Steel Construction. 1980. *Manual of Steel Construction*, 8th ed. Chicago.

Boresi, A. P., O. M. Sidebottom, F. B. Seely, and J. O. Smith. 1978. *Advanced Mechanics of Materials*, Wiley, New York.

Cowper, G. R. 1966. "The Shear Coefficient in Timoshenko's Beam Theory," *Journal of Applied Mechanics, ASME*, 33, 335–340.

Makowski, Z. S. 1962. "Braced Domes, Their History, Modern Trends and Recent Developments," *Architectural Science Review*, 5, No. 2, 62–79.

Makowski, Z. S. 1967. "Space Structures: A Short Review of Their Development," in *Space Structures*, R. M. Davies, ed., pp. 1–8. Wiley, New York.

Nickel, R. E., and G. A. Secor. 1972. "Convergence of Consistently Derived Timoshenko Beam Finite Elements," *International Journal for Numerical Methods in Engineering*, 5, No. 2, 243–252.

Sherman, D. R., et al. 1976. "Latticed Structures: State-of-the-Art Report," *Journal of the Structural Division, ASCE*, 102, No. ST11, 2197–2230.

Timoshenko, S. P., and J. M. Gere. 1972. *Mechanics of Materials.* Van Nostrand, New York.

Weaver, W., Jr., and J. M. Gere. 1980. *Matrix Analysis of Framed Structures*, 2nd ed. Van Nostrand, New York.

Yoo, C. H., and J. P. Fehrenbach. 1981. "Natural Frequencies of Curved Girders," *Journal of the Engineering Mechanics Division, ASCE*, 107, No. EM2, 339–354.

Additional References

Gjelsvik, A. 1981. *The Theory of Thin Walled Bars.* Wiley, New York.

Holzer, S. M., R. H. Plaut, A. E. Somers, Jr., and W. S. White. 1980. "Stability of Lattice Structures under Combined Loads," *Journal of the Engineering Mechanics Division, ASCE*, 106, No. EM2, 289–305.

Holzer, S. M., L. T. Watson, and P. Vu. 1981. "Stability Analyses of Lamella Domes," *Long Span Roof Structures*, Proceedings of a Symposium Held at the 1981 Annual Convention and Exhibit, ASCE, St. Louis, MO, October 26–30, pp. 179–209.

Johnston, B. G., F. J. Lin, and T. V. Galambos. 1980. *Basic Steel Design*, 2nd ed. Prentice-Hall, Englewood Cliffs, NJ.

Shanley, F. R. 1957. *Strength of Materials.* McGraw-Hill, New York.

SIX

SOLUTION OF SYSTEM EQUATIONS

6.1 INTRODUCTION

The accurate and efficient solution of linear algebraic equations is the central task in most structural analysis programs. It has been estimated (Meyer, 1973) that in linear static analysis programs, 20–50 percent of the execution time is expended in the solution of linear equations. This portion may increase to 80 percent in dynamic or nonlinear analysis programs. Accordingly, extensive research has been conducted to improve the efficiency of equation solvers by exploiting special features, such as the symmetry and sparsity of the coefficient matrix, and to estimate and control solution errors.

In this chapter, we review the literature, present solution techniques, and study solution errors relative to our system equation $\mathbf{Kq} = \mathbf{Q}$: The distribution of the coefficients of $\mathbf{K}$ in band form and appropriate storage schemes are addressed in Section 6.2. A general discussion of direct solution methods with reference to special structural analysis techniques for symmetric, positive definite band matrices is presented in Section 6.3. Algorithms for fixed and variable band solvers are presented and illustrated in Sections 6.4 and 6.5. The frontal solution, a special solution technique for large systems, is introduced in Section 6.6. Finally, measures of solution errors and methods for solution refinement are considered in Section 6.7.

6.2 BAND MATRICES

Stiffness matrices are generally sparsely populated; that is, they have a relatively small number of nonzero entries. Although the number of nonzero entries is independent of the order in which the joints are numbered, their distribution is not. In particular, if the joints are numbered such that the maximum difference in the joint numbers of each element is small relative to the total number of joints, the nonzero coefficients cluster in a *band* along the main diagonal. A general guideline for achieving a narrow band is to number the joints in sequence across the smaller dimension of the structure (Cook, 1981; Meyer, 1973).

A symmetric band matrix **K** has the property

$$K_{ij} = 0 \quad \text{if} \quad |j - i| > m \tag{6.1}$$

where m is the *half bandwidth*[1] (Bathe and Wilson, 1976), and $2m + 1$ is the *bandwidth*. The half bandwidth can be computed from the member code matrix as follows:

$$m = \max\{\Delta_i\}, \quad i = 1, \ldots, \text{NE} \tag{6.2}$$

where Δ_i is the difference between the largest and smallest nonzero integer in the ith column of the member code matrix.

For example, consider the frame in Figure 6.1a, whose elements are assumed to have inextensible centroidal axes (Section 4.3). The member code matrix, **M**, and the difference vector, **Δ**, are

$$
\mathbf{M}^T = \begin{bmatrix}
0 & 2 & 0 & 3 \\
1 & 2 & 4 & 5 \\
1 & 3 & 4 & 6 \\
0 & 5 & 0 & 6 \\
4 & 5 & 7 & 8 \\
4 & 6 & 7 & 9 \\
0 & 8 & 0 & 9 \\
7 & 8 & 10 & 11 \\
7 & 9 & 10 & 12 \\
0 & 11 & 0 & 12 \\
10 & 11 & 0 & 0 \\
10 & 12 & 0 & 0
\end{bmatrix}, \quad
\mathbf{\Delta}^T = \begin{bmatrix}
1 \\
4 \\
5 \\
1 \\
4 \\
5 \\
1 \\
4 \\
5 \\
1 \\
1 \\
2
\end{bmatrix} \tag{6.3}
$$

Thus, $m = 5$ and $K_{ij} = 0$ if $|j - i| > 5$. This is reflected by the stiffness matrix in Figure 6.2, whose nonzero entries are marked by $\times$'s; the locations that receive contributions from element 3 are shaded.

Note that the matrix in Figure 6.2 has zeros within the band.[2] By excluding all zeros above the first nonzero entry in each column, we obtain a variable

---

[1] Other definitions are also used in the literature. For example, in some publications (Cook, 1981; Felippa and Tocher, 1970; Meyer, 1973) the half bandwidth is defined as $m + 1$.

[2] In general, $K_{ij} = 0$ if no element connects q_i and q_j.

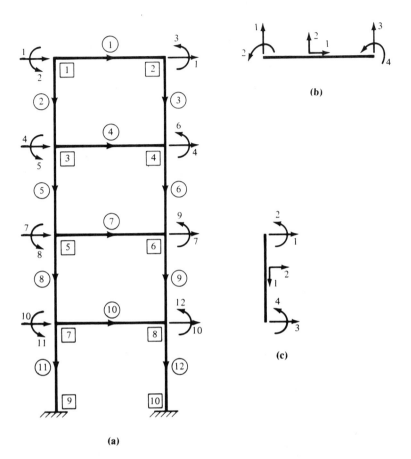

FIGURE 6.1 Plane frame. **(a)** Assembly of beam elements; **(b)** beams; **(c)** columns.

band that is contained in the fixed band. Specifically, if we denote (Bathe and Wilson, 1976)

$$m_j = \text{row number of the first nonzero entry of column } j$$

then

$$K_{ij} = 0 \quad \text{if} \quad i < m_j \tag{6.4}$$

and the half bandwidth

$$m = \max\{j - m_j\}, \quad j = 1, \ldots, n \tag{6.5}$$

The integers m_j define the *variable band* (Jennings, 1977; Tewarson, 1973), which is also called the *skyline* (Bathe and Wilson, 1976; Felippa, 1975), the *profile* (Zienkiewicz, 1977), or the *envelope* (George and Liu, 1981;

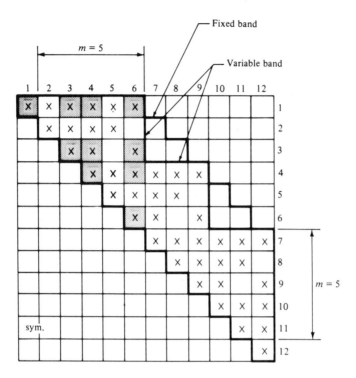

FIGURE 6.2 Stiffness matrix

Melosh and Bamford, 1969), and $(j - m_j + 1)$ is the number of elements in the jth *active column* (Bathe and Wilson, 1976), that is, the elements in the jth column from the main diagonal up to the first nonzero entry. The matrix in Figure 6.2 has the following variable band data:

j	1	2	3	4	5	6	7	8	9	10	11	12
m_j	1	1	1	1	1	1	4	4	4	7	7	7
$j - m_j$	0	1	2	3	4	5	3	4	5	3	4	5

and $m = \max\{j - m_j\} = 5$.

Band matrices are desirable because the zeros outside the band (fixed or variable) need not be stored nor processed in the solution of a system of equations (Section 6.3). There are several storage schemes for symmetric band matrices. In the simplest one, called *band storage*, all elements inside the fixed band in the upper half of **K** are stored. In a more sophisticated one, called *variable band*, *skyline*, *profile*, or *envelope storage*, only the elements in the active columns are stored.

Band Storage

The elements inside the fixed band in the upper half of the symmetric matrix
K of order $n \times n$ (Figure 6.3a) can be stored by rows in the matrix **A** of
order $n \times b$ (Figure 6.3b) or by columns in the matrix **A** of order $b \times n$
(Figure 6.3c), where $b = m + 1$. The choice depends on the order of decomposition of **K** in the solution of the band equations (Section 6.3).

The elements of **K** and the band storage arrays **A** are related as follows:
For row storage (Figure 6.3b)

$$K_{ij} = A_{ik}, \quad k = j - i + 1 \tag{6.6}$$

and for column storage (Figure 6.3c)

$$K_{ij} = A_{kj}, \quad k = i - j + m + 1 \tag{6.7}$$

where

$$K_{ij} = 0 \quad \text{if} \quad j > m + i \tag{6.8}$$

Note that the main diagonal elements of **K** are stored in the first column of **A**
in row storage (Figure 6.3b) and in the last row of **A** in column storage
(Figure 6.3c).

FIGURE 6.3 Band storage

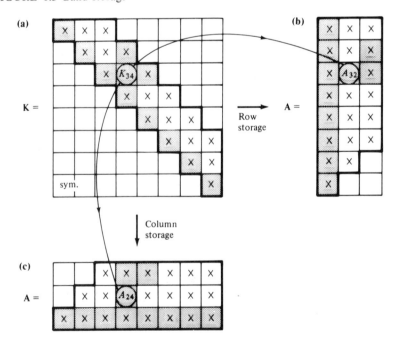

Skyline Storage

The elements in the active columns of **K** are stored consecutively in the one-dimensional array **A**. The relations between the elements of **K** and **A** can be defined by the array **MAXA** (Bathe and Wilson, 1976), which stores the addresses of the diagonal elements of **K** (Figures 6.4a–d). Specifically,

$$K_{jj} = A_r, \quad r = \textbf{MAXA}(j)$$
$$K_{ij} = A_s, \quad s = r + j - i$$

(6.9)

The number of elements in the *j*th *active column* of **K** is $\textbf{MAXA}(j + 1) - \textbf{MAXA}(j)$, $j = 1, \ldots, n$. **MAXA** can be generated from the *member code matrix*, which is called the *connectivity array* in (Bathe and Wilson, 1976).

In the following examples, stiffness matrices are constructed for simple structures to illustrate the effect of joint numbering on the distribution of stiffness coefficients.

Example 1. The continuous beams in Figure 6.5a are subjected to axial forces and are represented as an assembly of truss elements (Figure 6.5b) with the properties *A*, *E*, *L*. The element and system models are expressed

FIGURE 6.4 Skyline storage.

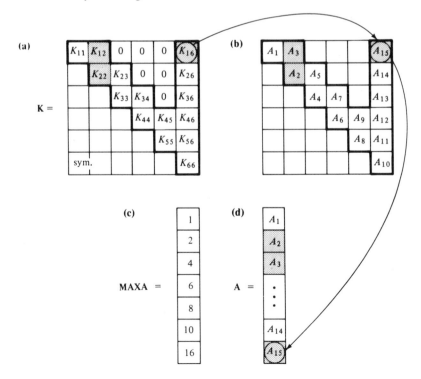

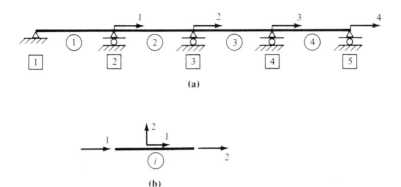

(a)

(b)

FIGURE 6.5 Assembly of truss elements. **(a)** Continuous beams; **(b)** truss element.

in nondimensional form by dividing the forces by EA and the deflections by L. The element stiffness matrix is

$$\mathbf{k} = \begin{bmatrix} 1 & -1 \\ \text{sym.} & 1 \end{bmatrix} \tag{6.10}$$

The member code matrix, the difference vector, and the half bandwidth [Eq. (6.2)] are

$$\mathbf{M}^T = \begin{bmatrix} 0 & 1 \\ 1 & 2 \\ 2 & 3 \\ 3 & 4 \end{bmatrix}, \quad \mathbf{\Delta}^T = \begin{bmatrix} 0 \\ 1 \\ 1 \\ 1 \end{bmatrix}, \quad m = 1 \tag{6.11}$$

The system stiffness matrix is

$$\mathbf{K} = \begin{bmatrix} 2 & -1 & 0 & 0 \\ & 2 & -1 & 0 \\ & & 2 & -1 \\ \text{sym.} & & & 1 \end{bmatrix} \tag{6.12}$$

which is called a tridiagonal matrix and can be stored in band form.

Example 2. The structure in Figure 6.6 is represented as an assembly of truss elements. Two joint numbering schemes are used. Element 6 connects joints 2 and 6 in scheme A and joints 4 and 5 in scheme B. The models are

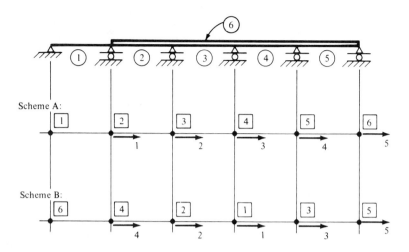

FIGURE 6.6 Assembly of truss elements

expressed in nondimensional form. All elements have the stiffness matrix in Eq, (6.10).

Scheme A

$$
\mathbf{M}^T = \begin{bmatrix} 0 & 1 \\ 1 & 2 \\ 2 & 3 \\ 3 & 4 \\ 4 & 5 \\ 1 & 5 \end{bmatrix}, \quad \mathbf{\Delta}^T = \begin{bmatrix} 0 \\ 1 \\ 1 \\ 1 \\ 1 \\ 4 \end{bmatrix}, \quad m = 4
\tag{6.13}
$$

The system stiffness matrix is

$$
\mathbf{K} = \begin{matrix}
& 4 & 2 & 1 & 3 & 5 & \\
& \begin{bmatrix} 3 & -1 & 0 & 0 & -1 \\ & 2 & -1 & 0 & 0 \\ & & 2 & -1 & 0 \\ & & & 2 & -1 \\ \text{sym.} & & & & 2 \end{bmatrix} & \begin{matrix} 4 \\ 2 \\ 1 \\ 3 \\ 5 \end{matrix}
\end{matrix}
\tag{6.14}
$$

In this case, we can only exclude zeros by using the skyline storage. The integers outside **K** represent the joint displacement numbers of scheme B.

Scheme B

$$\mathbf{M}^T = \begin{bmatrix} 0 & 4 \\ 4 & 2 \\ 2 & 1 \\ 1 & 3 \\ 3 & 5 \\ 4 & 5 \end{bmatrix}, \quad \mathbf{\Delta}^T = \begin{bmatrix} 0 \\ 2 \\ 1 \\ 2 \\ 2 \\ 1 \end{bmatrix}, \quad m = 2 \tag{6.15}$$

The system stiffness matrix is

$$\mathbf{K} = \begin{bmatrix} \overset{1}{2} & \overset{2}{-1} & \overset{3}{-1} & \overset{4}{0} & \overset{5}{0} \\[2pt] & 2 & 0 & -1 & 0 \\[2pt] & & 2 & 0 & -1 \\[2pt] & & & 3 & -1 \\[2pt] \text{sym.} & & & & 2 \end{bmatrix} \begin{matrix} 1 \\ 2 \\ 3 \\ 4 \\ 5 \end{matrix} \tag{6.16}$$

which can be stored as a band matrix. Note that Eq. (6.16) can be obtained directly from Eq. (6.14) by the code number technique.

For complex structures it may be difficult to determine by inspection a joint-numbering scheme that results in a narrow band. For this reason automatic reordering programs have been developed (Cook, 1981; George and Liu, 1981; Meyer, 1973). However, the best joint-numbering scheme is not necessarily the one that results in the smallest bandwidth. Of importance are also the effort required to produce the scheme and the number of operations required to formulate and solve the equations (Meyer, 1973).

6.3 SOLUTION OF LINEAR EQUATIONS

We are concerned with the solution of the system of equations

$$\mathbf{Kq} = \mathbf{Q} \tag{6.17}$$

Since $\mathbf{K}$ is a symmetric, positive definite matrix (Section 3.2), $\mathbf{K}$ is nonsingular and Eq. (6.17) has a unique solution (Dahlquist and Björk, 1974; Forsythe and Moler, 1967; Ralston and Rabinowitz, 1978).

Although we can obtain the solution to Eq. (6.17) by Cramer's rule (Thomas, 1956), the large amount of computation required to evaluate determinants makes the use of Cramer's rule impractical (Ralston and Rabinowitz, 1978). Therefore, we need to select a suitable numerical method.

In general, there are two types of numerical methods: *direct* and *iterative* methods. Direct methods give the exact solution (neglecting roundoff errors) in a finite number of arithmetic operations. Iterative methods give a sequence of approximate solutions that generally converge to the exact solution as the number of iterations tends to infinity.

Advantages and disadvantages of direct and iterative methods are discussed, for example, by Bathe and Wilson (1976), Cook (1981), and Meyer (1973). Direct methods are nearly always most effective (Bathe and Wilson, 1976), and they are widely used in structural analysis, particularly in the application of the finite element method. For these reasons, we will confine ourselves to direct methods.

Most direct methods are variations of Gaussian elimination. Commonly used direct methods include Gaussian elimination, the Cholesky method, and the modified Cholesky method (Felippa, 1975). In these methods, the system of equations is reduced to triangular systems, which are easy to solve. We consider first the solution of triangular systems.

Triangular Systems

We will encounter lower and upper triangular systems. Lower triangular systems are of the form

$$\mathbf{L}\mathbf{y} = \mathbf{Q} \tag{6.18}$$

where $\mathbf{L}$ is *lower triangular matrix*. Specifically, for four unknowns, Eq. (6.18) becomes

$$
\begin{bmatrix}
l_{11} & 0 & 0 & 0 \\
l_{21} & l_{22} & 0 & 0 \\
l_{31} & l_{32} & l_{33} & 0 \\
l_{41} & l_{42} & l_{43} & l_{44}
\end{bmatrix}
\begin{bmatrix}
y_1 \\ y_2 \\ y_3 \\ y_4
\end{bmatrix}
=
\begin{bmatrix}
Q_1 \\ Q_2 \\ Q_3 \\ Q_4
\end{bmatrix}
\tag{6.19}
$$

If $l_{ii} \neq 0$, the unknowns can be computed by *forward substitution* in the sequence

$$y_1 = \frac{Q_1}{l_{11}}$$

$$y_2 = \frac{Q_2 - l_{21} y_1}{l_{22}}$$

$$\vdots$$

$$y_4 = \frac{Q_4 - \sum_{k=1}^{3} l_{4k} y_k}{l_{44}}$$

In general,[3]

$$y_i = \frac{Q_i - \sum_{k=1}^{i-1} l_{ik} y_k}{l_{ii}}, \quad i = 1, \ldots, n \tag{6.20}$$

---

[3] The following convention is used: $\sum_{k=n}^{m} = 0$ if $m < n$.

Upper triangular systems have the form

$$\mathbf{U}\mathbf{q} = \mathbf{y} \tag{6.21}$$

where $\mathbf{U}$ is an *upper triangular matrix*.

For four unknowns, Eq. (6.21) becomes

$$\begin{bmatrix} u_{11} & u_{12} & u_{13} & u_{14} \\ 0 & u_{22} & u_{23} & u_{24} \\ 0 & 0 & u_{33} & u_{34} \\ 0 & 0 & 0 & u_{44} \end{bmatrix} \begin{bmatrix} q_1 \\ q_2 \\ q_3 \\ q_4 \end{bmatrix} = \begin{bmatrix} y_1 \\ y_2 \\ y_3 \\ y_4 \end{bmatrix} \tag{6.22}$$

If $u_{ii} \neq 0$, the unknowns can be solved by *back substitution* in the sequence

$$q_4 = \frac{y_4}{u_{44}}$$

$$q_3 = \frac{y_3 - u_{34}q_4}{u_{33}}$$

$$\vdots$$

$$q_1 = \frac{y_1 - \sum_{k=2}^{4} u_{1k}q_k}{u_{11}}$$

In general,

$$q_i = \frac{y_i - \sum_{k=i+1}^{n} u_{ik}q_k}{u_{ii}}, \quad i = n, \ldots, 1 \tag{6.23}$$

Gaussian Elimination

There are two basic forms of Gaussian elimination: the *standard* form and the *compact* form.

In the standard form, Eq. (6.17) is reduced to the upper triangular system

$$\mathbf{U}\mathbf{q} = \mathbf{y} \tag{6.24}$$

which is solved by back substitution.

In the compact form $\mathbf{K}$ is first decomposed (factorized) into the product

$$\mathbf{K} = \mathbf{L}\mathbf{U} \tag{6.25}$$

where $\mathbf{L}$ is a *unit lower triangular matrix* [Eq. (6.47)] and $\mathbf{U}$ is an *upper triangular matrix*.[4] Thus, Eq. (6.17) becomes

$$\mathbf{L}\mathbf{U}\mathbf{q} = \mathbf{Q} \tag{6.26}$$

---

[4] This is known as the *Doolittle* factorization. There is another version, called the *Crout* factorization, in which $\mathbf{L}$ is a *lower triangular matrix* and $\mathbf{U}$ is a *unit upper triangular matrix* (Ralston and Rabinowitz, 1978).

By Eq. (6.24), we can divide Eq. (6.26) into two triangular systems

$$\mathbf{Ly} = \mathbf{Q} \tag{6.27}$$

and

$$\mathbf{Uq} = \mathbf{y} \tag{6.28}$$

Thus, Eq. (6.27) can be solved for $\mathbf{y}$ by forward substitution, and Eq. (6.28) can be solved for $\mathbf{q}$ by back substitution.

Once we have obtained the $\mathbf{LU}$ decomposition of $\mathbf{K}$, we can solve Eq. (6.17) for any load vector. In particular, $\mathbf{Q}$ need not be known at the time of the $\mathbf{LU}$ decomposition. This is in contrast to standard Gaussian elimination where $\mathbf{K}$ and $\mathbf{Q}$ are reduced simultaneously to $\mathbf{U}$ and $\mathbf{y}$.

Standard Gaussian Elimination

Equation (6.17) is reduced to Eq. (6.24) by eliminating the unknowns in a systematic way (Atkinson, 1978; Dahlquist and Björk, 1974; Ralston and Rabinowitz, 1978). This procedure is illustrated for the system of equations $\mathbf{Kq} = \mathbf{Q}$ [Eq. (6.12)]:

$$\begin{bmatrix} 2 & -1 & 0 & 0 \\ -1 & 2 & -1 & 0 \\ 0 & -1 & 2 & -1 \\ 0 & 0 & -1 & 1 \end{bmatrix} \begin{bmatrix} q_1 \\ q_2 \\ q_3 \\ q_4 \end{bmatrix} = \begin{bmatrix} 0 \\ 1 \\ 0 \\ 0 \end{bmatrix} \tag{6.29}$$

STEP 1. q_1 is eliminated from the last three equations by subtracting the multiple

$$l_{i1} = \frac{K_{i1}}{K_{11}}, \quad i = 2, \ldots, 4$$

of the first equation from the ith equation. Thus,

$$l_{21} = -\tfrac{1}{2}, \quad l_{31} = 0, \quad l_{41} = 0$$

and we obtain the system $\mathbf{K}^{(1)}\mathbf{q} = \mathbf{Q}^{(1)}$:

$$\begin{bmatrix} 2 & -1 & 0 & 0 \\ 0 & \tfrac{3}{2} & -1 & 0 \\ 0 & -1 & 2 & -1 \\ 0 & 0 & -1 & 1 \end{bmatrix} \begin{bmatrix} q_1 \\ q_2 \\ q_3 \\ q_4 \end{bmatrix} = \begin{bmatrix} 0 \\ 1 \\ 0 \\ 0 \end{bmatrix} \tag{6.30}$$

STEP 2. q_2 is eliminated from the last two equations by subtracting the multiple

$$l_{i2} = \frac{K_{i2}^{(1)}}{K_{22}^{(1)}}, \quad i = 3, 4$$

of the second equation from the ith equation. Thus,

$$l_{32} = -\tfrac{2}{3}, \quad l_{42} = 0$$

and we obtain the system $\mathbf{K}^{(2)}\mathbf{q} = \mathbf{Q}^{(2)}$:

$$\left[\begin{array}{cc:cc} 2 & -1 & 0 & 0 \\ 0 & \tfrac{3}{2} & -1 & 0 \\ \hdashline 0 & 0 & \tfrac{4}{3} & -1 \\ 0 & 0 & -1 & 1 \end{array}\right] \begin{bmatrix} q_1 \\ q_2 \\ q_3 \\ q_4 \end{bmatrix} = \begin{bmatrix} 0 \\ 1 \\ \tfrac{2}{3} \\ 0 \end{bmatrix} \tag{6.31}$$

STEP 3. q_3 is eliminated from the last equation by subtracting the multiple

$$l_{43} = \frac{K_{43}^{(2)}}{K_{33}^{(2)}} = -\frac{3}{4}$$

of the third equation from the fourth equation. This yields the triangular system $\mathbf{K}^{(3)}\mathbf{q} = \mathbf{Q}^{(3)}$:

$$\left[\begin{array}{ccc:c} 2 & -1 & 0 & 0 \\ 0 & \tfrac{3}{2} & -1 & 0 \\ 0 & 0 & \tfrac{4}{3} & -1 \\ \hdashline 0 & 0 & 0 & \tfrac{1}{4} \end{array}\right] \begin{bmatrix} q_1 \\ q_2 \\ q_3 \\ q_4 \end{bmatrix} = \begin{bmatrix} 0 \\ 1 \\ \tfrac{2}{3} \\ \tfrac{1}{2} \end{bmatrix} \tag{6.32}$$

The solution of Eq. (6.32) by back substitution yields

$$q_4 = q_3 = q_2 = 2, \quad q_1 = 1 \tag{6.33}$$

Standard Gaussian elimination can be expressed by the following algorithm.

Algorithm. To eliminate q_k, we modify the coefficients of the last $n - k$ equations as follows (Ralston and Rabinowitz, 1978)

$$\begin{aligned} K_{ij}^{(k)} &= K_{ij}^{(k-1)} - l_{ik} K_{kj}^{(k-1)}, \quad k = 1, \ldots, n-1; \quad i, j = k+1, \ldots, n; \\ Q_i^{(k)} &= Q_i^{(k-1)} - l_{ik} Q_k^{(k-1)}, \end{aligned} \tag{6.34}$$

where

$$l_{ik} = \frac{K_{ik}^{(k-1)}}{K_{kk}^{(k-1)}}, \quad K_{ij}^{(0)} = K_{ij}, \quad Q_i^{(0)} = Q_i \tag{6.35}$$

By this example we can identify important properties of Gaussian elimination relative to the structural system in Eq. (6.17).

Pivoting

The pivotal elements $K_{kk}^{(k-1)}$ that appear during elimination must be non-zero, otherwise the process breaks down [see Eq. (6.35)]. If a zero pivotal element is encountered, row or column interchanges must be performed in order to place a nonzero element in the pivot position. This process is called pivoting. Pivoting may also be necessary when pivotal elements are nearly zero in order to ensure numerical stability (Dahlquist and Björk, 1974; Forsythe and Moler, 1967; Ralston and Rabinowitz, 1978).

For symmetric, positive definite matrices, the pivotal elements are always positive and pivoting may be omitted without serious increase in roundoff error (Forsythe and Moler, 1967). A solution for indefinite matrices is contained in LINPACK (Dongarra et al., 1979).

Symmetry

At any stage in the elimination of a symmetric matrix, the unreduced submatrix is symmetric [see Eqs. (6.30) and (6.31)]. This can be verified by Eqs. (6.34) and (6.35). Specifically, the coefficients of the kth unreduced submatrix are

$$K_{ij}^{(k)} = K_{ij}^{(k-1)} - \frac{K_{ik}^{(k-1)}}{K_{kk}^{(k-1)}} K_{kj}^{(k-1)},$$
$$\qquad k = 1, \ldots, n-1; i, j = k+1, \ldots, n$$
$$K_{ji}^{(k)} = K_{ji}^{(k-1)} - \frac{K_{jk}^{(k-1)}}{K_{kk}^{(k-1)}} K_{ki}^{(k-1)},$$

$$(6.36)$$

It follows that

$$K_{ij}^{(k)} = K_{ji}^{(k)} \quad \text{if} \quad K_{ij}^{(k-1)} = K_{ji}^{(k-1)}, \quad k = 1, \ldots, n-1 \qquad (6.37)$$

Thus, symmetry is preserved during Gaussian elimination.

We can take advantage of symmetry by modifying only the coefficients in the upper triangle of **K**. This requires the following changes in the algorithm, Eqs. (6.34) and (6.35): l_{ik} is replaced by

$$l_{ki} = \frac{K_{ki}^{(k-1)}}{K_{kk}^{(k-1)}} \qquad (6.38)$$

and $j = i, \ldots, n$.

With the modified algorithm, the reduction of the symmetric matrix

$$\mathbf{K} = \begin{bmatrix} 2 & -1 & 0 & 0 \\ & 2 & -1 & 0 \\ & & 2 & -1 \\ \text{sym.} & & & 1 \end{bmatrix} \qquad (6.39)$$

of Eq. (6.29) proceeds as follows.

STEP 1

$$l_{12} = -\tfrac{1}{2}, \quad l_{13} = 0, \quad l_{14} = 0$$

and

$$\mathbf{K}^{(1)} = \begin{bmatrix} 2 & -1 & 0 & 0 \\ 0 & \tfrac{3}{2} & -1 & 0 \\ 0 & & 2 & -1 \\ 0 & \text{sym.} & & 1 \end{bmatrix} \tag{6.40}$$

STEP 2

$$l_{23} = -\tfrac{2}{3}, \quad l_{24} = 0$$

and

$$\mathbf{K}^{(2)} = \begin{bmatrix} 2 & -1 & 0 & 0 \\ 0 & \tfrac{3}{2} & -1 & 0 \\ 0 & 0 & \tfrac{4}{3} & -1 \\ 0 & 0 & \text{sym.} & 1 \end{bmatrix} \tag{6.41}$$

STEP 3

$$l_{34} = -\tfrac{3}{4}$$

and

$$\mathbf{U} = \mathbf{K}^{(3)} = \begin{bmatrix} 2 & -1 & 0 & 0 \\ 0 & \tfrac{3}{2} & -1 & 0 \\ 0 & 0 & \tfrac{4}{3} & -1 \\ 0 & 0 & 0 & \tfrac{1}{4} \end{bmatrix} \tag{6.42}$$

Symmetry and Bandedness

The effort required in the reduction of a symmetric band matrix can be further reduced by noting that (1) the zeros outside the band are not affected [compare Eqs. (6.39)–(6.42)], and (2) in the elimination of q_k, the coefficients in any column below a zero in the kth row are not affected [compare the third and fourth columns in Eqs. (6.39)–(6.41)].

Thus, in the elimination of q_k in the symmetric band matrix in Figure 6.7 only the coefficients in the shaded triangle need to be modified.

Condensation

Gaussian elimination admits a physical interpretation. Specifically, the relations between the unreduced degrees of freedom and the corresponding forces in Eqs. (6.29)–(6.32) are reflected in Figures 6.8a–d, respectively.

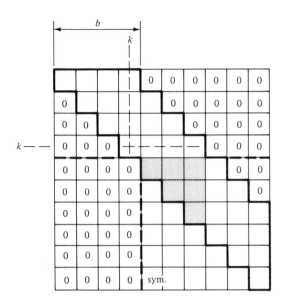

FIGURE 6.7 Reduction of symmetric band matrix

FIGURE 6.8 Physical interpretation of Gaussian elimination. **(a)** Start of solution; **(b)** elimination of q_1; **(c)** elimination of q_2; **(d)** elimination of q_3.

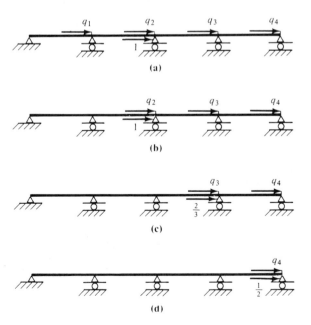

Moreover, at any stage in the elimination, the unreduced submatrix represents the condensed stiffness matrix (Section 4.6) corresponding to the unreduced degrees of freedom (Bathe and Wilson, 1976; Felippa and Tocher, 1970).

For illustration consider Example 4 of Section 4.6. The model of complex element 1 (Figures 4.22a–c) is according to Eqs. (4.163) and (4.167)

$$\frac{\gamma}{4}\begin{bmatrix} 5 & \sqrt{3} & -4 & 0 \\ \sqrt{3} & 3 & 0 & 0 \\ -4 & 0 & 5 & -\sqrt{3} \\ 0 & 0 & -\sqrt{3} & 3 \end{bmatrix}\begin{bmatrix} D_1 \\ D_2 \\ D_3 \\ D_4 \end{bmatrix} = \begin{bmatrix} \bar{F}_1 \\ \bar{F}_2 \\ 0 \\ -4\sqrt{3} \end{bmatrix} \qquad (6.43)$$

By eliminating D_4 and D_3 we obtain the partially reduced model

$$\frac{\gamma}{4}\left[\begin{array}{cc:cc} 1 & \sqrt{3} & 0 & 0 \\ \sqrt{3} & 3 & 0 & 0 \\ \hdashline -4 & 0 & 4 & 0 \\ 0 & 0 & -\sqrt{3} & 3 \end{array}\right]\begin{bmatrix} D_1 \\ D_2 \\ \hdashline D_3 \\ D_4 \end{bmatrix} = \begin{bmatrix} \bar{F}_1 - 4 \\ \bar{F}_2 + 0 \\ \hdashline -4 \\ -4\sqrt{3} \end{bmatrix} \qquad (6.44)$$

which yields the condensed model [Eqs. (4.159) and (4.169)]

$$\begin{bmatrix} \bar{F}_1 \\ \bar{F}_2 \end{bmatrix} = \frac{\gamma}{4}\begin{bmatrix} 1 & \sqrt{3} \\ \sqrt{3} & 3 \end{bmatrix}\begin{bmatrix} D_1 \\ D_2 \end{bmatrix} + \begin{bmatrix} 4 \\ 0 \end{bmatrix} \qquad (6.45)$$

and the equation

$$\frac{\gamma}{4}\begin{bmatrix} 4 & 0 \\ -\sqrt{3} & 3 \end{bmatrix}\begin{bmatrix} D_3 \\ D_4 \end{bmatrix} = \begin{bmatrix} -4 \\ -4\sqrt{3} \end{bmatrix} + \gamma\begin{bmatrix} 1 & 0 \\ 0 & 0 \end{bmatrix}\begin{bmatrix} D_1 \\ D_2 \end{bmatrix} \qquad (6.46)$$

from which D_3 and D_4 can be computed by forward substitution after D_1 and D_2 have been evaluated.

Compact Gaussian Elimination

By expressing Gaussian elimination in matrix form, we arrive at the compact form, Eq. (6.25). The equivalence of standard Gaussian elimination and the **LU** decomposition is demonstrated in most textbooks that deal with the solution of linear equations; see for example Bathe and Wilson (1976), Dahlquist and Björk (1974), Forsythe and Moler (1967), Jennings (1977), Ralston and Rabinowitz (1978). For our purpose it is sufficient to note that for the symmetric, positive definite matrix **K**, a unique **LU** decomposition exists and pivoting is not necessary (Forsythe and Moler, 1967).

The **LU** decomposition of the matrix **K** of order 4 has the form

$$\begin{bmatrix} K_{11} & K_{12} & K_{13} & K_{14} \\ K_{21} & K_{22} & K_{23} & K_{24} \\ K_{31} & K_{32} & K_{33} & K_{34} \\ K_{41} & K_{42} & K_{43} & K_{44} \end{bmatrix} = \begin{bmatrix} 1 & 0 & 0 & 0 \\ l_{21} & 1 & 0 & 0 \\ l_{31} & l_{32} & 1 & 0 \\ l_{41} & l_{42} & l_{43} & 1 \end{bmatrix} \begin{bmatrix} u_{11} & u_{12} & u_{13} & u_{14} \\ 0 & u_{22} & u_{23} & u_{24} \\ 0 & 0 & u_{33} & u_{34} \\ 0 & 0 & 0 & u_{44} \end{bmatrix}$$

$$(6.47)$$

The elements of **L** and **U** can be computed in succession, one row of **U** followed by the corresponding column of **L**, by equating the elements of **K** and **LU**. For example, the elements of the third row of **U** and the third column of **L** are determined as follows

$$K_{3j} = \sum_{k=1}^{2} l_{3k} u_{kj} + u_{3j} \rightarrow u_{3j} = K_{3j} - \sum_{k=1}^{2} l_{3k} u_{kj}, \quad j = 3, 4 \quad (6.48a)$$

$$K_{i3} = \sum_{k=1}^{2} l_{ik} u_{k3} + l_{i3} u_{33} \rightarrow l_{i3} = \frac{K_{i3} - \sum_{k=1}^{2} l_{ik} u_{k3}}{u_{33}}, \quad i = 4 \quad (6.48b)$$

In general, the elements of the rth row of **U** and the rth column of **L** can be computed as

$$u_{rj} = K_{rj} - \sum_{k=1}^{r-1} l_{rk} u_{kj}, \quad j = r, \dots, n \quad (6.49a)$$

$$l_{ir} = \frac{K_{ir} - \sum_{k=1}^{r-1} l_{ik} u_{kr}}{u_{rr}}, \quad i = r+1, \dots, n \quad (6.49b)$$

The elements of **L** and **U** can be stored in the space originally occupied by **K**, that is, by overwriting **K**. The resulting matrix is

$$\hat{\mathbf{K}} = \mathbf{L} + \mathbf{U} - \mathbf{I} \quad (6.50)$$

Specifically, the matrix **K** of order 4 becomes

$$\hat{\mathbf{K}} = \begin{bmatrix} u_{11} & u_{12} & u_{13} & u_{14} \\ l_{21} & u_{22} & u_{23} & u_{24} \\ l_{31} & l_{32} & u_{33} & u_{34} \\ l_{41} & l_{42} & l_{43} & u_{44} \end{bmatrix} \quad (6.51)$$

With the **LU** decomposition defined, we obtain by forward substitution [Eq. (6.20) for $l_{ii} = 1$]

$$y_r = Q_r - \sum_{k=1}^{r-1} l_{rk} y_k, \quad r = 1, \dots, n \quad (6.52)$$

and by back substitution [Eq. (6.23)]

$$q_r = \frac{y_r - \sum_{k=r+1}^{n} u_{rk} q_k}{u_{rr}}, \quad r = n, \dots, 1 \quad (6.53)$$

Example 1. Equation (6.29) is solved by compact Gaussian elimination.

1. *Decomposition* [Eqs. (6.49)]
To illustrate the overwriting of **K** during decomposition, the unreduced elements are separated by dashed lines from the reduced elements.

$r = 1$:

$$u_{1j} = K_{1j}, \quad j = 1, \ldots, 4$$

$$l_{i1} = \frac{K_{i1}}{K_{11}}, \quad i = 2, \ldots, 4$$

$$l_{21} = -\tfrac{1}{2}, \quad l_{31} = 0, \quad l_{41} = 0$$

$$\begin{bmatrix} 2 & -1 & 0 & 0 \\ -\tfrac{1}{2} & 2 & -1 & 0 \\ 0 & -1 & 2 & -1 \\ 0 & 0 & -1 & 1 \end{bmatrix} \tag{6.54}$$

$r = 2$:

$$u_{2j} = K_{2j} - l_{21}u_{1j}, \quad j = 2, \ldots, 4$$

$$u_{22} = \tfrac{3}{2}, \quad u_{23} = -1, \quad u_{24} = 0$$

$$l_{i2} = \frac{K_{i2} - l_{i1}u_{12}}{u_{22}}, \quad i = 3, 4$$

$$l_{32} = -\tfrac{2}{3}, \quad l_{42} = 0$$

$$\begin{bmatrix} 2 & -1 & 0 & 0 \\ -\tfrac{1}{2} & \tfrac{3}{2} & -1 & 0 \\ 0 & -\tfrac{2}{3} & 2 & -1 \\ 0 & 0 & -1 & 1 \end{bmatrix} \tag{6.55}$$

$r = 3$:

$$u_{3j} = K_{3j} - \sum_{k=1}^{2} l_{3k}u_{kj}, \quad j = 3, 4$$

$$u_{33} = \tfrac{4}{3}, \quad u_{34} = -1$$

$$l_{43} = \frac{K_{43} - \sum_{k=1}^{2} l_{4k}u_{k3}}{u_{33}} = -\frac{3}{4}$$

$$\begin{bmatrix} 2 & -1 & 0 & 0 \\ -\tfrac{1}{2} & \tfrac{3}{2} & -1 & 0 \\ 0 & -\tfrac{2}{3} & \tfrac{4}{3} & -1 \\ 0 & 0 & -\tfrac{3}{4} & 1 \end{bmatrix} \tag{6.56}$$

$r = 4$:

$$u_{44} = K_{44} - \sum_{k=1}^{3} l_{4k} u_{k4} = \frac{1}{4}$$

$$\hat{\mathbf{K}} = \begin{bmatrix} 2 & -1 & 0 & 0 \\ -\frac{1}{2} & \frac{3}{2} & -1 & 0 \\ 0 & -\frac{2}{3} & \frac{4}{3} & -1 \\ 0 & 0 & -\frac{3}{4} & \frac{1}{4} \end{bmatrix} = \mathbf{L} + \mathbf{U} - \mathbf{I} \qquad (6.57)$$

Note that the elements l_{ij} below the main diagonal of $\mathbf{L}$ are the multipliers of the standard Gaussian elimination.

2. *Forward Substitution* [Eqs. (6.52)]

$$y_1 = Q_1 = 0, \qquad\qquad y_2 = Q_2 - l_{21} y_1 = 1$$

$$y_3 = Q_3 - \sum_{k=1}^{2} l_{3k} y_k = \frac{2}{3}, \quad y_4 = Q_4 - \sum_{k=1}^{3} l_{4k} y_k = \frac{1}{2} \qquad (6.58)$$

3. *Back Substitution* [Eqs. (6.53)]

$$q_4 = \frac{y_4}{u_{44}} = 2, \qquad\qquad q_3 = \frac{y_3 - u_{34} q_4}{u_{33}} = 2$$

$$\qquad\qquad (6.59)$$

$$q_2 = \frac{y_2 - \sum_{k=3}^{4} u_{2k} q_k}{u_{22}} = 2, \quad q_1 = \frac{y_1 - \sum_{k=2}^{4} u_{1k} q_k}{u_{11}} = 1$$

Order of Decomposition

It is not necessary to compute the elements of $\hat{\mathbf{K}}$ in the row–column order of Eqs. (6.49). For example, the elements of $\hat{\mathbf{K}}$ can be computed in the column order

$$u_{ij} = K_{ij} - \sum_{k=1}^{i-1} l_{ik} u_{kj}, \quad i = 1, \ldots, j \qquad (6.60a)$$

$$l_{ij} = \frac{K_{ij} - \sum_{k=1}^{j-1} l_{ik} u_{kj}}{u_{jj}}, \quad i = j+1, \ldots, n \qquad (6.60b)$$

The only restriction on the order of computation is that to compute element (i, j) of $\hat{\mathbf{K}}$ all other elements with a row number less than or equal to i and a column number less than or equal to j must have been computed (Jennings, 1977).

Symmetry

The decomposition of **K**, Eq. (6.25), can also be expressed as (Forsythe and Moler, 1967; Jennings, 1977; Ralston and Rabinowitz, 1978)

$$\mathbf{K} = \mathbf{L}\mathbf{D}\hat{\mathbf{U}} \tag{6.61}$$

where

$$\mathbf{U} = \mathbf{D}\hat{\mathbf{U}} \tag{6.62}$$

D is a *diagonal matrix* with the diagonal elements

$$d_i = u_{ii} \tag{6.63}$$

and $\hat{\mathbf{U}}$ is a *unit upper triangular matrix*. By Eqs. (6.62) and (6.63)

$$\hat{u}_{ij} = \frac{u_{ij}}{d_i} \tag{6.64}$$

and since **K** is symmetric, $\hat{\mathbf{U}} = \mathbf{L}^T$, that is

$$l_{ik} = \hat{u}_{ki} \tag{6.65}$$

Thus, we need only compute the elements of **D** and $\hat{\mathbf{U}}$ and store them by overwriting the elements in the upper triangle of **K**. This is illustrated for symmetric band matrices in Section 6.4 and 6.5.

Determinants

The most efficient way to compute the value of the determinant of **K** is by the **LU** decomposition of **K** (Dahlquist and Björk, 1974; Ralston and Rabinowitz, 1978). Specifically, from Eq. (6.61) we obtain

$$\det(\mathbf{K}) = \det(\mathbf{L}) \det(\mathbf{D}) \det(\hat{\mathbf{U}}) = \prod_{i=1}^{n} d_i > 0 \tag{6.66}$$

if **K** is positive definite.

Since the value of the determinant can be very large or very small, a special product subroutine may be required to avoid exponent overflow or underflow (Forsythe and Moler, 1967).

Cholesky Method

In the Cholesky, or square-root, method, the symmetric positive definite matrix **K** is decomposed in the form[5] (Dahlquist and Björk, 1974; Forsythe and Moler, 1967; Wilkinson and Reinsch, 1971)

$$\mathbf{K} = \mathbf{U}^T\mathbf{U} \tag{6.67}$$

---

[5] If the elements in the lower triangle of **K** are stored, the matrix is decomposed in the form $\mathbf{K} = \mathbf{L}\mathbf{L}^T$.

where U is an *upper triangular matrix* with positive diagonal elements. The elements of U may be computed in row or column order by equating corresponding elements in Eq. (6.67). In particular, the elements of the jth column are

$$u_{ij} = \frac{K_{ij} - \sum_{k=1}^{i-1} u_{ki} u_{kj}}{u_{ii}}, \quad i = 1, \ldots, j-1$$

$$u_{jj} = \left(K_{jj} - \sum_{k=1}^{j-1} u_{kj}^2 \right)^{1/2}$$

(6.68)

The solution of Eq. (6.17) can be completed by Eqs. (6.20) and (6.23), where in Eq. (6.20) $l_{ik} = u_{ki}$.

Example 2. In the solution of Eq. (6.29) by the Cholesky method we obtain

$$U = \begin{bmatrix} \sqrt{2} & -\dfrac{1}{\sqrt{2}} & 0 & 0 \\ 0 & \sqrt{\dfrac{3}{2}} & -\sqrt{\dfrac{2}{3}} & 0 \\ 0 & 0 & \dfrac{2}{\sqrt{3}} & -\dfrac{\sqrt{3}}{2} \\ 0 & 0 & 0 & \dfrac{1}{2} \end{bmatrix}, \quad y = \begin{bmatrix} 0 \\ \sqrt{\dfrac{2}{3}} \\ \dfrac{1}{\sqrt{3}} \\ 1 \end{bmatrix}, \quad q = \begin{bmatrix} 1 \\ 2 \\ 2 \\ 2 \end{bmatrix}$$

(6.69)

Modified Cholesky Method

The square roots in the Cholesky method can be avoided by letting

$$U = \hat{D}\hat{U}$$

(6.70)

where $\hat{D}$ is a *diagonal matrix* with the diagonal elements $\hat{d}_i = u_{ii}$, and $\hat{U}$ is a *unit upper triangular matrix*. Thus,

$$u_{ij} = u_{ii} \hat{u}_{ij}$$

(6.71)

Equations (6.67) and (6.70) yield the alternative decomposition

$$K = \hat{U}^T D \hat{U}$$

(6.72)

where D is a *diagonal matrix* with

$$d_i = u_{ii}^2$$

(6.73)

Implementation details for the modified Cholesky method are contained in Wilkinson and Reinsch (1971).

Comparison of Methods

Following an error analysis, Felippa and Tocher (1970) state, "There are often good reasons for using Gauss in preference to Cholesky: programming simplicity, ... physical interpretation, and simple determination of condensed stiffness or flexibility matrices. However, solution accuracy is certainly not one of them."

6.4 BAND SOLVERS

If $\mathbf{K}$ is a band matrix, we can reduce the effort required in the solution of Eq. (6.17) by confining the operations to the elements in the band. This reduces the range of the indices in the inner product computations during decomposition of $\mathbf{K}$ [Eqs. (6.49), (6.60), (6.68)] and in the forward and back substitutions [Eqs. (6.52), (6.53)].

For illustration consider the $\mathbf{LU}$ decomposition of a band matrix of order 6 and with half bandwidth 3.

$$
\mathbf{K} = \begin{bmatrix}
1 & 0 & 0 & 0 & 0 & 0 \\
l_{21} & 1 & 0 & 0 & 0 & 0 \\
l_{31} & l_{32} & 1 & 0 & 0 & 0 \\
l_{41} & l_{42} & l_{43} & 1 & 0 & 0 \\
0 & l_{52} & l_{53} & l_{54} & 1 & 0 \\
0 & 0 & l_{63} & l_{64} & l_{65} & 1
\end{bmatrix}
\begin{bmatrix}
u_{11} & u_{12} & u_{13} & u_{14} & 0 & 0 \\
0 & u_{22} & u_{23} & u_{24} & u_{25} & 0 \\
0 & 0 & u_{33} & u_{34} & u_{35} & u_{36} \\
0 & 0 & 0 & u_{44} & u_{45} & u_{46} \\
0 & 0 & 0 & 0 & u_{55} & u_{56} \\
0 & 0 & 0 & 0 & 0 & u_{66}
\end{bmatrix}
\tag{6.74}
$$

From Eq. (6.74) we can compute, for example, the elements in the fifth column of $\mathbf{U}$ as follows:

$$
K_{25} = u_{25} \qquad\qquad\qquad \to u_{25} = K_{25}
$$
$$
\vdots \qquad\qquad\qquad\qquad\qquad \vdots
\tag{6.75}
$$
$$
K_{45} = \sum_{k=2}^{3} l_{4k} u_{k5} + u_{45} \to u_{45} = K_{45} - \sum_{k=2}^{3} l_{4k} u_{k5}
$$

The elements involved in the inner product computation for u_{45} are shaded in Eq. (6.74).

In general, the elements of the jth column of $\mathbf{U}$ can be computed as

$$
u_{tj} = K_{tj}
$$
$$
u_{ij} = K_{ij} - \sum_{k=t}^{i-1} l_{ik} u_{kj}, \quad i = t+1, \ldots, j
\tag{6.76}
$$

where

$$
t = \max\{1, j - m\}, \quad m = \text{half bandwidth}
$$

By taking advantage of symmetry and bandedness, we can express the solution of Eq. (6.17) by Gaussian elimination as follows.

1. Decomposition

$\mathbf{K}$ is decomposed into the product $\mathbf{K} = \mathbf{L}\mathbf{D}\hat{\mathbf{U}}$. The computation of $\mathbf{D}$ and $\hat{\mathbf{U}}$ by Eqs. (6.63)–(6.65) and (6.76) is based on the algorithm for variable band matrices by Bathe and Wilson (1976).

$j = 1$:

$$d_1 = K_{11} \qquad (6.77)$$

$j = 2, \ldots, n$:

Compute

$$u_{tj} = K_{tj}$$

$$u_{ij} = K_{ij} - \sum_{k=t}^{i-1} \hat{u}_{ki} u_{kj}, \quad i = t + 1, \ldots, j - 1 \qquad (6.78)$$

$$t = \max\{1, j - m\}$$

Compute and store

$$\hat{u}_{ij} = \frac{u_{ij}}{d_i}, \quad i = t, \ldots, j - 1$$

$$d_j = K_{jj} - \sum_{k=t}^{j-1} \hat{u}_{kj} u_{kj} \qquad (6.79)$$

The elements of $\mathbf{D}$ and $\hat{\mathbf{U}}$ can be stored by overwriting $\mathbf{K}$. Actually K_{ij} is replaced by u_{ij}, which in turn is replaced by $\hat{u}_{ij}$, and K_{jj} is replaced by d_j.

2. Forward Substitution

By specializing Eq. (6.52) for a band matrix and using Eq. (6.65) we obtain

$$y_i = Q_i - \sum_{k=i-m}^{i-1} \hat{u}_{ki} y_k, \quad i = 1, \ldots, n; k \geq 1 \qquad (6.80)$$

3. Back Substitution

Similarly for a band matrix, Eq. (6.53) subject to Eq. (6.64) becomes

$$q_i = \frac{y_i}{d_i} - \sum_{k=i+1}^{i+m} \hat{u}_{ik} q_k, \quad i = n, \ldots, 1; k \leq n \qquad (6.81)$$

This algorithm can be adapted to the Cholesky method (Problem 6.5) and the modified Cholesky method.

Example 1. Equation (6.29), with $m = 1$, is solved by Eqs. (6.77)–(6.81).

1. *Decomposition.* To mark the progress of decomposition, the unreduced columns are separated by dashed lines from the reduced columns.

$j = 1$:

$$d_1 = K_{11} = 2$$

$$\begin{bmatrix} 2 & \vdots & -1 & & \\ & \vdots & 2 & -1 & \\ & \vdots & & 2 & -1 \\ & \vdots & & & 1 \end{bmatrix} \tag{6.82}$$

$j = 2$: $t = j - m = 1$

$$u_{12} = K_{12} = -1, \quad \hat{u}_{12} = \frac{u_{12}}{d_1} = -\frac{1}{2}$$

$$d_2 = K_{22} - \hat{u}_{12}u_{12} = \tfrac{3}{2}$$

$$\begin{bmatrix} 2 & -\frac{1}{2} & \vdots & & \\ & \frac{3}{2} & \vdots & -1 & \\ & & \vdots & 2 & -1 \\ & & \vdots & & 1 \end{bmatrix} \tag{6.83}$$

$j = 3$: $t = j - m = 2$

$$u_{23} = K_{23} = -1, \quad \hat{u}_{23} = \frac{K_{23}}{d_2} = -\frac{2}{3}$$

$$d_3 = K_{33} - \hat{u}_{23}u_{23} = \tfrac{4}{3}$$

$$\begin{bmatrix} 2 & -\frac{1}{2} & & \vdots & \\ & \frac{3}{2} & -\frac{2}{3} & \vdots & \\ & & \frac{4}{3} & \vdots & -1 \\ & & & \vdots & 1 \end{bmatrix} \tag{6.84}$$

$j = 4$: $t = j - m = 3$

$$u_{34} = K_{34} = -1, \quad \hat{u}_{34} = \frac{u_{34}}{d_3} = -\frac{3}{4}$$

$$d_4 = K_{44} - \hat{u}_{34}u_{34} = \tfrac{1}{4}$$

$$\begin{bmatrix} 2 & -\frac{1}{2} & & \\ & \frac{3}{2} & -\frac{2}{3} & \\ & & \frac{4}{3} & -\frac{3}{4} \\ & & & \frac{1}{4} \end{bmatrix} = \begin{bmatrix} d_1 & \hat{u}_{12} & & \\ & d_2 & \hat{u}_{23} & \\ & & d_3 & \hat{u}_{34} \\ & & & d_4 \end{bmatrix} \tag{6.85}$$

2. *Forward Substitution*

$$y_1 = Q_1 = 0, \qquad y_2 = Q_2 - \hat{u}_{12} y_1 = 1$$

$$y_3 = Q_3 - \hat{u}_{23} y_2 = \tfrac{2}{3}, \quad y_4 = Q_4 - \hat{u}_{34} y_3 = \tfrac{1}{2} \tag{6.86}$$

3. *Back Substitution*

$$q_4 = \frac{y_4}{d_4} = 2, \qquad q_3 = \frac{y_3}{d_3} - \hat{u}_{34} q_4 = 2$$

$$q_2 = \frac{y_2}{d_2} - \hat{u}_{23} q_3 = 2, \quad q_1 = \frac{y_1}{d_1} - \hat{u}_{12} q_2 = 1 \tag{6.87}$$

Example 2. Equations (6.77)–(6.81) are applied to solve the system of equations (Bathe and Wilson, 1976; Wilkinson and Reinsch, 1971)

$$\begin{bmatrix} 5 & -4 & 1 & 0 \\ -4 & 6 & -4 & 1 \\ 1 & -4 & 6 & -4 \\ 0 & 1 & -4 & 5 \end{bmatrix} \begin{bmatrix} q_1 \\ q_2 \\ q_3 \\ q_4 \end{bmatrix} = \begin{bmatrix} 0 \\ 1 \\ 0 \\ 0 \end{bmatrix} \tag{6.88}$$

$\mathbf{K}$ is symmetric and $m = 2$.

1. *Decomposition*
$j = 1$:

$$d_1 = K_{11} = 5$$

$$\begin{bmatrix} 5 & -4 & 1 & \\ & 6 & -4 & 1 \\ & & 6 & -4 \\ & & & 5 \end{bmatrix} \tag{6.89}$$

$j = 2$: $\ j - m = 0, \quad t = 1$

$$u_{12} = K_{12} = -4, \quad \hat{u}_{12} = \frac{u_{12}}{d_1} = -\frac{4}{5}$$

$$d_2 = K_{22} - \hat{u}_{12} u_{12} = \tfrac{14}{5}$$

$$\begin{bmatrix} 5 & -\tfrac{4}{5} & 1 & \\ & \tfrac{14}{5} & -4 & 1 \\ & & 6 & -4 \\ & & & 5 \end{bmatrix} \tag{6.90}$$

$j = 3:$ $t = j - m = 1$

$$u_{13} = K_{13} = 1, \quad u_{23} = K_{23} - \hat{u}_{12}u_{13} = -\tfrac{16}{5}$$

$$\hat{u}_{13} = \frac{u_{13}}{d_1} = \frac{1}{5}, \quad \hat{u}_{23} = \frac{u_{23}}{d_2} = -\frac{8}{7}$$

$$d_3 = K_{33} - \sum_{k=1}^{2} \hat{u}_{k3}u_{k3} = \tfrac{15}{7}$$

$$\begin{bmatrix} 5 & -\tfrac{4}{5} & \tfrac{1}{5} & \vdots & \\ & \tfrac{14}{5} & -\tfrac{8}{7} & \vdots & 1 \\ & & \tfrac{15}{7} & \vdots & -4 \\ & & & & 5 \end{bmatrix} \tag{6.91}$$

$j = 4:$ $t = j - m = 2$

$$u_{24} = K_{24} = 1, \quad u_{34} = K_{34} - \hat{u}_{23}u_{24} = -\tfrac{20}{7}$$

$$\hat{u}_{24} = \frac{u_{24}}{d_2} = \frac{5}{14}, \quad \hat{u}_{34} = \frac{u_{34}}{d_3} = -\frac{4}{3}$$

$$d_4 = K_{44} - \sum_{k=2}^{3} \hat{u}_{k4}u_{k4} = \frac{5}{6}$$

$$\begin{bmatrix} 5 & -\tfrac{4}{5} & \tfrac{1}{5} & \\ & \tfrac{14}{5} & -\tfrac{8}{7} & \tfrac{5}{14} \\ & & \tfrac{15}{7} & -\tfrac{4}{3} \\ & & & \tfrac{5}{6} \end{bmatrix} = \begin{bmatrix} d_1 & \hat{u}_{12} & \hat{u}_{13} & \\ & d_2 & \hat{u}_{23} & \hat{u}_{24} \\ & & d_3 & \hat{u}_{34} \\ & & & d_4 \end{bmatrix} \tag{6.92}$$

2. *Forward Substitution*

$$y_1 = Q_1 = 0, \qquad\qquad y_2 = Q_2 - \hat{u}_{12}y_1 = 1$$

$$y_3 = Q_3 - \sum_{k=1}^{2} \hat{u}_{k3}y_k = \frac{8}{7}, \quad y_4 = Q_4 - \sum_{k=2}^{3} \hat{u}_{k4}y_k = \frac{7}{6} \tag{6.93}$$

3. *Back Substitution*

$$q_4 = \frac{y_4}{d_4} = \frac{7}{5}, \qquad\qquad q_3 = \frac{y_3}{d_3} - \hat{u}_{34}q_4 = \frac{12}{5}$$

$$q_2 = \frac{y_2}{d_2} - \sum_{k=3}^{4} \hat{u}_{2k}q_k = \frac{13}{5}, \quad q_1 = \frac{y_1}{d_1} - \sum_{k=2}^{3} \hat{u}_{1k}q_k = \frac{8}{5} \tag{6.94}$$

Operation Count

The number of arithmetic operations in the solution of a system of equations is usually measured by the number of multiplications (a division counts as a multiplication) since it is approximately equal to the number of additions (Dhalquist and Björk, 1974; Forsythe and Moler, 1967).

The **LU** decomposition of a large full matrix of order n requires approximately $\frac{1}{3}n^3$ operations, while only n^2 operations are required in the solution by forward and back substitutions. The number of operations required in the **LU** decomposition is reduced to $\frac{1}{6}n^3$ for a symmetric matrix and to $\frac{1}{2}nm^2$ for a symmetric band matrix if the half bandwidth $m \ll n$.

Computer Programs

Computer programs are available for the solution of symmetric band equations by Gaussian elimination (Cook, 1981; Becker et al., 1981; Brebbia and Ferrante, 1978; Felippa and Tocher, 1970; Wilson, 1968), the Cholesky method (Brebbia and Ferrante, 1978; Dongarra et al., 1979; Wilkinson and Reinsch, 1971), and the modified Cholesky method (Schwarz et al., 1973). See also the Additional References at the end of this chapter.

6.5 VARIABLE BAND SOLVERS

For a variable band matrix (Section 6.2), the solution to Eq. (6.17) by Gaussian elimination can be expressed as follows (Bathe and Wilson, 1976).

1. Decomposition

K is decomposed into the product $\mathbf{K} = \mathbf{L}\mathbf{D}\hat{\mathbf{U}}$, where **D** is a diagonal matrix, $\hat{\mathbf{U}}$ is a unit upper triangular matrix, and $\mathbf{L} = \hat{\mathbf{U}}^T$ [Eqs. (6.61)–(6.65)]. The elements of **D** and $\hat{\mathbf{U}}$ are computed in column order and stored by overwriting **K**:

$j = 1$:

$$d_1 = K_{11} \tag{6.95}$$

$j = 2, \ldots, n$: $t = m_j, s = \max\{m_i, m_j\}$

$$u_{tj} = K_{tj}$$

$$u_{ij} = K_{ij} - \sum_{k=s}^{i-1} \hat{u}_{ki} u_{kj}, \quad i = t + 1, \ldots, j - 1 \tag{6.96}$$

m_j = row number of the first nonzero entry of column j

$$\hat{u}_{ij} = \frac{u_{ij}}{d_i}, \quad i = t, \ldots, j - 1$$

$$d_j = K_{jj} - \sum_{k=t}^{j-1} \hat{u}_{kj} u_{kj} \tag{6.97}$$

2. Forward Substitution

The solution of equation $\mathbf{Ly} = \mathbf{Q}$ is obtained, analogous to Eq. (6.80), as

$$y_i = Q_i - \sum_{k=m_i}^{i=1} \hat{u}_{ki} y_k \tag{6.98}$$

3. Back Substitution

Consistent with the skyline form of $\hat{\mathbf{U}}$, the equation

$$\hat{\mathbf{U}}\mathbf{q} = \bar{\mathbf{y}}, \quad \bar{y}_i = \frac{y_i}{d_i} \tag{6.99}$$

is solved in column order by using standard Gaussian elimination (Section 6.3) to eliminate all the elements above the main diagonal of $\hat{\mathbf{U}}$ (Jennings, 1966). In the process, $\hat{\mathbf{U}}$ becomes a unit matrix and $\mathbf{q}$ becomes equal to the reduced right-hand side vector. This procedure can be expressed by the algorithm

$$\bar{y}^{(n)} = \bar{y}$$

where

$$q_n = \bar{y}_n^{(n)} \tag{6.100}$$

$i = n, \ldots, 2$:

$$\bar{y}_k^{(i-1)} = \bar{y}_k^{(i)} - \hat{u}_{ki} q_i, \quad k = m_i, \ldots, i-1$$

$$q_{i-1} = \bar{y}_{i-1}^{(i-1)} \tag{6.101}$$

Storage procedure: Q_k is replaced in sequence by y_k, $\bar{y}_k^{(n)} = \bar{y}_k, \ldots, \bar{y}_k^{(1)} = q_k$.
This algorithm can be deduced from the solution of the equation $\hat{\mathbf{U}}\mathbf{q} = \bar{\mathbf{y}}^{(5)}$:

$$\begin{bmatrix} 1 & \hat{u}_{12} & 0 & 0 & \hat{u}_{15} \\ 0 & 1 & \hat{u}_{23} & 0 & \hat{u}_{25} \\ 0 & 0 & 1 & \hat{u}_{34} & \hat{u}_{35} \\ 0 & 0 & 0 & 1 & \hat{u}_{45} \\ 0 & 0 & 0 & 0 & 1 \end{bmatrix} \begin{bmatrix} q_1 \\ q_2 \\ q_3 \\ q_4 \\ q_5 \end{bmatrix} = \bar{\mathbf{y}}^{(5)} \tag{6.102}$$

where $q_5 = \bar{y}_5^{(5)}$. By eliminating q_5 from the first four equations we obtain

$$\begin{bmatrix} 1 & \hat{u}_{12} & 0 & 0 & 0 \\ 0 & 1 & \hat{u}_{23} & 0 & 0 \\ 0 & 0 & 1 & \hat{u}_{34} & 0 \\ 0 & 0 & 0 & 1 & 0 \\ 0 & 0 & 0 & 0 & 1 \end{bmatrix} \begin{bmatrix} q_1 \\ q_2 \\ q_3 \\ q_4 \\ q_5 \end{bmatrix} = \begin{bmatrix} \bar{y}_1^{(5)} - \hat{u}_{15} q_5 \\ \bar{y}_2^{(5)} - \hat{u}_{25} q_5 \\ \bar{y}_3^{(5)} - \hat{u}_{35} q_5 \\ \bar{y}_4^{(5)} - \hat{u}_{45} q_5 \\ q_5 \end{bmatrix} = \bar{\mathbf{y}}^{(4)} \tag{6.103}$$

where $q_4 = \bar{y}_4^{(4)} = \bar{y}_4^{(5)} - \hat{u}_{45} q_5$. Eliminating q_4, q_3, and q_2 we find, respectively,

$$
\bar{\mathbf{y}}^{(3)} = \begin{bmatrix} \bar{y}_1^{(4)} \\ \bar{y}_2^{(4)} \\ \bar{y}_3^{(4)} - \hat{u}_{34} q_4 \\ q_4 \\ q_5 \end{bmatrix}, \quad \bar{\mathbf{y}}^{(2)} = \begin{bmatrix} \bar{y}_1^{(3)} \\ \bar{y}_2^{(3)} - \hat{u}_{23} q_3 \\ q_3 \\ q_4 \\ q_5 \end{bmatrix}
$$

$$
\bar{\mathbf{y}}^{(1)} = \begin{bmatrix} \bar{y}_1^{(2)} - \hat{u}_{12} q_2 \\ q_2 \\ q_3 \\ q_4 \\ q_5 \end{bmatrix} = \mathbf{q}
$$
(6.104)

Example. The variable band algorithm [Eqs. (6.95)–(6.101)] is applied to the equations

$$
\begin{bmatrix} 3 & -1 & 0 & 0 & -1 \\ & 2 & -1 & 0 & 0 \\ & & 2 & -1 & 0 \\ & & & 2 & -1 \\ \text{sym.} & & & & 2 \end{bmatrix} \begin{bmatrix} q_1 \\ q_2 \\ q_3 \\ q_4 \\ q_5 \end{bmatrix} = \begin{bmatrix} 0 \\ 0 \\ 1 \\ 0 \\ 0 \end{bmatrix}
$$
(6.105)

where $\mathbf{K}$ is constructed in Example 2 of Section 6.2.

1. *Decomposition.*
The unreduced columns are separated from the reduced columns by dashed lines.

$j = 1$:

$$
d_1 = K_{11} = 3
$$
(6.106)

$j = 2$: $\quad t = m_2 = 1$

$$
u_{12} = K_{12} = -1, \quad \hat{u}_{12} = \frac{u_{12}}{d_1} = -\frac{1}{3}
$$

$$
d_2 = K_{22} - \hat{u}_{12} u_{12} = \tfrac{5}{3}
$$

$$
\begin{bmatrix} 3 & -\frac{1}{3} & \vdots & & -1 \\ & \frac{5}{3} & \vdots & -1 & 0 \\ & & \vdots & 2 & -1 & 0 \\ & & \vdots & & 2 & -1 \\ & & \vdots & & & 2 \end{bmatrix}
$$
(6.107)

$j = 3: \quad t = m_3 = 2$

$$u_{23} = K_{23} = -1, \quad \hat{u}_{23} = \frac{u_{23}}{d_2} = -\frac{3}{5}$$

$$d_3 = K_{33} - \hat{u}_{23}u_{23} = \tfrac{7}{5}$$

$$\begin{bmatrix} 3 & -\frac{1}{3} & & & -1 \\ & \frac{5}{3} & -\frac{3}{5} & & 0 \\ & & \frac{7}{5} & -1 & 0 \\ & & & 2 & -1 \\ & & & & 2 \end{bmatrix} \tag{6.108}$$

$j = 4: \quad t = m_4 = 3$

$$u_{34} = K_{34} = -1, \quad \hat{u}_{34} = \frac{u_{34}}{d_3} = -\frac{5}{7}$$

$$d_4 = K_{44} - \hat{u}_{34}u_{34} = \tfrac{9}{7}$$

$$\begin{bmatrix} 3 & -\frac{1}{3} & & & -1 \\ & \frac{5}{3} & -\frac{3}{5} & & 0 \\ & & \frac{7}{5} & -\frac{5}{7} & 0 \\ & & & \frac{9}{7} & -1 \\ & & & & 2 \end{bmatrix} \tag{6.109}$$

$j = 5: \quad t = m_5 = 1; \quad s = 1, 2, 3$

$$u_{15} = K_{15} = -1, \qquad \hat{u}_{15} = \frac{u_{15}}{d_1} = -\frac{1}{3}$$

$$u_{25} = K_{25} - \hat{u}_{12}u_{15} = -\frac{1}{3}, \quad \hat{u}_{25} = \frac{u_{25}}{d_2} = -\frac{1}{5}$$

$$u_{35} = K_{35} - \hat{u}_{23}u_{25} = -\frac{1}{5}, \quad \hat{u}_{35} = \frac{u_{35}}{d_3} = -\frac{1}{7}$$

$$u_{45} = K_{45} - \hat{u}_{34}u_{35} = -\frac{8}{7}, \quad \hat{u}_{45} = \frac{u_{45}}{d_4} = -\frac{8}{9}$$

$$d_5 = K_{55} - \sum_{k=1}^{4} \hat{u}_{k5}u_{k5} = \frac{5}{9}$$

$$\begin{bmatrix} 3 & -\frac{1}{3} & & & -\frac{1}{3} \\ & \frac{5}{3} & -\frac{3}{5} & & -\frac{1}{5} \\ & & \frac{7}{5} & -\frac{5}{7} & -\frac{1}{7} \\ & & & \frac{9}{7} & -\frac{8}{9} \\ & & & & \frac{5}{9} \end{bmatrix} \tag{6.110}$$

2. *Forward Substitution*

$$y_1 = Q_1 = 0$$

$$y_2 = Q_2 - \hat{u}_{12} y_1 = 0$$

$$y_3 = Q_3 - \hat{u}_{23} y_2 = 1 \qquad (6.111)$$

$$y_4 = Q_4 - \hat{u}_{34} y_3 = \tfrac{5}{7}$$

$$y_5 = Q_5 - \sum_{k=1}^{4} \hat{u}_{k5} y_k = \frac{7}{9}$$

3. *Back Substitution*

$$\bar{y}^{(5)} = \bar{y} = \begin{bmatrix} 0 \\ 0 \\ \frac{5}{7} \\ \frac{5}{9} \\ \frac{7}{5} \end{bmatrix}, \quad q_5 = \bar{y}_5^{(5)} = \tfrac{7}{5} \qquad (6.112)$$

$i = 5: \quad k = 1, \ldots, 4$

$$\bar{y}_k^{(4)} = \bar{y}_k^{(5)} - \hat{u}_{k5} q_5$$

$$\bar{y}^{(4)} = \begin{bmatrix} \frac{7}{15} \\ \frac{7}{25} \\ \frac{32}{35} \\ \frac{9}{5} \\ \frac{7}{5} \end{bmatrix}, \quad q_4 = \bar{y}_4^{(4)} = \tfrac{9}{5} \qquad (6.113)$$

$i = 4: \quad k = 3$

$$\bar{y}_3^{(3)} = \bar{y}_3^{(4)} - \hat{u}_{34} q_4$$

$$\bar{y}^{(3)} = \begin{bmatrix} \frac{7}{15} \\ \frac{7}{25} \\ \frac{11}{5} \\ \frac{9}{5} \\ \frac{7}{5} \end{bmatrix}, \quad q_3 = \bar{y}_3^{(3)} = \tfrac{11}{5} \qquad (6.114)$$

$i = 3: \quad k = 2$

$$\bar{y}_2^{(2)} = \bar{y}_2^{(3)} - \hat{u}_{23} q_3$$

$$\bar{y}^{(2)} = \begin{bmatrix} \frac{7}{15} \\ \frac{8}{5} \\ \frac{11}{5} \\ \frac{9}{5} \\ \frac{7}{5} \end{bmatrix}, \quad q_2 = \bar{y}_2^{(2)} = \tfrac{8}{5} \qquad (6.115)$$

$i = 2$: $k = 1$

$$\bar{y}_1^{(1)} = \bar{y}_1^{(2)} - \hat{u}_{12}q_2 = 1 = q_1$$

$$\mathbf{q} = \bar{\mathbf{y}}^{(1)} = \begin{bmatrix} 1 \\ \frac{8}{5} \\ \frac{11}{5} \\ \frac{9}{5} \\ \frac{7}{5} \end{bmatrix} \qquad (6.116)$$

Computer Programs

Computer programs for variable band solvers have been written based on Gaussian elimination (Bathe and Wilson, 1976; Zienkiewicz, 1977; Mondkar and Powell, 1974a, 1974b; Wiberg, 1974; Wilson et al., 1974), the Cholesky method (George and Liu, 1981; Jennings, 1977), and the modified Cholesky method (Felippa, 1975). See also the Additional References at the end of this chapter.

6.6 FRONTAL SOLUTIONS

A system of equations is *large* if it cannot be solved directly in central memory of a given computer and peripheral storage is required (Meyer, 1975). Various methods have been developed for the solution of large sparse matrices. Among them are (Meyer, 1975) *out-of-core band solvers*, which reduce equations by blocks in high-speed storage;[6] *partitioning methods*, which include substructuring; and *frontal solvers*.

The *frontal solution technique* (Irons and Ahmad, 1980; Irons, 1970; Melosh and Bamford, 1969) is a special form of *symmetric standard Gaussian elimination* in which the assembly of elements alternates with the elimination of variables. It is based on two properties peculiar to the assembly process and symmetric standard Gaussian elimination:

1. The coefficients in the kth row of the system stiffness matrix and the load vector are fully summed after the last element that contains q_k has been assembled; this corresponds to the last column of the member code matrix with an entry k.
2. When the coefficients of the kth row are fully summed, q_k can be eliminated.

---

[6] For example, Figure 6.7 indicates that in the elimination of q_k, only the stiffness coefficients of the *active triangle* (the elements in the kth row and in the shaded triangle below) need to be in central memory.

This follows from the algorithm for symmetric Gaussian elimination [Eqs. (6.34), (6.38)]:

$$K_{ij}^{(k)} = K_{ij}^{(k-1)} - l_{ki}K_{kj}^{(k-1)}$$
$$Q_i^{(k)} = Q_i^{(k-1)} - l_{ki}Q_k^{(k-1)}$$

(6.117)

where

$$l_{ki} = \frac{K_{ki}^{(k-1)}}{K_{kk}^{(k-1)}}$$

Equations (6.117) indicate that the coefficients in the ith row need not be fully summed in order to eliminate q_k. Specifically, the order of additions and subtractions does not matter provided the second terms on the right-hand side of Eqs. (6.117) are correct, that is, the coefficients of the kth row are fully summed.

The frontal solution proceeds element by element, analogous to a wave passing over the structure. A variable becomes *active* on its first appearance and is immediately eliminated on its last. The number of active variables is called the *wave front* or simply the *front*. The front may change continually during the solution, similar to the active column heights in the skyline solution.

Example. The frontal solution is illustrated for the structure in Figure 6.9a, whose properties are defined in Example 2 of Section 6.2. The stiffness matrix of each element is

$$\mathbf{k} = \begin{bmatrix} 1 & -1 \\ -1 & 1 \end{bmatrix}$$

(6.118)

The assembly is performed with the member code matrix

$$\mathbf{M}^T = \begin{bmatrix} 0 & 1 \\ 1 & 2 \\ 2 & 3 \\ 3 & 4 \\ 4 & 5 \\ 1 & 5 \end{bmatrix}$$

(6.119)

The variation of the front during the assembly–elimination process of the frontal reduction is depicted graphically (Irons, 1970) in Figure 6.9b. For clarity, symmetry is not exploited in the solution.

Element 1
Assemble: write equation in q_1.

$$[1]q_1 = [0]$$

(6.120)

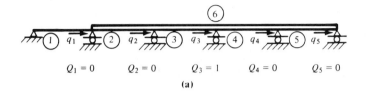

$Q_1 = 0$ $Q_2 = 0$ $Q_3 = 1$ $Q_4 = 0$ $Q_5 = 0$

(a)

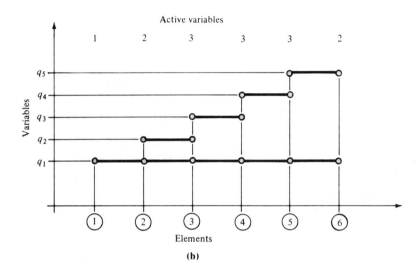

FIGURE 6.9 Frontal solution. (a) Assembly of truss elements; (b) variation of front.

Element 2
Assemble: write equations in q_1 and q_2.

$$\begin{bmatrix} 2 & -1 \\ -1 & 1 \end{bmatrix} \begin{bmatrix} q_1 \\ q_2 \end{bmatrix} = \begin{bmatrix} 0 \\ 0 \end{bmatrix} \tag{6.121}$$

Element 3
Assemble: write equations in q_1–q_3.

$$\begin{bmatrix} 2 & -1 & 0 \\ -1 & 2 & -1 \\ 0 & -1 & 1 \end{bmatrix} \begin{bmatrix} q_1 \\ q_2 \\ q_3 \end{bmatrix} = \begin{bmatrix} 0 \\ 0 \\ 1 \end{bmatrix} \tag{6.122}$$

Eliminate q_2 from the first and third equations.

$$\begin{bmatrix} \frac{3}{2} & -\frac{1}{2} \\ -\frac{1}{2} & \frac{1}{2} \end{bmatrix} \begin{bmatrix} q_1 \\ q_3 \end{bmatrix} = \begin{bmatrix} 0 \\ 1 \end{bmatrix} \tag{6.123}$$

Element 4

Assemble: write equations in q_1, q_3, q_4.

$$\begin{bmatrix} \frac{3}{2} & -\frac{1}{2} & 0 \\ -\frac{1}{2} & \frac{3}{2} & -1 \\ 0 & -1 & 1 \end{bmatrix} \begin{bmatrix} q_1 \\ q_3 \\ q_4 \end{bmatrix} = \begin{bmatrix} 0 \\ 1 \\ 0 \end{bmatrix} \tag{6.124}$$

Eliminate q_3 from the first and third equations.

$$\begin{bmatrix} \frac{4}{3} & -\frac{1}{3} \\ -\frac{1}{3} & \frac{1}{3} \end{bmatrix} \begin{bmatrix} q_1 \\ q_4 \end{bmatrix} = \begin{bmatrix} \frac{1}{3} \\ \frac{2}{3} \end{bmatrix} \tag{6.125}$$

Element 5

Assemble: write equations in q_1, q_4, q_5.

$$\begin{bmatrix} \frac{4}{3} & -\frac{1}{3} & 0 \\ -\frac{1}{3} & \frac{4}{3} & -1 \\ 0 & -1 & 1 \end{bmatrix} \begin{bmatrix} q_1 \\ q_4 \\ q_5 \end{bmatrix} = \begin{bmatrix} \frac{1}{3} \\ \frac{2}{3} \\ 0 \end{bmatrix} \tag{6.126}$$

Eliminate q_4 from the first and third equations.

$$\begin{bmatrix} \frac{5}{4} & -\frac{1}{4} \\ -\frac{1}{4} & \frac{1}{4} \end{bmatrix} \begin{bmatrix} q_1 \\ q_5 \end{bmatrix} = \begin{bmatrix} \frac{1}{2} \\ \frac{1}{2} \end{bmatrix} \tag{6.127}$$

Element 6

Assemble: write equations in q_1, q_5.

$$\begin{bmatrix} \frac{9}{4} & -\frac{5}{4} \\ -\frac{5}{4} & \frac{5}{4} \end{bmatrix} \begin{bmatrix} q_1 \\ q_5 \end{bmatrix} = \begin{bmatrix} \frac{1}{2} \\ \frac{1}{2} \end{bmatrix} \tag{6.128}$$

Eliminate q_1 from the second equation.

$$\begin{bmatrix} \frac{5}{9} \end{bmatrix} q_5 = \begin{bmatrix} \frac{7}{9} \end{bmatrix} \tag{6.129}$$

The shaded coefficients in Eqs. (6.122), (6.124), (6.126), (6.128), and (6.129) define the reduced system of equations. (The equations are rewritten only in order to display the triangular form.)

$$\begin{bmatrix} 2 & -1 & 0 & -1 & 0 \\ 0 & \frac{3}{2} & -1 & -\frac{1}{2} & 0 \\ 0 & 0 & \frac{4}{3} & -\frac{1}{3} & -1 \\ 0 & 0 & 0 & \frac{9}{4} & -\frac{5}{4} \\ 0 & 0 & 0 & 0 & \frac{5}{9} \end{bmatrix} \begin{bmatrix} q_2 \\ q_3 \\ q_4 \\ q_1 \\ q_5 \end{bmatrix} = \begin{bmatrix} 0 \\ 1 \\ \frac{2}{3} \\ \frac{1}{2} \\ \frac{7}{9} \end{bmatrix} \tag{6.130}$$

The solution of Eq. (6.130) by back substitution yields

$$q_5 = \tfrac{7}{5}, \quad q_1 = 1, \quad q_4 = \tfrac{9}{5}, \quad q_3 = \tfrac{11}{5}, \quad q_2 = \tfrac{8}{5}$$

This example reveals some important properties of the frontal technique:
(1) The front depends on the order in which the elements are assembled,
but the ordering of variables, which is crucial for band solvers, is irrelevant.
For example, the two joint-numbering schemes in Figure 6.6 result in
different variable numbers but not in different fronts. (2) The variables may
be activated in a different order from that in which they are eliminated. This
can be an advantage over band solvers because only the submatrix of
coefficients corresponding to active variables (see footnote 6, p. 312) must
be in core. For example, in the frontal solution of the structure in Figure 6.9a,
the front never exceeds the value 3, whereas in the variable band solution
of Section 6.5, the maximum active column height is 5.

Additional advantages of the frontal technique are cited in Irons and
Ahmad (1980), Irons (1970), Melosh and Bamford (1969), and Meyer (1973).
However, they can only be realized for large systems that cannot be solved
in core. "The main disadvantage of frontal solutions is the formidable
amount of bookkeeping required to keep track of the unknowns and the
reduced stiffness coefficients, while band solvers can be set up with a mini-
mum amount of indexing and data manipulation" (Meyer, 1973). According
to Felippa (1979), "The skyline storage scheme offers a reasonable compro-
mise between organizational simplicity and computational efficiency."
Another disadvantage of the frontal solution is inherent in *standard* Gaussian
elimination [Section 6.2]: In the solution of $\mathbf{Kq} = \mathbf{Q}$ for several load vectors
that are not known at the same time, the reduction of $\mathbf{K}$ has to be repeated,
at a considerable cost in the number of operations. This becomes a factor in
nonlinear structural analysis (Agrawal et al., 1980).

Computer Programs

Computer programs based on the frontal technique are listed or cited in
Irons and Ahmad (1980), Agrawal et al. (1980), Hinton and Owen (1977),
Irons (1970), and Melosh and Bamford (1969) and in the Additional
References at the end of this chapter.

6.7 SOLUTION ERRORS

Numerical errors in the solution of linear equations have been studied
within the contexts of numerical analysis and structural analysis. Summaries
of these investigations are contained in papers (Felippa and Tocher, 1970;

Melosh, 1971; Meyer, 1975; Roy, 1971) and textbooks (Bathe and Wilson, 1976; Cook, 1981; McGuire and Gallagher, 1979). They identify possible sources of errors and describe methods of error detection and error control. An introduction to this important topic is presented in this section.

Computational Errors

In the formulation and solution of the equation $\mathbf{Kq} = \mathbf{Q}$, two computational errors are introduced: *Truncation errors*, which are caused by the representation of the coefficients of $\mathbf{K}$ and $\mathbf{Q}$ by a finite number of digits as floating point numbers, and *roundoff errors*, which are caused by floating point arithmetic in the solution for $\mathbf{q}$ (Atkinson, 1978; Dahlquist and Björk, 1974; Forsythe and Moler, 1967; Fox, 1965; Hamming, 1973; Johnson and Riess, 1977; Ralston and Rabinowitz, 1978; Wilkinson, 1963).

The truncation error should not be confused with the discretization error of the finite element method (Cook, 1981) or the error in the data defining structural properties (Roy, 1971). The truncation error is simply a roundoff error contained in $\mathbf{K}$ and $\mathbf{Q}$ at the start of the solution. It is caused by floating point computation in the formulation of element models and their assembly into the system model.

Condition Number

The matrix $\mathbf{K}$ is said to be *ill conditioned* if small variations in $\mathbf{K}$ and $\mathbf{Q}$ can cause large changes in the computed solution. Variations in $\mathbf{K}$ and $\mathbf{Q}$ are introduced through truncation and roundoff errors. A measure of the condition of $\mathbf{K}$ is given by the *condition number*

$$\kappa = \frac{\lambda_n}{\lambda_1} \tag{6.131}$$

where λ_n and λ_1 are the largest and smallest eigenvalues of $\mathbf{K}$, that is, they are the largest and smallest roots of the *characteristic determinant*

$$\det(\mathbf{K} - \lambda\mathbf{I}) = 0 \tag{6.132}$$

The sensitivity of the computed solution to computational errors increases with the condition number. If the condition number is large, $\mathbf{K}$ is ill-conditioned.

Procedures for computing or estimating the condition number are described, for example, by Cook (1981). It may be necessary to scale $\mathbf{K}$ in order to avoid artificial ill-conditioning (Cook, 1981; McGuire and Gallagher, 1979)

Error Detection

"Tests can detect errors but cannot prove their absence" (Cook, 1981).

Roundoff Errors

The effect of roundoff errors on the solution can be determined as follows (Meyer, 1975; Roy, 1971; Wilkinson, 1963): (1) **K** and **Q** are computed in single precision and the solution is obtained in single precision. (2) The single precision arrays **K** and **Q** are stored in double precision and the solution is performed in double precision. The difference in the two solutions is due to roundoff errors.

Roy (1971) found in his error analysis of structures by the finite element displacement method that the effect of roundoff errors on the solution accuracy is not significant. This is consistent with experience in finite element analysis (Bathe and Wilson, 1976; Cook, 1981). However, the round-off error depends on the word length and may be important for micro-computers.

Truncation Errors

The effect of truncation errors on the solution can be estimated as (Bathe and Wilson, 1976; Cook, 1981; Roy, 1971)

$$s \geq t - \log \kappa \qquad (6.133)$$

where s is the number of significant decimal digits in the solution and t is the number of decimal digits to which coefficients of **K** are represented in the computer. For example, if $t = 7$ and $\kappa = 1000$, $s \geq 4$.

Roy (1971) found that the truncation error is the decisive error influencing the accuracy of the solution.

Diagonal Decay Test

In this simple test, which may reveal local ill-conditioning during Gaussian elimination, the decay of diagonal coefficients is monitored (Cook, 1981; Irons and Ahmad, 1980; Felippa and Tocher, 1970; Melosh and Bamford, 1969). Specifically, before the elimination of the kth variable, the ratio $K_{kk}^{(k-1)}/K_{kk}$ is computed, where K_{kk} is the initial diagonal coefficient and $K_{kk}^{(k-1)}$ is the pivot (Section 6.3).

If this ratio becomes unacceptably small, the execution can be terminated (Cook, 1981). For example, if the ratio is 10^{-6}, six leading digits of K_{kk} have been lost during elimination, which is unacceptable for a six-digit computer precision.

Equilibrium Check

Let $\bar{q}$ denote the computed solution to $Kq = Q$; then the solution error is

$$q - \bar{q} = K^{-1} \Delta Q \qquad (6.134)$$

and the residual

$$\Delta Q = Q - K\bar{q} \qquad (6.135)$$

The residual, which should be computed in double precision, represents unbalanced forces.

Gross errors in the solution may be detected by computing the residual. If ΔQ is large, then there is clearly something wrong with the solution. However, a small residual does not guarantee an accurate solution because relatively large values in K^{-1}, a symptom of ill-conditioning, can cause a large solution error even for small residual [Eq. (6.134)] (Bathe and Wilson, 1976; Cook, 1981; McGuire and Gallagher, 1979; Meyer, 1975).

Improvement of Solution

Various procedures are available for reducing truncation and roundoff errors in the solution (Meyer, 1975; Roy, 1971).

Iterative Improvement

Let us denote the *initial solution* to $Kq = Q$ as $q^{(1)}$ and the corresponding *residual*

$$\Delta Q^{(1)} = Q - Kq^{(1)} \qquad (6.136)$$

If K, Q, and $q^{(1)}$ are given to t digits, $\Delta Q^{(1)}$ is computed to $2t$ digits and rounded to t digits (Dahlquist and Björke, 1974; Felippa and Tocher, 1970; Forsythe and Moler, 1967). Using the decomposition of K from the initial solution, we compute the *correction* $\Delta q^{(1)}$ (in single precision) from

$$K \Delta q^{(1)} = \Delta Q^{(1)} \qquad (6.137)$$

and obtain the *improved solution*

$$q^{(2)} = q^{(1)} + \Delta q^{(1)} \qquad (6.138)$$

This procedure can be continued until the residual is sufficiently small.

The relative error in the initial solution is approximately

$$e = \frac{|\Delta q^{(1)}|}{|q^{(1)}|} \qquad (6.139)$$

If $e \ll 1$, the improved solution has approximately twice as many significant digits as the initial solution. If e is close to or larger than unity, the initial solution has no correct digits and cannot be improved (Felippa and Tocher, 1970). In that case full double precision arithmetic is unavoidable.

Double Precision Arithmetic

If vital information is lost by the single precision representation of $\mathbf{K}$, the solution can only be improved by performing the computation of $\mathbf{K}$ and the solution for $\mathbf{q}$ in double precision.

Modification of Model

Procedures for improving the solution accuracy through model changes include substructuring and the selection of special degrees of freedom (Bathe and Wilson, 1976; Cook, 1981; McGuire and Gallagher, 1979; Roy, 1971) of special degrees of freedom.

Error Analysis

The model of the structure in Figure 6.10 is used to reveal a possible source of ill-conditioning, to apply error tests, to show the effects of truncation and roundoff errors on the solution accuracy, and to compute improved solutions.

The extensional stiffness of elements 1 and 2 are γ^* and γ, respectively. The system stiffness matrix and its inverse are

$$\mathbf{K} = \begin{bmatrix} \gamma^* + \gamma & -\gamma \\ -\gamma & \gamma \end{bmatrix}, \quad \mathbf{K}^{-1} = \begin{bmatrix} \dfrac{1}{\gamma^*} & \dfrac{1}{\gamma^*} \\ \dfrac{1}{\gamma^*} & \dfrac{1}{\gamma^*} + \dfrac{1}{\gamma} \end{bmatrix} \tag{6.140}$$

and the eigenvalues of $\mathbf{K}$ are

$$\lambda_1 = \gamma + \frac{\gamma^*}{2} - \left[\gamma^2 + \left(\frac{\gamma^*}{2} \right)^2 \right]^{1/2}$$

$$\lambda_2 = \gamma + \frac{\gamma^*}{2} + \left[\gamma^2 + \left(\frac{\gamma^*}{2} \right)^2 \right]^{1/2} \tag{6.141}$$

If we let

$$\gamma^* = 1.17, \quad \gamma = 100\gamma^*, \quad Q_1 = Q_2 = 1.17 \tag{6.142}$$

the exact solution of $\mathbf{Kq} = \mathbf{Q}$ is

$$q_1 = 2, \quad q_2 = 2.01 \tag{6.143}$$

and the condition number

$$\kappa = \frac{\lambda_2}{\lambda_1} = 402 \tag{6.144}$$

Equations (6.143) can be verified by Figure 6.10: Since the structure is statically determinate, we can compute directly the element forces, the element elongations, and the joint deflections.

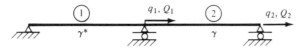

FIGURE 6.10 Assembly of truss elements

Error Source

Roy (1971) identified four types of structures that may cause ill-conditioning. Among them are structures containing adjacent elements with widely varying stiffnesses. This can be illustrated by the structure in Figure 6.10. Specifically, if $\gamma \gg \gamma^*$, γ dominates $\mathbf{K}$ but γ^* dominates $\mathbf{K}^{-1}$ [Eqs. (6.140)]. Thus, the solution $\mathbf{q} = \mathbf{K}^{-1}\mathbf{Q}$ is sensitive to a truncation error in the coefficient $\gamma + \gamma^*$. In the event that γ^* is lost in the addition, that is, $\gamma + \gamma^*$ is represented in the computer as γ, $\mathbf{K}$ becomes singular.

This type of ill-conditioning may arise in unbraced frames (Livesley, 1975) where the extensional stiffness of elements are much larger than the flexural stiffness. This problem may be eliminated, for example, by excluding the extensional stiffnesses in the formulation of the model (Section 4.3).

Truncation Errors

To illustrate the effect of truncation errors, the coefficients of the system model are represented (a) to three digits and (b) to five digits. Then Gaussian elimination is performed by ten-digit floating point arithmetic to exclude roundoff errors from the computed solution.

(a) *Three-Digit Arithmetic*

$$\begin{bmatrix} 118. & -117. \\ -117. & 117. \end{bmatrix} \begin{bmatrix} \bar{q}_1 \\ \bar{q}_2 \end{bmatrix} = \begin{bmatrix} 1.17 \\ 1.17 \end{bmatrix} \tag{6.145}$$

The computed solution is

$$\bar{q}_1 = 2.34, \quad \bar{q}_2 = 2.35 \tag{6.146}$$

(b) *Five-Digit Arithmetic*

$$\begin{bmatrix} 118.17 & -117.00 \\ -117.00 & 117.00 \end{bmatrix} \begin{bmatrix} \bar{q}_1 \\ \bar{q}_2 \end{bmatrix} = \begin{bmatrix} 1.1700 \\ 1.1700 \end{bmatrix} \tag{6.147}$$

The computed solution is

$$\bar{q}_1 = 2.0000, \quad \bar{q}_2 = 2.0100 \tag{6.148}$$

Equations (6.143), (6.146), and (6.148) indicate that the three-digit representation causes a significant truncation error but the five-digit representation is sufficient. The inadequacy of the three-digit representation is reflected by Eq. (6.133),

$$s \geq 3 - \log 402 = 0.4 \tag{6.149}$$

which shows that truncation may cause the loss of all significant digits. Note that the prediction of s is not sensitive to the accuracy of κ. For example, if we double the value of κ in Eq. (6.149), we obtain $s \geq 0.1$.

Roundoff Errors

To illustrate the effect of roundoff errors, Eqs. (6.145) and (6.147) are solved by Gaussian elimination using three-digit and five-digit floating point arithmetic, respectively; that is, after each arithmetic operation, the resulting number is rounded to three digits in the solution of Eq. (6.145) and to five digits in the solution of Eq. (6.147).

(a) *Three-Digit Arithmetic* By standard Gaussian elimination, Eq. (6.145) is reduced to the triangular system

$$\begin{bmatrix} 118. & -117. \\ 0 & 1. \end{bmatrix}\begin{bmatrix} \bar{q}_1 \\ \bar{q}_2 \end{bmatrix} = \begin{bmatrix} 1.17 \\ 2.33 \end{bmatrix} \tag{6.150}$$

which yields

$$\bar{q}_1 = 2.32, \quad \bar{q}_2 = 2.33 \tag{6.151}$$

(b) *Five-Digit Arithmetic* The reduced form of Eq. (6.147) is

$$\begin{bmatrix} 118.17 & -117.00 \\ 0 & 1.1600 \end{bmatrix}\begin{bmatrix} \bar{q}_1 \\ \bar{q}_2 \end{bmatrix} = \begin{bmatrix} 1.1700 \\ 2.3284 \end{bmatrix} \tag{6.152}$$

and the solution is

$$\bar{q}_1 = 1.9972, \quad \bar{q}_2 = 2.0072 \tag{6.153}$$

The differences in Eqs. (6.146) and (6.151) and in Eqs. (6.148) and (6.153) are due to roundoff errors.

If we apply the diagonal decay test to Eqs. (6.150), we obtain $K_{22}^{(1)}/K_{22}$ $= 0.009$, which signals a serious loss of accuracy in the three-digit solution.

Iterative Improvements

Through iterative improvements we can recover the roundoff errors in the three-digit and five-digit solutions, but we cannot recover the truncation error in the three-digit solution.

(a) *Three-Digit Solution* By Eqs. (6.136), (6.145), and (6.151)

$$\mathbf{q}^{(1)} = \begin{bmatrix} 2.32 \\ 2.33 \end{bmatrix}, \quad \Delta\mathbf{Q}^{(1)} = \mathbf{Q} - \mathbf{Kq} = \begin{bmatrix} 0.02 \\ 0.00 \end{bmatrix} \tag{6.154}$$

The solution to

$$\mathbf{K}\,\Delta\mathbf{q}^{(1)} = \Delta\mathbf{Q}^{(1)} \tag{6.155}$$

is

$$\Delta \mathbf{q}^{(1)} = \begin{bmatrix} 0.020 \\ 0.020 \end{bmatrix} \tag{6.156}$$

and the improved solution is

$$\mathbf{q}^{(2)} = \mathbf{q}^{(1)} + \Delta \mathbf{q}^{(1)} = \begin{bmatrix} 2.34 \\ 2.35 \end{bmatrix} \tag{6.157}$$

which agrees with the solution in Eqs. (6.146). To avoid repeating the decomposition of $\mathbf{K}$ in the solution of Eq. (6.155), we have to use the compact form of Gaussian elimination (or the Cholesky method).

(b) *Five-Digit Solution* By Eqs. (6.136)–(6.138), (6.147), and (6.153) we obtain

$$\mathbf{q}^{(1)} = \begin{bmatrix} 1.9972 \\ 2.0072 \end{bmatrix}, \quad \Delta \mathbf{Q}^{(1)} = \begin{bmatrix} 0.00328 \\ 0.00000 \end{bmatrix}, \quad \Delta \mathbf{q}^{(1)} = \begin{bmatrix} 0.00280 \\ 0.00280 \end{bmatrix}$$

and the improved solution

$$\mathbf{q}^{(2)} = \begin{bmatrix} 2.0000 \\ 2.0100 \end{bmatrix} \tag{6.158}$$

which corresponds to the exact solution.

PROBLEMS

6.1 Consider the frame in Figure 6.1a: Number the joints in sequence across the height of the frame.
   a. Determine the member code matrix $\mathbf{M}$, the difference vector $\boldsymbol{\Delta}$, and the half bandwidth m.
   b. Analogous to Figure 6.2, show the fixed band and the variable band, and identify the nonzero entries by $\times$'s.

6.2 Determine the half bandwidths for the two joint-numbering schemes of the unconstrained plane truss. For each system stiffness matrix identify the fixed

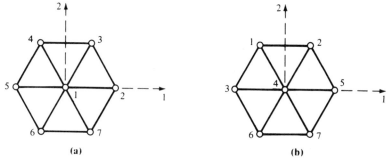

(a)   (b)

PROBLEM 6.2

band and the variable band (skyline), and mark the nonzero entries in the upper triangle by $\times$'s.

6.3   Use standard Gaussian elimination to solve the system of equations $\mathbf{Kq} = \mathbf{Q}$, where $\mathbf{K}$ is defined in Eq. (6.12) and the joint loads are $Q_1 = 1, Q_2 = Q_3 = Q_4 = 0$ (you may wish to exploit symmetry). The solution can be verified by Figure 6.5a: $q_1 = Q_1/\gamma = 1$, and since elements 2–4 undergo rigid-body displacements, $q_4 = q_3 = q_2 = q_1$.

6.4   Use the Gaussian band solver algorithm [Eqs. (6.77)–(6.81)] to solve the system of equations $\mathbf{Kq} = \mathbf{Q}$, where $\mathbf{K}$ is defined in Eq. (6.16), which corresponds to scheme B in Figure 6.6, and $\mathbf{Q}^T = [1 \quad 0 \quad 0 \quad 0 \quad 0]$. The solution for scheme A in Figure 6.6 is given by Eq. (6.116).

6.5   Use the procedure illustrated by Eqs. (6.74)–(6.76) to verify that for a symmetric band matrix $\mathbf{K}$ of order n and half bandwidth m the solution of $\mathbf{Kq} = \mathbf{Q}$ by the *Cholesky method* can be expressed as follows.
1. *Decomposition.* $\mathbf{K}$ is decomposed into the product $\mathbf{K} = \mathbf{U}^T\mathbf{U}$ [Eq. (6.67)] and the elements of $\mathbf{U}$ are computed in column order:

$j = 1$:

$$u_{11} = (K_{11})^{1/2}$$

$j = 2, \ldots, n: \quad t = \max\{1, j - m\}$

$$u_{tj} = \frac{K_{tj}}{u_{tt}}$$

$$u_{ij} = \frac{K_{ij} - \sum_{k=t}^{i-1} u_{ki}u_{kj}}{u_{ii}}, \quad i = t + 1, \ldots, j - 1$$

$$u_{jj} = \left(K_{jj} - \sum_{k=t}^{j-1} u_{kj}^2\right)^{1/2}$$

2. *Forward Substitution*

$$y_i = \frac{Q_i - \sum_{k=i-m}^{i-1} u_{ki}y_k}{u_{ii}}, \quad i = 1, \ldots, n; k \geq 1$$

3. *Back Substitution*

$$q_i = \frac{y_i - \sum_{k=i+1}^{i+m} u_{ik}q_k}{u_{ii}}, \quad i = n, \ldots, 1; k \leq n$$

6.6   Solve the equations of Problem 6.4 by the Cholesky algorithm formulated in Problem 6.5.

6.7   Solve the equations of Problem 6.4 by the variable band solver algorithm of Section 6.5.

6.8   Perform the frontal solution for the structure in Figure 6.5a (Example 1 of Section 6.2) with $Q_2 = 1$, $Q_1 = Q_3 = Q_4 = 0$. Depict frontal solution process graphically analogous to Figure 6.9.

6.9 Apply the frontal solution method to the structure in Figure 4.22a (Example 4 of Section 4.6). Depict active variables analogous to Figure 6.9.

6.10 Repeat the error analysis of the structure in Figure 6.10 (Section 6.7) by using compact Gaussian elimination.

6.11 Work Problem 6.10 with the values of γ and γ^* reversed; that is, let $\gamma = 1.17$ and $\gamma^* = 100\gamma$.

REFERENCES

Agrawal, A. B., A. A. Mufti, and L. G. Jaeger. 1980. "Band-Schemes vs. Frontal-Routines in Nonlinear Structural Analysis," *International Journal for Numerical Methods in Engineering*, 15, 753–766.

Atkinson, K. E. 1978. *An Introduction to Numerical Analysis*. Wiley, New York.

Bathe, K. J., and E. L. Wilson, 1976. *Numerical Methods in Finite Element Analysis*. Prentice-Hall, Englewood Cliffs, NJ.

Becker, E. B., G. F. Carey, and J. T. Oden. 1981. *Finite Elements: An Introduction*, Vol. 1. Prentice-Hall, Englewood Cliffs, NJ.

Brebbia, C. A., and A. J. Ferrante. 1978. *Computational Methods for the Solution of Engineering Problems*. Pentech Press, London

Cook, R. D. 1981. *Concepts and Applications of Finite Element Analysis*, 2nd ed. Wiley, New York.

Dahlquist, G., and A. Björk. 1974. *Numerical Methods* (translated by N. Anderson). Prentice-Hall, Englewood Cliffs, NJ.

Dongarra, J. J., C. B. Moler, J. R. Bunch, and G. W. Stewart. 1979. *LINPACK Users' Guide*. SIAM, Philadelphia

Felippa, C. A. 1975. "Solution of Linear Equations with Skyline-Stored Symmetric Matrix," *Computers and Structures*, 5, 13–29.

Felippa, C. A., and J. L. Tocher, 1970. Discussion of "Efficient Solution to Load-Deflection Equations," by R. J. Melosh and R. M. Bamford, *Journal of the Structural Division, ASCE*, 96, No. ST2, 422–426.

Forsythe, G. E., and C. B. Moler. 1967. *Computer Solution of Linear Algebraic Systems*. Prentice-Hall, Englewood Cliffs, NJ.

Fox, L. 1965. *An Introduction to Numerical Linear Algebra*. Oxford University Press, New York.

George, A., and J. W. Liu. 1981. *Computer Solution of Large Sparse Positive Definite Systems*. Prentice-Hall, Englewood Cliffs. NJ.

Hamming, R. W. 1973. *Numerical Methods for Scientists and Engineers*, 2nd ed. McGraw-Hill, New York.

Hinton, E., and D. R. J. Owen. 1977. *Finite Element Programming*. Academic Press, New York.

Irons, B. M. 1970. "A Frontal Solution Program for Finite Element Analysis," *International Journal for Numerical Methods in Engineering*, 2, 5–32.

Irons, B., and S. Ahmad. 1980. *Techniques of Finite Elements*. Wiley, New York.

Jennings, A. 1966. "A Compact Storage Scheme for the Solution of Symmetric Linear Simultaneous Equations," *Computer Journal*, 9, 281–285.

Jennings, A. 1977. *Matrix Computation for Engineers and Scientists.* Wiley, New York.

Johnson, L. W., and R. D. Riess. 1977. *Numerical Analysis.* Addison-Wesley, Reading, MA.

McGuire, W., and R. H. Gallagher. 1979. *Matrix Structural Analysis.* Wiley, New York.

Melosh, R. J. 1971. "Manipulation Errors in Finite Element Analysis," *Recent Advances in Matrix Methods of Structural Analysis and Design,* R. H. Gallagher, Y. Yamada, and J. T. Oden, eds., pp. 857–877. University of Alabama Press, Huntsville.

Melosh, R. J., and R. M. Bamford. 1969. "Efficient Solution of Load–Deflection Equations," *Journal of the Structural Division, ASCE,* 95, No. ST4, 661–676.

Meyer, C. 1973. "Solution of Linear Equations—State-of-the-Art," *Journal of the Structural Division, ASCE,* 99, No. ST7, 1507–1526.

Meyer, C. 1975. "Special Problems Related to Linear Equation Solvers," *Journal of the Structural Division, ASCE,* 101, No. ST4, 869–890.

Mondkar, D. P., and G. H. Powell. 1974a. "Towards Optimal In-Core Equation Solving," *Computers and Structures,* 4, 531–548.

Mondkar, D. P., and G. H. Powell. 1974b. "Large Capacity Equation Solver for Structural Analysis," *Computers and Structures,* 4, 699–728.

Ralston, A., and P. Rabinowitz. 1978. *A First Course in Numerical Analysis,* 2nd ed. McGraw-Hill, New York.

Rosanoff, R. A., J. F. Gloudeman, and S. Levy. 1968. "Numerical Conditioning of Stiffness Matrix Formulations for Frame Structures," *Proceedings, 2nd Conference on Matrix Methods in Structural Mechanics,* AFFDL-TR-68-150, Wright-Patterson Air Force Base, Ohio, pp. 1029–1060.

Roy, J. R. 1971. "Numerical Error in Structural Solutions," *Journal of the Structural Division, ASCE,* 97, No. ST4, 1039–1054.

Schwarz, H. R., H. Rutishauser, and E. Stiefel. 1973. *Numerical Analysis of Symmetric Matrices* (translated by P. Hertelendy). Prentice-Hall, Englewood Cliffs, NJ.

Tewarson, R. P. 1973. *Sparse Matrices.* Academic Press, New York.

Thomas, G. B., Jr. 1956. *Calculus and Analytic Geometry,* 2nd ed. Addison-Wesley, Reading, MA.

Wiberg, N. E. 1974. "Matrix Structural Analysis with Mixed Variables," *International Journal for Numerical Methods in Engineering,* 8, 167–194.

Wilkinson, J. H. 1963. *Rounding Errors in Algebraic Processes.* Prentice-Hall, Englewood Cliffs, NJ.

Wilkinson, J. H. 1965. *The Algebraic Eigenvalue Problem.* Oxford (Clarendon Press), New York and London.

Wilkinson, J. H., and C. Reinsch. *Handbook for Automatic Computation,* Vol. II: *Linear Algebra.* Springer-Verlag, New York.

Wilson, E. L. 1968. "A Computer Program for the Dynamic Analysis of Underground Structures," Report 68-1, Civil Engineering Department, University of California, Berkley.

Wilson, E. L., K. J. Bathe, and W. P. Doherty. 1974. "Direct Solution of Large Systems of Linear Equations," *Computers and Structures,* 4, 363–372.

Zienkiewicz, O. C. 1977. *The Finite Element Method,* 3d ed. McGraw-Hill, New York.

Additional References

Akhras, G., and G. Dhatt. 1976. "An Automatic Node Relabelling Scheme for Minimizing a Matrix or Network Bandwidth," *International Journal for Numerical Methods in Engineering*, 10, 787–797.

Argyris, J. H., Th. L. Johnsen, R. A. Rosanoff, and J. R. Roy. 1976. "On Numerical Error in the Finite Element Method," *Computer Methods in Applied Mechanics and Engineering*, 7, 261–282.

Bossavit, A., and M. Fremond. 1976. "The Frontal Method Based on Mechanics and Dynamic Programming," *Computer Methods in Applied Mechanics and Engineering*, 8, 153–178.

Collins, R. J. 1973. "Bandwidth Reduction by Automatic Renumbering," *International Journal for Numerical Methods in Engineering*, 6, 345–356.

Eisenstat, S. C., M. C. Gursky, M. H. Schultz, and A. H. Sherman. 1982. "Yale Sparse Matrix Package I: The Symmetric Codes," *International Journal for Numerical Methods in Engineering*, 18, 1145–1151.

Everstine, G. C. 1979. "A Comparison of Three Resequencing Algorithms for the Reduction of Matrix Profile and Wavefront," *International Journal for Numerical Methods in Engineering*, 14, 837–853.

Fuchs, G. von, J. R. Roy, and E. Schrem. 1972. "Hypermatrix Solution of Large Sets of Symmetric Positive-Definite Linear Equations," *Computer Methods in Applied Mechanics and Engineering*, 1, 197–216.

Irons, B. M., and E. Elkamshoshy. 1976. "A Small Frontal Package for Finite Elements," in *Proceedings of the International Symposium on Large Engineering Systems*, A. Wexler, ed., pp. 121–128. Pergamon Press, Elmsford, NY.

Kamel, H. A., and M. W. McCabe. 1978. "Direct Numerical Solution of Large Sets of Simultaneous Equations," *Computers and Structures*, 9, 113–123.

Pao, Y. C. 1978. "Algorithms for Direct-Access Gaussian Solution of Structural Stiffness Matrix Equation," *International Journal for Numerical Methods in Engineering*, 12, 751–764.

Recuero, A., and J. P. Gutierrez. 1979. "A Direct Linear System Solver with Small Core Requirements," *International Journal for Numerical Methods in Engineering*, 14, 633–645.

Rose, D. J., G. G. Whitten, A. H. Sherman, and R. E. Tarjan. 1980. "Algorithms and Software for In-Core Factorization of Sparse Symmetric Positive Definite Matrices," *Computers and Structures*, 11, 597–608.

Taylor, R. L., E. L. Wilson, and S. J. Sackett. 1981. "Direct Solution of Equations by Frontal and Variable Band, Active Column Methods," in *Nonlinear Finite Element Analysis in Structural Mechanics*, W. Wunderlich, E. Stein, and K. J. Bathe, eds., pp. 521–552. Springer-Verlag, New York.

Wilson, E. L., and H. H. Dovey. 1978. "Solution or Reduction of Equilibrium Equations for Large Complex Structural Systems," *Advances in Engineering Software*, 1, 19–25.

PROGRAM DEVELOPMENT

7.1 INTRODUCTION

We are concerned with the development of good computer programs for the analysis of structures by the matrix displacement method. Good programming, like good writing, is based on structure "that allows the reader to execute your reasoning, maintain your perspective, and be convinced of your conclusions" (Linger et al., 1979).

The importance of structure in composition is conveyed by Strunk and White (1979, p. 15):

> A basic structural design underlies every kind of writing. The writer will in part follow this design, in part deviate from it, according to his skill, his needs, and the unexpected events that accompany the act of composition. Writing, to be effective, must follow closely the thoughts of the writer, but not necessarily in the order in which those thoughts occur. This calls for a scheme of procedure.

The recognition of the importance of structure in programming has led to a powerful method of programming, called *structured programming*.

Structured programming is introduced in Section 7.2, and its principles are applied in Section 7.3 in the design of a frame program. The frame program forms the basis of a variety of programming problems, Problems 7.1–7.18.

7.2 STRUCTURED PROGRAMMING

Structured programming (Bates, 1976; Dahl et al., 1972; Denning, 1976; Linger et al., 1979; Wirth, 1974) is a philosophy of programming that focuses on the limited ability of the human mind to process information. For example, to Wirth (1974), structured programming is the expression of a conviction that the programmer's knowledge must not consist of a bag of tricks and trade secrets, but a general intellectual ability to tackle problems systematically, and that particular techniques should be replaced (or augmented) by a method. At its heart lies an attitude rather than a recipe: the admission of the limitations of our minds.

The key principle of structured programming is the importance of structure for the intellectual manageability and reliability of programs. The main concerns are the reduction of complexity and program correctness (Bates, 1976).

Reduction of Complexity

Structured programming proposes *rules* (control structures) to master complexity, in the programming process and the resulting program, by decomposing the problem into manageable *subproblems* (modules). This decomposition is guided by *separation of concerns* of what is to be done from how it is to be done (Dijkstra, 1976; Woodger, 1976). Its dual purpose is "to parcel out the necessary detailed reasoning into portions of manageable size and, more important still, to reduce the total amount of detailed reasoning that remains necessary" (Dijkstra, 1976).

Control Structures

Control structures proposed for the decomposition of problems, the composition of programs, are based on well-known patterns of reasoning. The basic control structures are (Bates, 1976; Böhm and Jacopini, 1966; Dahl et al., 1972):

Sequence structure: sequential decomposition (Figure 7.1)

Alternative structure: If-then-else selection (Figure 7.2)

Loop structure: While-do repetition (Figure 7.3)

The basic control structures are represented by standard flow charts in Figures 7.1a-7.3a and by structured flow charts, called Nassi-Schneiderman diagrams (Nassi and Schneiderman, 1973), in Figures 7.1b-7.3b.

The control structures define the order in which *program blocks* are to be executed. A program block may be empty, it may consist of a single instruction, or it may contain a sequence of instructions (for example, it may represent a subprogram). A program block has a single entry point and a single exit point. The line of flow through a program block is indicated by arrows in standard flow charts, and it is from *top to bottom* in the structured flow charts. Note that the control structures themselves form program blocks; thus, they can be interpreted as single actions in a sequential computation.

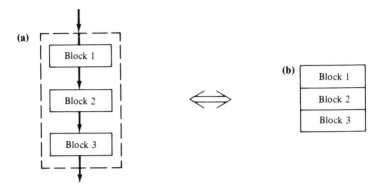

FIGURE 7.1 Sequence structure

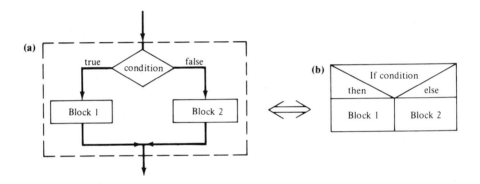

FIGURE 7.2 Alternative structure

FIGURE 7.3 Loop structure

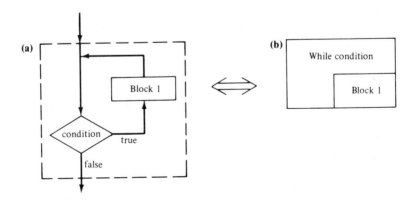

Although the basic control structures reflect recurring thought patterns in problem solving, the programmer need not confine himself to them. For example, the *case structure*, a generalization of the alternative structure (Dahl et al., 1972), provides a choice among several program blocks (Figure 7.4). Denning (1976) asserts that new forms are acceptable if they are simple, understandable, and have a well-defined proof schema: "To hold that only certain forms are admissible, is to hold that structured programming is fixed and inflexible."

Methods of Modularization

In a review of the principle of modularity, Ross et al. (1975) proposed the following general, unifying definition: "Modularity deals with how the *structure* of an object can make the attainment of some *purpose* easier. Modularity is *purposeful structuring*." Our principal goals, the reduction of complexity and program correctness, encompass the fundamental goals of software engineering (Ross et al., 1975): modifiability, efficiency, reliability, and understandability.

FIGURE 7.4 Case structure

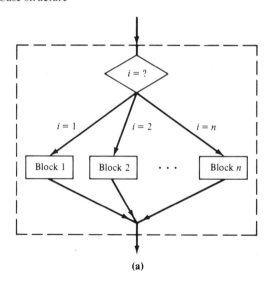

The two most widely discussed methods of modularization in structured programming are the top-down approach and the bottom-up approach. The two terms refer to the directions of *successive decomposition* and *successive composition* in the design process (Bates, 1976).

The *top-down approach*, also called the *method of successive refinements*, is characterized by a sequence of expansion steps (Denning, 1976; Linger et al., 1979). It starts with a precise description of the program function, which is expanded (decomposed or refined) into simpler functions. The expansion of functions is continued until all functions are sufficiently simple that *correct programs* can be written from the beginning. This surprising expectation reflects the new reality of programming expressed by Linger et al. (1979, p. 1): "The new reality is that you can learn to consistently design and write programs that are correct from the beginning and that prove to be error free in their testing and subsequent use."

Denning (1976) has formalized the top-down design process in terms of *marked* and *unmarked functions*. A function is marked if and only if it has been refined or programmed. Consistent with the principle of *separation of concerns*, unmarked functions define what is to be done, whereas marked functions define how it is to be done. The function expansion (modularization) proceeds as follows:

First step: The function of the entire program is defined.

Each step: The first *unmarked function* in the sequence is selected. If it is sufficiently simple, a program is written; otherwise it is expanded into one or more simpler functions, which are added to the sequence.

Result: A sequence of *marked functions* linked by control statements corresponding to the expansion steps is obtained.

The resulting program structure can be represented by a tree chart (Chapin, 1976; Hughes and Michtom, 1977; Tocher, 1976; Yourdon, 1975; Yourdon and Constantine, 1978). For example, the functions in Figure 7.5 are marked in the sequence A, B, C, D, E, F, G, H, I. Specifically, A is expanded into B and C; B is expanded into D and E; C is expanded into F, G, and H; D is programmed; E is expanded into I; F is programmed; G is programmed; H is programmed; and I is programmed. Accordingly, the functions in Figure 7.5 are marked level by level, from left to right.

The *bottom-up approach*, also called the *method of successive compositions*, is the inverse of the top-down approach (Denning, 1976; Bates, 1976). Thus, we can regard the last function decomposition in the top-down design as the first function composition in the bottom-up design. Specifically, a set of base functions is composed into a more complex function. The composition of subsets of previously defined functions into more complex functions is continued until a single function is obtained that implements the desired program.

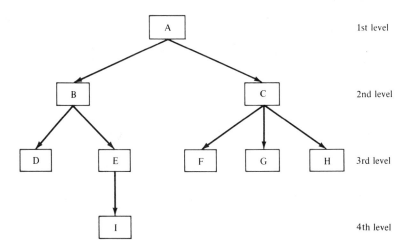

FIGURE 7.5 Program structure

The relative merits of the top-down and bottom-up approaches of modularization are discussed in Bates (1976). It is noted that although the top-down approach seems to be almost universally accepted, there are strong arguments against the exclusive use of either technique. For example, McClure (1975) states

> Neither approach really claims that it can be followed rigidly or exclusively throughout the program composition process. When designing essentially top-down, one will always peek below into lower layers using previous experience and knowledge of the program language attributes to influence design decisions. When following a bottom-up approach, the designer always will keep in mind the general control structure and the program objective ultimately to be achieved. But whatever approach is used, the design is not likely to be successful unless the program composition proceeds in well thought-out steps.

The principal criterion in structured programming is simplicity. Good programming means finding deep simplicities in a complex logical task (Linger et al., 1979). More specific criteria for effective modularization and desirable characteristics of modules, such as independence, are discussed, for example, in Hughes and Michtom (1977), Bates (1976), Kernighan and Plauger (1976), Yourdon (1975), Yourdon and Constantine (1978).

Data Structure

The selection of data types and data structuring forms an important part of program design Bates (1976). In fact, "many programmers claim that the very first aspect of a problem solution is the organization of the data; only after they have settled at least on the rough outlines of the data structure do they turn attention to developing the control structures" (Denning, 1976). Jackson has taken this approach one step further by basing the program structure on the data structure (Jackson, 1976).

Languages for Structured Programming

"Programs must be read and understood by people, as part of the creation process and so that they can be trusted or modified to meet changing needs" (Linger et al., 1979, p. 15). The implications of structured programming for language design are discussed in Bates (1976). It is stressed that the distinction between programming and coding must be reflected in the languages used for each activity. Laski (1976) offers the following definitions.

Programming language: The language the programmer uses to express his thinking about his program.

Coding language: The language in which the program is expressed for machine input.

An example of a standardized programming language is the process design language (PDL) proposed by IBM (Linger et al., 1979; Nanney, 1981). Combinations of graphical and narrative means for expressing program design are used, for example, in IBM's HIPO (hierarchical-input–process–output) charts (Chapin, 1976; Yourdon, 1975; Yourdon and Constantine, 1978), in Chapin charts (Chapin, 1976), and in Nassi–Schneiderman charts (Merchant, 1981; Nassi and Schneiderman, 1973).

FORTRAN was not considered to be well suited for coding structured programs (Bates, 1976). However, the alternative structure (Figure 7.2) has been implented in standard FORTRAN 77 (Merchant, 1981; Nanney, 1981), the loop structure (Figure 7.3) has been added in some FORTRAN compilers (Digital Equipment, 1980), and the FORTRAN version processed by the WATFIV compiler (Moore, 1975; Moore and Makela, 1978) contains the basic control structures and the case structure (Figures 7.1–7.4).

Program Correctness

Following Denning (1976), we define program correctness relative to the function

$$y = f(x) \tag{7.1}$$

where f is the program function, x is the input data set, and y is the output data set.

In the computer program, P, f is represented by some function f_P. The program is correct if

$$f = f_P \tag{7.2}$$

Verification and Testing

There is a spectrum of methods concerned with correctness proving (Bates, 1976; Denning, 1976). The two extremes are *program verification* and *program testing*. Verification seeks to prove correctness by a systematic study of the program's specifications and structure. Testing seeks to prove correctness by applying test sets x to the program and checking that

$$f_P(x) = f(x) \tag{7.3}$$

Note that verification can be done without running the program, whereas testing can be done without knowing the program's structure. However, "only by taking the program's structure into account will it be possible to construct small test sets that, if passed, produce high confidence in the program's correctness" (Denning, 1976). An example is the *minimal test* where a test set is constructed to assure that every statement of the program is executed at least once.

Incremental Testing

To reduce the difficulty of detecting errors and of making corrections, program testing should proceed in an incremental fashion with program development. Yourdon and Constantine (1978) provide an illuminating argument for this approach: Suppose we have constructed a system of N modules that apparently works. We add a new module and test the combination of $N + 1$ modules. If the new combination does not work, the bug may or may not be located in the most recently added module (although frequently it is); what is important is that something about the new module has aggravated the system to the point where a bug has exposed itself. In contrast, if each module is tested separately and then all modules are combined at once, the process of tracking down bugs in that system may be like looking for a needle in a haystack.

Analogous to program modularization, testing may proceed top-down or bottom-up (Denning, 1976; Yourdon and Constantine, 1978).

Top-down testing is frequently integrated with top-down design (Bates, 1976; Linger et al., 1979; McGowan and Kelly, 1975; Yourdon and Constantine, 1978). This requires *dummy programs*, also called *stubs*, to represent unmarked functions in the test programs. A dummy program may simply exist without performing any function, or it may print a message to indicate that it was invoked. In conjunction with top-down design, top-down testing proceeds as follows:

First step: The main program is coded and tested.

Each step: A stub is replaced by a completed program and the new system is tested.

Result: All stubs have been replaced by completed programs and the final program has been tested.

For example, in top-down testing of the program in Figure 7.5, the following sequence of nine test programs is constructed:

1. The main program A is composed of stubs B and C;
2. stub B is replaced by a program containing stubs D and E;
3. stub C is replaced by a program containing stubs F, G, and H;
4. stub D is replaced by a program;
5. stub E is replaced by a program containing stub I;
6. stub F is replaced by a program;
7. stub G is replaced by a program;
8. stub H is replaced by a program; and
9. stub I is replaced by a program.

Note that in top-down testing, the modules are tested in their system environment. That is, the control programs that integrate the modules are written and tested first, and the functional code is added progressively.

Bottom-up testing is characterized by the sequence of activities (Yourdon and Constantine, 1978): module testing, subsystem testing, system testing. For example, in bottom-up testing of the program in Figure 7.5, the modules I, E, D, and B can be tested, combined, and tested in incremental fashion to produce the subprogram B. Similarly, the subprogram C can be constructed, combined with B, and tested to yield the final program A. In most cases, bottom-up testing requires *test drivers* that exercise the test programs.

The relative merits of top-down and bottom-up testing are discussed, for example, in Bates (1976); Denning (1976); Linger et al. (1979), and Yourdon and Constantine (1978).

Program Efficiency

It is perhaps fitting to address program efficiency last. As Knuth (1974) states,

Premature emphasis on efficiency is a big mistake, which may well be the source of most programming complexity and grief. We should ordinarily keep efficiency considerations in the background when we formulate our programs. We need to be subconsciously aware of the data processing tools available to us, but we should strive most of all for a program that is easy to understand and almost sure to work.

Moreover, in view of the increasing ratio of software costs to hardware costs, it is not sensible to measure efficiency merely in terms of the object program's speed of execution and size. What is important is the efficiency of the complete program life cycle, from the initial design through maintenance (Bates, 1976; Yourdon and Constantine, 1978).

7.3 FRAME PROGRAM

A design of a computer program for the matrix displacement analysis of plane frames is presented in this section. It forms the basis for the programming Problems 7.1–7.18.

Program Function

The program is to perform the matrix displacement analysis (Figure 3.9) of plane frames composed of prismatic members, which may have distinct geometric and material properties. It must be able to handle multiple load conditions. Each load condition may consist of joint loads and combinations of the following member actions: A concentrated load at any point of a member (Figure A.9), a uniformly distributed load over a portion of a member (Figure A.10), and a uniform temperature change [Eq. (4.238) with $\Delta t = 0$].

Program Design

The program structure is presented in Figure 7.6. The subprograms are described by Nassi–Schneiderman diagrams (Section 7.2) with lists of input and output arguments.

Although it is desirable to check input data for validity and plausibility (Kernighan and Plauger, 1974), only the limits of the control variables NE and NJ are tested to keep the program development manageable as a class project. The dimensions of all arrays that depend on the size of the structure can be controlled in the main program by the parameter MX. Adjustable arrays (Dongarra et al., 1979; Merchant, 1981; Digital Equipment, 1980)

FIGURE 7.6 Program structure

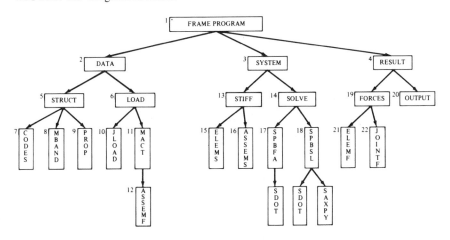

are used in the subprograms. Since FORTRAN stores matrices by columns, all matrices are processed by columns to generate sequential access to memory (Dongarra et al., 1979).

Coding is illustrated in standard FORTRAN 77 and the FORTRAN version of the WATFIV compiler. However, any suitable coding language can be used.

Symbolic Names

The names of program parameters (symbolic constants) and variables are defined. The FORTRAN default-type declaration for variables is used. The words element and member are used interchangeably.

PARAMETERS

MX   Maximum number of elements, joints

MXNEQ   Maximum number of equations (degrees of freedom)

VARIABLES

ACT   Action imposed on member (consistent with conventions in Sections 1.5 and 4.8):

$$ACT = \begin{cases} P & \text{for MAT} = 1 \text{ (Figure A.9)} \\ p & \text{for MAT} = 2 \text{ (Figure A.10)} \\ t_0 & \text{for MAT} = 3 \text{ (Figure 4.27 with } \Delta t = 0) \end{cases}$$

AREA(I)   Cross-sectional area of element I

C1(I),C2(I)   Direction cosines of element I

CTE(I)   Coefficient of thermal expansion of element I

D(6)   Global element displacement vector

DIST   Distance for member actions:

$$DIST = \begin{cases} aL & \text{for MAT} = 1, 2 \text{ (Figures A.9, A.10)} \\ \text{not applicable} & \text{for MAT} = 3 \end{cases}$$

ELENG(I)   Length of element I

EMOD(I)   Modulus of elasticity of element I

F(6,MX)   Local element force matrix:

F(L,I) = Lth local force of element I (Figure 2.4c)

FORCE   Applied joint force

G(7)   Global element stiffness coefficients defined in Eqs. (3.64)

INDEX(6,6)   Index matrix defined in Eq. (7.4)

JCODE(3,MX)   Joint code matrix:

$$JCODE(L,J) = \begin{cases} K & \text{if } U_{lj} = q_k \\ 0 & \text{otherwise} \end{cases}$$

where U_{lj} is the displacement in the l direction of joint j and q_k is the kth joint displacement (Section 2.3).

JDIR   Joint direction

JNUM   Joint number

LC   Load condition

MAT   Member action type:

$$MAT = \begin{cases} 1 & \text{for load in Figure A.9} \\ 2 & \text{for load in Figure A.10} \\ 3 & \text{for uniform temperature change} \\ & \text{(Figure 4.27 with } \Delta t = 0) \end{cases}$$

MBD   Half bandwidth [Eq. (6.2)]

MCODE(6,MX)   Member code matrix:

$$MCODE(L,I) = \begin{cases} K & \text{if } D_l^i = q_k \\ 0 & \text{otherwise} \end{cases}$$

where D_l^i is the lth global displacement of element i (Figure 2.4c) and q_k is the kth joint displacement (Sections 2.3, 2.4).

MINC(2,MX)   Member incidence matrix:

$$MINC(1,I) = \text{joint at the } a \text{ end of element I}$$

$$MINC(2,I) = \text{joint at the } b \text{ end of element I}$$

MN   Member number

NA(I)   Number of actions on member I

NE   Number of elements

NEQ   Number of equations (degrees of freedom)

NJ   Number of joints

NLC   Number of load conditions

P(3,MX)   Joint force matrix:

$$P(L,J) = L\text{th force (applied or reactive) at joint J}$$

Q(MXNEQ)   Joint load, or displacement, vector

$$Q(K) = \begin{cases} K\text{th joint load (Section 2.3) on entry to subroutines} \\ \text{SYSTEM and SOLVE} \\ K\text{th joint displacement (Section 2.3) on return from} \\ \text{subroutines SYSTEM and SOLVE} \end{cases}$$

SS(MXNEQ,MXNEQ)   System stiffness band matrix stored in the form of Figure 6.3c in the region SS(MBD + 1,NEQ)

X(1,J),X(2,J)   Global 1, 2 coordinates of joint J stored temporarily in the matrix P(3,MX)

ZI(I)   Moment of inertia about the local z (3) axis of element I

Input Data

The required data can be entered as follows.

```
C     LIST-DIRECTED INPUT
C     INPUT UNITS: KIP, INCH, RADIAN, FAHRENHEIT 1
C
C     1. ENTER DATE IN THE FORM '02/24/84' (IN MAIN)
C          DATE
C
C     2. ENTER NUMBER OF ELEMENTS, NUMBER OF JOINTS,
C        AND NUMBER OF LOAD CONDITIONS (IN MAIN)
C          NE, NJ, NLC
C
C     3. ENTER MEMBER INCIDENCES (IN STRUCT)
C          MINC(1,I), MINC(2,I)     I = 1 to NE
C
C     4. ENTER FOR EACH JOINT CONSTRAINT(IN STRUCT)
C          JNUM, JDIR
C          AFTER LAST JOINT CONSTRAINT ENTER
C          0, 0
C
C     5. ENTER JOINT COORDINATES(IN PROP)
C          X(1,J), X(2,J)     J = 1 to NJ
C
C     6. ENTER MEMBER PROPERTIES(IN PROP)2
C          AREA(I), ZI(I), EMOD(I), CTE(I)     I = 1 to NE
C
C     7. DO FOR EACH LOAD CONDITION
C          IF THERE ARE JOINT LOADS ENTER(IN JLOAD)
C          JNUM, JDIR, FORCE
C          AFTER LAST JOINT LOAD ENTER
C          0, 0, 0
C          ELSE ENTER
C          0, 0, 0
C          END IF
C
C          IF THERE ARE MEMBER ACTIONS ENTER(IN MACT)3
C          MN, MAT, ACT, DIST
C          AFTER LAST MEMBER ACTION ENTER
C          0, 0, 0, 0
C          ELSE ENTER
C          0, 0, 0, 0
C          END IF
C        END DO
```

Main Program

The main program consists of the subroutines DATA, SYSTEM, and RESULT (Figure 7.6). DATA reads and echos input data and generates

---

[1] See Problem 7.2.
[2] See Problem 7.18.
[3] See Problem 7.17.

the data required in the matrix displacement analysis. SYSTEM constructs the system stiffness matrix and solves the system equations for the joint displacements. RESULT computes element and joint forces and produces output.

Function: Initialize parameters MX, MXNEQ; read and echo NE, NJ, NLC; if NE and NJ are less than or equal to MX, call for each load condition DATA, SYSTEM, and RESULT; else print error message and stop.

NS diagram:

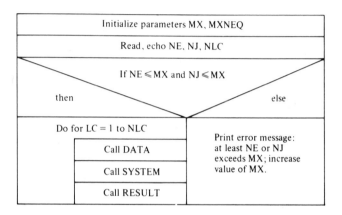

FORTRAN 77 code:

```
          CHARACTER * (*) TITLE, UNITS, DATE*8
          PARAMETER (MX = 30, MXNEQ = 3 * (MX − 1),
     $       TITLE = 'PLANE FRAME ANALYSIS',
     $       UNITS ='UNITS: KIP, INCH, RADIAN, FAHRENHEIT')
          DIMENSION F(6,MX),P(3,MX),SS(MXNEQ,MXNEQ),Q(MXNEQ),
     $       AREA(MX),ZI(MX),EMOD(MX),CTE(MX),ELENG(MX),C1(MX),
     $       C2(MX),MCODE(6,MX),JCODE(3,MX),MINC(2,MX),NA(MX)
C
C         INITIALIZE PARAMETERS MX, MXNEQ; READ AND ECHO NE, NJ,
C         NLC; IF NE AND NJ ARE LESS THAN OR EQUAL TO MX, CALL FOR
C         EACH LOAD CONDITION DATA, SYSTEM, AND RESULT; ELSE
C         PRINT ERROR MESSAGE AND STOP.
C
          READ *, DATE
          PRINT 10, TITLE, 'DATE: ', DATE, UNITS
       10 FORMAT('1', T10, 68('*')/ T10, '*', T34, A, T77, '*'/
     $             T10, 68('*')// T64, 2(A)// T10, A/ )
          READ *, NE, NJ, NLC
          PRINT 20, 'NUMBER OF ELEMENTS', NE, 'NUMBER OF JOINTS', NJ,
     $             'NUMBER OF LOAD CONDITIONS', NLC
```

(*Continued* over page)

```
   20 FORMAT(3(T10, A, T35, I4/ ))
C
      IF (NE .LE. MX .AND. NJ .LE. MX) THEN
         DO 30 LC = 1, NLC
            CALL DATA(F,P,Q,AREA,ZI,EMOD,CTE,ELENG,C1,C2,
     $                MCODE,JCODE,MINC,NA,NE,NJ,NEQ,MBD,LC)
            CALL SYSTEMS(SS,Q,AREA,ZI,EMOD,ELENG,C1,C2,
     $                MCODE,NE,NEQ,MBD,LC,MXNEQ)
            CALL RESULT(F,P,Q,AREA,ZI,EMOD,ELENG,C1,C2,
     $                MCODE,JCODE,MINC,NE,NJ)
   30    CONTINUE
      ELSE
         PRINT 40, 'ERROR MESSAGE: AT LEAST NE OR NJ EXCEEDS MX:'.
     $                             'INCREASE VALUE OF MX.'
   40    FORMAT(T10, A/ T25, A/ )
      END IF
C
      STOP
      END
```

Subprogram Data

DATA is expanded into the subroutines STRUCT and LOAD (Figure 7.6). STRUCT reads, echos, and processes structural data: It calls the subroutine CODES to generate the joint and member codes, it invokes the function MBAND to compute the half bandwidth, and it calls the subroutine PROP to compute the element lengths and direction cosines. LOAD reads, echos, and processes load data: It calls the subroutine JLOAD to store joint loads, and it calls the subroutine MACT to compute fixed-end forces induced by member actions and to assemble them into the equivalent joint load vector.

Subroutine DATA

Function: For the first load condition $LC = 1$, call STRUCT and LOAD; for subsequent load conditions, $LC > 1$, call LOAD.

Input arguments: LC, NE, NJ

Output arguments: F, X, Q, AREA, ZI, EMOD, CTE, ELENG, C1, C2, MCODE, JCODE, MINC, NA, NEQ, MBD

NS diagram:

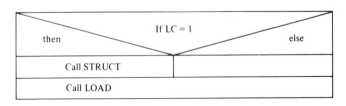

Subroutine STRUCT

Function: Read and echo the member incidences, MINC(L, I); initialize the elements of the joint code matrix, JCODE, to unity, read and echo for each joint constraint the joint number, JNUM, and joint direction, JDIR, and store a zero in the corresponding location of JCODE (end of data marker JNUM = 0); call CODES, MBAND, and PROP.

Input arguments: NE, NJ

Output arguments: X, AREA, ZI, EMOD, CTE, ELENG, C1, C2, MCODE, JCODE, MINC, NEQ, MBD

NS diagram:

```
┌────────────────────────────────────────────────────────────┐
│ Read, echo                                                   │
│ MINC(1, I), MINC(2, I)  I = 1 to NE                          │
├────────────────────────────────────────────────────────────┤
│ Initialize                                                   │
│ JCODE(L, J) = 1   J = 1 to NJ, L = 1 to 3                    │
├────────────────────────────────────────────────────────────┤
│ Read joint constraint                                        │
│ JNUM, JDIR                                                   │
├────────────────────────────────────────────────────────────┤
│ While JNUM ≠ 0                                               │
│        ┌─────────────────────────────────────────────────┐  │
│        │ Print JNUM, JDIR                                 │  │
│        ├─────────────────────────────────────────────────┤  │
│        │ JCODE(JDIR, JNUM) = 0                            │  │
│        ├─────────────────────────────────────────────────┤  │
│        │ Read joint constraint                            │  │
│        │ JNUM, JDIR                                       │  │
│        └─────────────────────────────────────────────────┘  │
├────────────────────────────────────────────────────────────┤
│ Call CODES                                                   │
├────────────────────────────────────────────────────────────┤
│ MBD = MBAND(MCODE, NE)                                       │
├────────────────────────────────────────────────────────────┤
│ Call PROP                                                    │
└────────────────────────────────────────────────────────────┘
```

Subroutine CODES

Function: Generate the joint code, JCODE, by assigning integers in sequence, by columns, to all nonzero elements of JCODE from 1 to NEQ; generate the member code, MCODE, by transferring via MINC columns of JCODE into columns of MCODE.

Input arguments: JCODE, MINC, NE, NJ

Output arguments: MCODE, JCODE, NEQ

NS diagram:

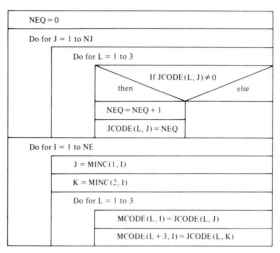

Function MBAND

Function: Compute the half bandwidth, MBAND, by Eq. (6.2): In each column of MCODE, the first and last nonzero integers are the smallest and largest nonzero integers, respectively, of that column. MBAND is the maximum difference of the nonzero integers in any column of MCODE.

Input arguments: MCODE, NE

Output arguments: MBAND

NS diagram:

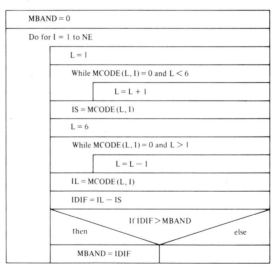

FORTRAN–WATFIV code:

```
      FUNCTION MBAND(MCODE,NE)
      DIMENSION MCODE(6,1)
C
C     COMPUTE THE HALF BANDWIDTH, MBAND, BY EQ. 6.2: IN EACH
C     COLUMN OF MCODE, THE FIRST AND LAST NONZERO INTEGERS ARE
C     THE SMALLEST AND LARGEST NONZERO INTEGERS, RESPECTIVELY.
C     OF THAT COLUMN. MBAND IS THE MAXIMUM DIFFERENCE OF THE
C     NONZERO INTEGERS IN ANY COLUMN OF MCODE.
C
      MBAND = 0
      DO 10 I = 1, NE
        L = 1
        WHILE (MCODE(L,I) .EQ. 0 .AND. L .LT. 6) DO
           L = L + 1
        END WHILE
        IS = MCODE(L,I)
        L = 6
        WHILE (MCODE(L,I) .EQ. 0 .AND. L .GT. 1) DO
           L = L - 1
        END WHILE
        IL = MCODE(L,I)
        IDIF = IL - IS
        IF (IDIF .GT. MBAND) THEN
          MBAND = IDIF
        END IF
   10 CONTINUE
      RETURN
      END
```

COMMENT: In standard FORTRAN 77, the While–do loop can be written as follows:

```
    5 IF (MCODE(L, I) .EQ. 0 .AND. L .LT. 6) THEN
        L = L + 1
        GO TO 5
      END IF
```

Subroutine PROP

Function: Read and echo the joint coordinates, $X(L,J)$; compute for each element by Eqs. (C.21) the length, $ELENG(I)$, and the direction cosines, $C1(I)$, $C2(I)$; read for each element the cross-sectional area, $AREA(I)$, the moment of inertia about the local z (3) axis, $ZI(I)$, the modulus of elasticity, $EMOD(I)$, and the coefficient of thermal expansion, $CTE(I)$: print element properties.

Input arguments: MINC, NE, NJ

Output arguments: X, AREA, ZI, EMOD, CTE, ELENG, C1, C2

NS diagram:

Read, echo joint coordinates $X(1, J), X(2, J)$ $J = 1$ to NJ		
Do·for $I = 1$ to NE		
	$J = MINC(1, I)$	
	$K = MINC(2, I)$	
	$EL1 = X(1, K) - X(1, J)$	
	$EL2 = X(2, K) - X(2, J)$	
	$ELENG(I) = SQRT(EL1**2 + EL2**2)$	
	$C1(I) = EL1/ELENG(I)$ $C2(I) = EL2/ELENG(I)$	
	Read, print element properties	

Subroutine LOAD

Function: Initialize to zero the joint load vector, Q, the local element (member) force vector, F, and the number of actions vector, NA; call JLOAD and MACT.

Input arguments: AREA, EMOD, CTE, ELENG, C1, C2, MCODE, JCODE, NE, NEQ

Output arguments: F, Q, NA

NS diagram:

Initialize $Q(K) = 0.$ $K = 1$ to NEQ $F(L, I) = 0.$ $NA(I) = 0$ $I = 1$ to NE, $L = 1$ to 6
Call JLOAD
Call MACT

Subroutine JLOAD

Function: Read the joint number, JNUM, the joint direction, JDIR, and the applied force, FORCE; while JNUM $\neq$ 0, print JNUM, JDIR, FORCE, store FORCE in Q, and read JNUM, JDIR, FORCE.

Input arguments: Q, JCODE

Output arguments: Q

NS diagram:

```
┌────────────────────────────────────────────────────────────────────┐
│  Read JNUM, JDIR, FORCE                                             │
├────────────────────────────────────────────────────────────────────┤
│                            If JNUM ≠ 0                               │
│  then                                            else               │
├──────────────────────────────────────┬─────────────────────────────┤
│  Print caption for joint loads        │  Print: no joint loads      │
├──────────────────────────────────────┤                             │
│  While JNUM ≠ 0                       │                             │
│  ┌──────────────────────────────────┐│                             │
│  │ Print JNUM, JDIR, FORCE          ││                             │
│  ├──────────────────────────────────┤│                             │
│  │ K = JCODE(JDIR, JNUM)            ││                             │
│  ├──────────────────────────────────┤│                             │
│  │ Q(K) = FORCE                     ││                             │
│  ├──────────────────────────────────┤│                             │
│  │ Read JNUM, JDIR, FORCE           ││                             │
│  └──────────────────────────────────┘│                             │
└──────────────────────────────────────┴─────────────────────────────┘
```

Subroutine MACT

Function: Read the member number, MN, the member action type, MAT, the action, ACT, and the distance, DIST; while MN ≠ 0, print MN, MAT, ACT, DIST, increment the action counter, compute and accumulate the fixed-end forces, and read MN, MAT, ACT, DIST; call ASSEMF.

Input arguments: F, Q, AREA, EMOD, CTE, ELENG, C1, C2, MCODE, NA, NE

Output arguments: F, Q, NA

NS diagram:

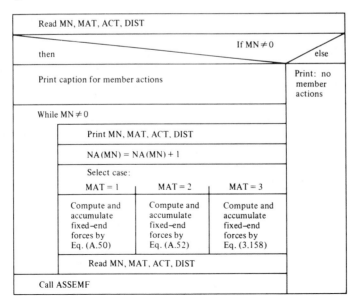

Subroutine ASSEMF

Function: Transform and assemble the local fixed-end forces, F(L,I), to produce the equivalent joint load vector, Q, by Eqs. (2.34), (2.37), (3.92)–(3.94), and the force transformation of Section 2.4.

Input arguments: F, Q, C1, C2, MCODE, NA, NE

Output arguments: Q

NS diagram:

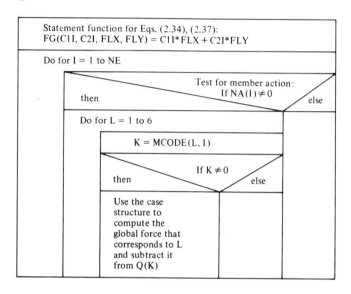

FORTRAN–WATFIV code:

```
      SUBROUTINE ASSEMF(F,Q,C1,C2,MCODE,NA,NE)
      DIMENSION F(6,1),Q(1),C1(1),C2(1),MCODE(6,1),NA(1)
C
C     TRANSFORM AND ASSEMBLE THE LOCAL FIXED-END FORCES, F(L.1),
C     TO PRODUCE THE EQUIVALENT JOINT LOAD VECTOR, Q, BY EQS. 2.34,
C     2.37, 3.92–3.94 AND THE FORCE TRANSFORMATION OF SECTION 2.4.
C
C     STATEMENT FUNCTION
      FG(C1I,C2I,FLX,FLY) = C1I * FLX + C2I * FLY
C
      DO 20 I = 1, NE
        IF (NA(I) .NE. 0) THEN
          DO 10 L = 1, 6
            K = MCODE(L,I)
            IF (K .NE. 0) THEN
              DO CASE L
              CASE
                Q(K) = Q(K) − FG(C1(I), − C2(I),F(1,I),F(2,I))
              CASE
                Q(K) = Q(K) − FG(C2(I), C1(I),F(1,I),F(2,I))
```

```
                CASE
                  Q(K) = Q(K) − F(3,I)
                CASE
                  Q(K) = Q(K) − FG(C1(I), − C2(I),F(4,I),F(5,I))
                CASE
                  Q(K) = Q(K) − FG(C2(I), C1(I),F(4,I),F(5,I))
                CASE
                  Q(K) = Q(K) − F(6,I)
              END CASE
              END IF
    10    CONTINUE
              END IF
    20 CONTINUE
          RETURN
          END
```

FORTRAN 77 code:

```
          SUBROUTINE ASSEMF(F,Q,C1,C2,MCODE,NA,NE)
          DIMENSION F(6,*),Q(*),C1(*),C2(*),MCODE(6,*),NA(*)
C
C     TRANSFORM AND ASSEMBLE THE LOCAL FIXED-END FORCES,
C     F(L,I), TO PRODUCE THE EQUIVALENT JOINT LOAD VECTOR, Q, BY
C     EQS. 2.34, 2.37, 3.92–3.94 AND THE FORCE TRANSFORMATION OF
C     SECTION 2.4.
C
C     STATEMENT FUNCTION
      FG(C1I,C2I,FLX,FLY) = C1I * FLX + C2I * FLY
C
          DO 20 I = 1, NE
            IF (NA(I) .NE. 0) THEN
              DO 10 L = 1, 6
                K = MCODE(L,I)
                IF (K .NE. 0) THEN
                  IF (L .EQ. 1) THEN
                    Q(K) = Q(K) − FG(C1(I), − C2(I),F(1,I),F(2,I))
                  ELSE IF (L .EQ. 2) THEN
                    Q(K) = Q(K) − FG(C2(I), C1(I),F(1,I),F(2,I))
                  ELSE IF (L .EQ. 3) THEN
                    Q(K) = Q(K) − F(3,I)
                  ELSE IF (L .EQ. 4) THEN
                    Q(K) = Q(K) − FG(C1(I), − C2(I),F(4,I),F(5,I))
                  ELSE IF (L .EQ. 5) THEN
                    Q(K) = Q(K) − FG(C2(I), C1(I),F(4,I),F(5,I))
                  ELSE
                    Q(K) = Q(K) − F(6,I)
                  END IF
                END IF
    10        CONTINUE
            END IF
    20 CONTINUE
          RETURN
          END
```

Subprogram System

System is expanded into the subroutines STIFF and SOLVE (Figure 7.6).
STIFF generates the symmetric, positive definite, system stiffness matrix,
Eq. (3.24), and stores it in the band form shown in Figure 6.3c. SOLVE
solves the system equations, Eq. (6.17), for the joint displacements.

For each element, STIFF calls the subroutines ELEMS and ASSEMS.
ELEMS computes the distinct global element stiffness coefficients of Eqs.
(3.64). Their locations in the global element stiffness matrix, Eq. (3.63), are
defined by the INDEX matrix, which stores the negative signs and subscripts
of these functions. Specifically,

$$
\text{INDEX} =
\begin{bmatrix}
1 & 2 & 4 & -1 & -2 & 4 \\
  & 3 & 5 & -2 & -3 & 5 \\
  &   & 6 & -4 & -5 & 7 \\
  &   &   & 1 & 2 & -4 \\
  &   &   &   & 3 & -5 \\
\text{sym.} &   &   &   &   & 6
\end{bmatrix}
\tag{7.4}
$$

ASSEMS assigns the contributions of the global element stiffness matrix
to the system stiffness band matrix by the MCODE and Eq. (6.7). For
example, the location of a global stiffness coefficient with subscripts (IE,JE)
of element N in the system stiffness band matrix can be determined as
shown in Figure 7.7: First the MCODE is used to compute the location
(I,J) in the system stiffness matrix: $J = \text{MCODE(JE,N)}, I = \text{MCODE(IE,N)}$;
then the appropriate row K of the column-stored system stiffness band matrix
is computed by Eq. (6.7): $K = I - J + MBD + 1$.

LINPACK subroutines (Dongarra et al., 1979) are used to solve the
system equations by the Cholesky method (Sections 6.3 and 6.4, and Problem
6.5). Specifically, SOLVE calls SPBFA to factorize the system stiffness band
matrix and SPBSL to compute the joint displacements for each load condition.

LINPACK also provides subroutines to estimate the condition number of
the system stiffness matrix (Section 6.7) and to calculate determinants
(Section 6.3). Other computer programs for the solution of symmetric,
positive definite band equations are cited in Section 6.4.

Subroutine SYSTEM

   Function: For the first load condition, LC = 1, call STIFF and SOLVE;
      for subsequent load conditions, LC > 1, call SOLVE.

   Input arguments: Q, AREA, ZI, EMOD, ELENG, C1, C2, MCODE,
      NE, NEQ, MBD, LC, MXNEQ
      Output arguments: SS, Q

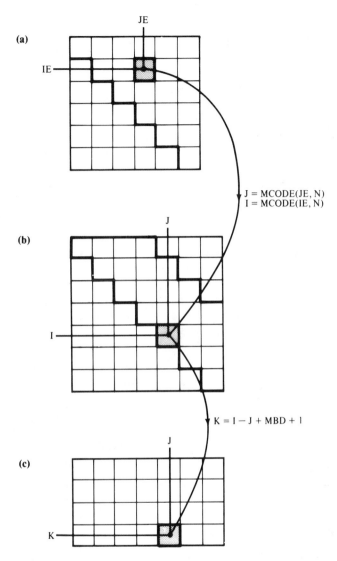

FIGURE 7.7 Band storage. **(a)** Stiffness matrix of element N; **(b)** system stiffness matrix; **(c)** system stiffness band matrix.

NS diagram:

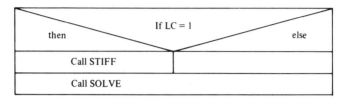

NOTE: The input argument Q stores the joint loads; the output argument Q stores the joint displacements.

Subroutine STIFF

> Function: Initialize the system stiffness matrix, SS, to zero; for each element call ELEMS and ASSEMS.

> Input arguments: AREA, ZI, EMOD, ELENG, C1, C2, MCODE, NE, NEQ, MBD, MXNEQ

> Output arguments: SS

> NS diagram:

Initialize $SS(I, J) = 0$. $J = 1$ to NEQ, $I = 1$ to MBD $+ 1$	
Do for N = 1 to NE	
	Call ELEMS
	Call ASSEMS

Subroutine ELEMS

> Function: For element N, compute the global stiffness coefficients, $G(7)$, defined in Eqs. (3.64).

> Input arguments: AREA, ZI, EMOD, ELENG, C1, C2, N

> Output arguments: G

> NS diagram:

Compute by Eqs. (3.64) $G(1)–G(7)$

Subroutine ASSEMS

Function: Initialize INDEX by Eq. (7.4); assign stiffness coefficients, G(L), of element N to the system stiffness band matrix, SS, by INDEX, MCODE, and Eq. (6.7).

Input arguments: SS, G, MCODE, MBD, MXNEQ, N

Output arguments: SS

NS diagram:

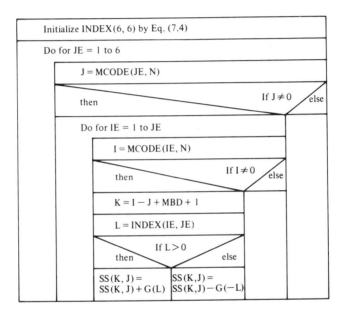

NOTE: IE, JE are the subscripts of a coefficient in the global element stiffness matrix, Eq. (3.63), and the corresponding INDEX matrix, Eq. (7.4); the subscripts K, J define the location of this coefficient in the system stiffness band matrix, SS(MBD + 1, NEQ).

FORTRAN–WATFIV code:

```
      SUBROUTINE ASSEMS(SS,G,MCODE,MBD,MXNEQ,N)
      DIMENSION SS(MXNEQ,1),G(7),MCODE(6,1)
      INTEGER INDEX(6,6)/1,2,4,-1,-2,4, 2,3,5,-2,-3,5, 4,5,6,-4,-5,7,
     $                   -1,-2,-4,1,2,-4, -2,-3,-5,2,3,-5, 4,5,7,-4,-5,6/
C
C     INITIALIZE INDEX BY EQ. 7.4; ASSIGN STIFFNESS COEFFICIENTS,
C     G(L), OF ELEMENT N TO THE SYSTEM STIFFNESS BAND MATRIX, SS,
C     BY INDEX, MCODE, AND EQ. 6.7.
```

(*Continued* over page)

```
C
      DO 20 JE = 1, 6
        J = MCODE(JE,N)
        IF (J .NE. 0) THEN
          DO 10 IE = 1, JE
            I = MCODE(IE,N)
            IF (I .NE. 0) THEN
              K = I − J + MBD + 1
              L = INDEX(IE,JE)
              IF (L .GT. 0) THEN
                SS(K,J) = SS(K,J) + G(L)
              ELSE
                SS(K,J) = SS(K,J) − G(−L)
              END IF
            END IF
10        CONTINUE
        END IF
20    CONTINUE
      RETURN
      END
```

Subroutine SOLVE

Function: For the first load condition, LC = 1, call SPBFA and SPBSL; for subsequent load conditions, LC > 1, call SPBSL.

Input arguments: SS, Q, NEQ, MBD, LC, MXNEQ

Output arguments: Q

NS diagram:

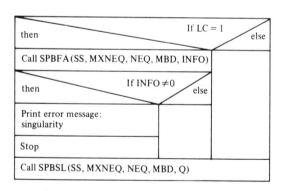

NOTE

$$INFO = \begin{cases} 0 & \text{for normal return} \\ K & \text{if the leading minor of order } k \text{ is not positive definite.} \end{cases}$$

The error condition INFO ≠ 0 is usually caused by improper subroutine arguments (Dongarra et al., 1979).

Subprogram Result

RESULT is expanded into the subroutines FORCES and OUTPUT (Figure 7.6). FORCES computes the local element forces and the joint forces. OUTPUT prints the joint displacements, the local element forces, and the joint forces.

Subroutine RESULT

   Function: Initialize the joint force matrix, P, to zero; call FORCES and OUTPUT.

   Input arguments: F, Q, AREA, ZI, EMOD, ELENG, C1, C2, MCODE, JCODE, MINC, NE, NJ

   Output arguments: P

   NS diagram:

Initialize P(L, J)   J = 1 to NJ, L = 1 to 3
Call FORCES
Call OUTPUT

Subroutine FORCES

   Function: For each element call ELEMF and JOINTF.

   Input arguments: F, P, Q, AREA, ZI, EMOD, ELENG, C1, C2, MCODE, MINC, NE

   Output arguments: F, P

   NS diagram:

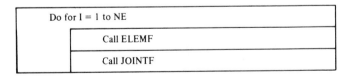

Subroutine ELEMF

   Function: Compute the local forces of element I, F(6,I): determine the global element displacements, D(6), from the joint displacement vector, Q, via MCODE; compute the local forces at the *a* end of the element by Eqs. (2.34), (2.36), (3.59); use equilibrium to compute the local forces at the b end of the element and Eq. (3.95) to compute the actual element forces.

Input arguments: F, Q, AREA, ZI, EMOD, ELENG, C1, C2, MCODE, I

Output arguments: F

NS diagram:

Use Q and MCODE to determine the global element displacements D(1)–D(6)
Use Eqs. (2.34), (2.36), (3.59) to compute the local forces at the a end of the element: F1, F2, F3
Use equilibrium and Eq. (3.95) to compute the actual element forces: F(1, I) = F(1, I) + F1 F(2, I) = F(2, I) + F2 F(3, I) = F(3, I) + F3 F(4, I) = F(4, I) − F1 F(5, I) = F(5, I) − F2 F(6, I) = F(6, I) + F2*ELENG(I) − F3

Subroutine JOINTF

Function: Transform the local forces of element I, F(6,I), to global forces and assign them to the joint force matrix, P, by Eqs. (2.29), (2.34), (2.37), and MINC.

Input arguments: F, P, C1, C2, MINC, I

Output arguments: P

NS diagram:

Statement function for Eqs. 2.34, 2.37: FG(C1I, C2I, FLX, FLY) = C1I*FLX + C2I*FLY
J = MINC(1, I)
K = MINC(2, I)
P(1, J) = P(1, J) + FG(C1(I), −C2(I), F(1, I), F(2, I)) P(2, J) = P(2, J) + FG(C2(I), C1(I), F(1, I), F(2, I)) P(3, J) = P(3, J) + F(3, I)
P(1, K) = P(1, K) + FG(C1 (I), −C2 (I), F(4, I), F(5, I)) P(2, K) = P(2, K) + FG(C2 (I), C1(I), F(4, I), F(5, I)) P(3, K) = P(3, K) + F(6, I)

Subroutine OUTPUT

Function: Use the joint displacement vector, Q, and JCODE to print the joint displacements (including joint constraints); print the local element forces, F(6,NE); print the joint forces, P(3,NJ).

Input arguments: F, P, Q, JCODE, NE, NJ

Output arguments: None

NS diagram:

Print joint displacements by Q and JCODE
Print local element forces F(L, I) I = 1 to NE, L = 1 to 6
Print joint forces P(L, J) J = 1 to NJ, L = 1 to 3

Test Problem

The symmetric frame in Figure 7.8 can serve as one of the test problems. The element properties are identical to those of Example 1 in Section 3.4. The two load conditions are symmetric (Figure 7.8b) and antisymmetric (Figure 7.8c) with respect to the line of symmetry of the frame. Thus, the substructures in Figures 7.8d and e are alternative test problems for load conditions 1 and 2, respectively (Section 4.2).

Result

The matrix displacement analysis (Figure 3.9) was performed with a FORTRAN 77 program in double precision. The output units are kilopound, inch, and radian.

Load Condition 1

	Joint Displacements		
Joint	Direction 1	Direction 2	Direction 3
1	0.0010	−0.0288	−0.0028
2	−0.0010	−0.0288	0.0028
3	−0.0007	−0.0192	−0.0014
4	0.0007	−0.0192	0.0014
5	0.0000	0.0000	0.0000
6	0.0000	0.0000	0.0000

358

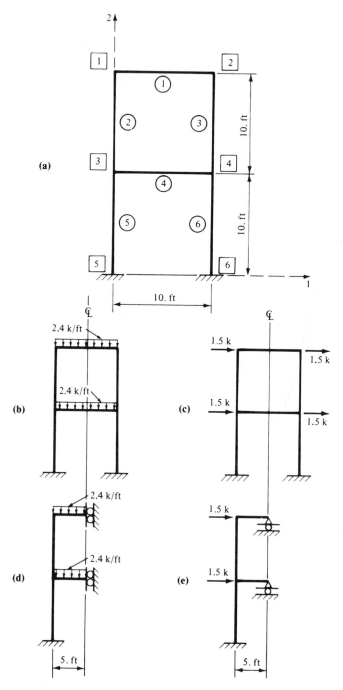

FIGURE 7.8 Test problem. **(a)** Symmetric frame; **(b)** load condition 1; **(c)** load condition 2; **(d)** subframe for load condition 1; **(e)** subframe for load condition 2.

| | Local Element Forces | | | | | |
Element	f_1	f_2	f_3	f_4	f_5	f_6
1	2.564	12.000	171.094	−2.564	12.000	−171.094
2	12.000	−2.564	−171.094	−12.000	2.564	−136.535
3	12.000	2.564	171.094	−12.000	−2.564	136.535
4	−1.698	12.000	205.653	1.698	12.000	−205.653
5	24.000	−0.866	−69.118	−24.000	0.866	−34.771
6	24.000	0.866	69.118	−24.000	−0.866	34.771

| | Joint Forces | | |
Joint	Direction 1	Direction 2	Direction 3
1	0.000	0.000	0.000
2	0.000	0.000	0.000
3	0.000	0.000	0.000
4	0.000	0.000	0.000
5	0.866	24.000	−34.771
6	−0.866	24.000	34.771

Load Condition 2

| | Joint Displacements | | |
Joint	Direction 1	Direction 2	Direction 3
1	0.8773	0.0057	−0.0015
2	0.8773	−0.0057	−0.0015
3	0.4647	0.0043	−0.0029
4	0.4647	−0.0043	−0.0029
5	0.0000	0.0000	0.0000
6	0.0000	0.0000	0.0000

| | Local Element Forces | | | | | |
Element	f_1	f_2	f_3	f_4	f_5	f_6
1	0.000	−1.795	−107.674	0.000	1.795	−107.674
2	−1.795	1.500	107.674	1.795	−1.500	72.326
3	1.795	1.500	107.674	−1.795	−1.500	72.326
4	0.000	−3.592	−215.510	0.000	3.592	−215.510
5	−5.386	3.000	143.184	5.386	−3.000	216.816
6	5.386	3.000	143.184	−5.386	−3.000	216.816

	Joint Forces		
Joint	Direction 1	Direction 2	Direction 3
1	1.500	0.000	0.000
2	1.500	0.000	0.000
3	1.500	0.000	0.000
4	1.500	0.000	0.000
5	−3.000	−5.386	216.816
6	−3.000	5.386	216.816

PROBLEMS

7.1 Complete the development of the frame program of Section 7.3 (coding and testing). The programming requirements are as follows.

1. *Coding Language*: Use a coding language that contains at least the alternative structure (Figure 7.2) of the basic control structures.

2. *Documentation*: Integrate documentation with program development (Kernighan and Plauger, 1974). Specify input requirements (for example, the list-directed input in Section 7.3), and define symbolic names and the functions of the main program and subprograms.

3. *Testing*: Use incremental testing (Section 7.2) in the modified top-down sequence indicated by the following program numbers (Figure 7.6); a starred number signifies a program stub. Submit the following program sequences at dates to be specified:

   a. 1, 2, 3*, 4*, 5, 6, 7, 8*, 9, 10, 11*;
   b. replace 3* by 3, 13, 14, 15, 16, 17, 18, and 8* by 8;
   c. replace 4* by 4, 19, 20, 21, 22;
   d. replace 11* by 11, 12.

   Test problems will be specified.

4. *Output*: Print captions, output units, echo input, print error messages, and for each load condition print joint displacements, local element forces, and joint forces. During testing print values of variables generated in each new subprogram.

7.2 Modify the frame program of Problem 7.1 to allow the user to select input and output units: For example, to enter joint coordinates in feet or to use SI units.

7.3 Develop an alternative design for the frame program in Figure 7.6. Consider, for example, alternative design strategies in Chapter 12 of Yourdon (1975): Present a tree chart of your design and describe the modified and new subroutines by Nassi–Schneiderman diagrams. Discuss the relative merits of the two designs.

7.4 Common blocks can be used in FORTRAN when argument lists become cumbersome (Kernighan and Plauger, 1974; Yourdon and Constantine, 1978). Define labeled common blocks for the subroutines DATA, SYSTEM, and RESULT in

Section 7.3 such that the subroutines are only coupled by data that must be shared. Discuss the relative merits of common blocks and argument lists.

7.5   Develop a matrix displacement analysis program for plane trusses (Section 3.5):
   a. Present your design, analogous to the frame program in Section 7.3, by a tree chart and Nassi–Schneiderman diagrams. Note that many subroutines of the frame program can be adapted with minor modifications. For example, the subroutines of SYSTEM require changes only in the G functions in ELEMS, the INDEX matrix in ASSEMS, and the upper limit of JE from 6 to 4 in ASSEMS.
   b. Complete the program similar to Problem 7.1.

7.6   Develop a matrix displacement analysis program for space trusses (Section 5.4): Carry out steps (a) and (b) similar to Problem 7.5.

7.7   Develop a matrix displacement analysis program for grids (Section 5.6): Carry out steps (a) and (b) similar to Problem 7.5.

7.8   Develop a matrix displacement analysis program for space frames with joint loads (Section 5.5): Carry out steps (a) and (b) similar to Problem 7.5.

7.9   Design a program to analyze plane frames and plane trusses by the matrix displacement method. Present your design by a tree chart and Nassi–Schneiderman diagrams; use or modify subroutines of Section 7.3 when practical. View this project as the framework for a general purpose program.

7.10   Replace the fixed band solver in the frame program (Problem 7.1) with the variable band solver (Sections 6.2 and 6.5) of Bathe and Wilson (1976). This can be accomplished as follows:
   1. Replace MBAND with one or two subroutines that generate the vector MAXA, which defines the skyline storage illustrated in Figure 6.4 (see subroutines COLHT and ADDRES in Bathe and Wilson, 1976);
   2. modify ASSEMS to store the system stiffness coefficients in vector form by Eqs. (6.9);
   3. replace the LINPACK subroutines in SOLVE with the variable band solver (subroutine COLSOL in Bathe and Wilson, 1976).

7.11   Modify the frame program of Problem 7.1 to analyze orthogonal frames with internal constraints (Section 4.3).

7.12   Modify the frame program of Problem 7.1 to analyze symmetric frames with asymmetric actions by decomposing the frames into substructures with symmetric and antisymmetric actions (Sections 4.2).

7.13   Modify the frame program of Problem 7.1 to accept one or more of the following element actions:
   a. additional member loads;
   b. additional temperature variations (Section 4.8);
   c. imperfect elements (Section 4.7).

7.14 Modify the frame program of Problem 7.1 to accept geometric imperfections in joint locations (Section 4.7).

7.15 Modify the frame program of Problem 7.1 to construct influence lines (Section 4.10).

7.16 Modify the frame program of Problem 7.1 to analyze structures with one or more of the following special features:
a. internal releases (Sections 4.4 and 4.6);
b. distinct elements (Section 4.5);
c. distinct joint reference frames (Section 4.9).

7.17 Modify the frame program of Problem 7.1 to compute internal forces, normal stresses, and transverse deflections of elements. These element responses can be expressed as (Sections 1.5 and 3.6)

$$\overline{N}(x) = -\overline{f}_1 + N_p(x)$$

$$\overline{V}(x) = \overline{f}_2 + V_p(x)$$

$$\overline{M}(x) = \overline{f}_2 x - \overline{f}_3 + M_p(x)$$

$$\overline{\sigma}(x, y) = \frac{\overline{N}(x)}{A} - \frac{\overline{M}(x)}{I} y$$

$$\overline{v}(x) = d_2 + d_3 x + \frac{1}{EI}\left(\frac{1}{6}\overline{f}_2 x^3 - \frac{1}{2}\overline{f}_3 x^2\right) + v_p(x)$$

The contributions $N_p(x)$, $V_p(x)$, $M_p(x)$ to the internal forces and $v_p(x)$ to the transverse deflection are caused by element loads. They can be formulated for each individual element load, as illustrated in Problems A.1–A.3 of Appendix A, and accumulated. For example, if $p_1(x)$ and $p_2(x)$ are axial and transverse load functions, respectively,

$$N_p = -\int_0^x p_1 \, dx$$

$$V_p = \int_0^x p_2 \, dx, \quad M_p = \int_0^x V_p \, dx$$

$$v_p = \int_0^x \theta_p \, dx, \quad \theta_p = \frac{1}{EI}\int_0^x M_p \, dx$$

7.18 Modify the frame program of Problem 7.1 to key section and material properties of elements to tables of possible properties (for example, a file of AISC section properties).

REFERENCES

Bathe, K. J., and E. L. Wilson. 1976. *Numerical Methods in Finite Element Analysis.* Prentice-Hall, Englewood Cliffs, NJ.

Bates, D. (ed.). 1976. *Structured Programming, Infotech State of the Art Report.* Infotech International, England.

Böhm, C., and G. Jacopini. 1966. "Flow Diagrams, Turing Machines, and Language with Only Two Formation Rules," *Communications of the ACM,* 9, 366–371.

Chapin, N. 1976. "Aids to Producing Comprehensible Software," in *Structured Programming, Infotech State of the Art Report,* pp. 165–181. Infotech International, England.

Dahl, O.-J., E. W. Dijkstra, and C. A. R. Hoare. 1972. *Structured Programming.* Academic Press, New York.

Denning, P. J. 1976. "A Hard Look at Structured Programming," in *Structured Programming, Infotech State of the Art Report,* pp. 183–202. Infotech International, England.

Digital Equipment Corporation. 1980. *VAX-11 FORTRAN Language Reference Manual,* Order No. AA-D034B-Te. Maynard, MA.

Dijkstra, E. W. 1976. "Programming Methodologies: Their Objectives and Their Nature," in *Structured Programming, Infotech State of the Art Report,* pp. 203–216. Infotech International, England.

Dongarra, J. J., C. B. Moler, J. R. Bunch, and G. W. Stewart, 1979. *LINPACK Users' Guide.* SIAM, Philadelphia.

Hughes, J. K., and J. I. Michtom. 1977. *A Structured Approach to Programming.* Prentice-Hall, Englewood Cliffs, NJ.

Jackson, M.A. 1976. "Data Structure as a Basis for Program Structure," in *Structured Programming, Infotech State of the Art Report,* pp. 279–291. Infotech International, England.

Kernighan, B. W., and P. J. Plauger. 1974. *The Elements of Programming Style.* McGraw-Hill, New York.

Kernighan, B. W., and P. J. Plauger, 1976. *Software Tools.* Addison-Wesley, Reading, MA.

Knuth, D. E. 1974. "Structured Programming with GO TO Statements," *Computing Surveys, ACM,* 6, 261–301.

Laski, J. G. 1976. "What, Why, and Wherefore of Structured Programming," in *Structured Programming, Infotech State of the Art Report,* pp. 293–307. Infotech International, England.

Linger, R. C., H. D. Mills, and B. I. Witt. 1979. *Structured Programming: Theory and Practice.* Addison-Wesley, Reading, MA.

McClure, C. L. 1975. "Top-Down, Bottom-Up, and Structured Programming," *Proceedings of the 1st IEEE National Conference on Software Engineering,* pp. 89–94.

McGowan, C., and J. Kelly. 1975. *Top-Down Structured Programming.* Petrocelli/Charter Books.

Merchant, M. J. 1981. *FORTRAN 77: Language and Style.* Wadsworth, Belmont, CA.

Moore, J. B. 1975. *WATFIV: Fortran Programming with the WATFIV Compiler.* Reston Publishing Company, Reston, VA.

Moore, J. B., and L. J. Makela. 1978. *Structured FORTRAN with WATFIV,* Reston Publishing Company, Reston, VA.

Nanney, T. R. 1981. *Computing: A Problem-Solving Approach with FORTRAN 77*. Prentice-Hall, Englewood Cliffs, NJ.

Nassi, I., and B. Schneiderman. 1973. "Flow Chart Techniques for Structured Programming," *ACM SIGPLAN Notices*, 8, No. 8, 12–26.

Ross, D. T., J. B. Goodenough, and C. A. Irvine. 1975. "Software Engineering: Process, Principles, and Goals," *Computer*, 8, No. 5, 17–25.

Strunk, B. Jr., and E. B. White. 1979. *The Elements of Style*, 3d ed. Macmillan, New York.

Tocher, J. L. 1976. "Specifications for Successful Programming Projects," *Methods of Structural Analysis*, Vol. II, W. E. Saul and A. H. Peyrot, eds., pp. 837–850. American Society of Civil Engineers, New York.

Wirth, N. 1974. "On the Composition of Well-Structured Programs," *Computing Surveys*, 6, 247–259.

Woodger, M. 1976. "The Aims of Structured Programming," in *Structured Programming, Infotech State of the Art Report*, pp. 395–420. Infotech International, England.

Yourdon, E. 1975. *Techniques of Program Structure and Design*. Prentice-Hall, Englewood Cliffs, NJ.

Yourdon, E., and L. L. Constantine. 1978. *Structured Design: Fundamentals of a Discipline of Computer Program and Systems Design*. Yourdon Press, New York.

Additional References

Baron, R. J., and L. G. Shapiro. 1980. *Data Structures and their Implementation*. Van Nostrand Reinhold, New York.

Champine, G. A., R. D. Coop, and R. C. Heinselman. 1980. *Distributed Computer Systems*. North-Holland, Amsterdam.

Dyck, V. A., J. D. Lawson, and J. A. Smith. 1984. *FORTRAN 77: An Introduction to Structured Problem Solving*. Reston Publishing Co., Reston, VA.

Ellis, T. M. R. 1982. *A Structured Approach to FORTRAN 77 Programming*. Addison-Wesley, London.

Felippa, C. A. 1981. "Trends in Computational Engineering," *Advances in Engineering Software*, 3, No. 2, 50–54.

Fenves, S. J. 1983. "Computers in the Future of Structural Engineering," *Proceedings of the Eighth Conference on Electronic Computation*, J. K. Nelson, Jr., ed., pp. 1–8. American Society of Civil Engineers, New York.

Lopez, L. A. 1977. "FINITE: An Approach to Structural Mechanics Systems," *International Journal for Numerical Methods in Engineering*, 11, 851–866.

McCracken, D. D. 1984. *Computing for Engineers and Scientists with FORTRAN 77*. Wiley, New York.

Pollack, S. V. 1982. *Structured FORTRAN 77 Programming*. Boyd & Fraser, San Francisco.

Wagener, J. L. 1980. *FORTRAN 77: Principles of Programming*. Wiley, New York.

Wilson, E. L. 1980. "The Use of Minicomputers in Structural Analysis," *Computers and Structures*, 12, 695–698.

Wilson, E. L. 1980. "SAP-80-Structural Analysis Programs for Small or Large Computer Systems," *CEPA 1980 Fall Conference and Annual Meeting*, October 13–15, Newport Beach, CA.

ELEMENT ACTIONS AND RESPONSES

Formulas are presented for the application of the matrix displacement method to structures with various element actions (Section 3.6). They include displacement–deformation relations, singularity functions for the representation of discontinuous loads, and general fixed-end force formulas. In addition it is illustrated how the displacement–deformation relations can be used to compute displacements at specific points of elements, analogous to the moment–area method.

Displacement–Deformation Relations

If we integrate the strain–displacement relation, Eq. (1.31), subject to the boundary conditions, Eqs. (1.66), and the curvature–displacement relation

$$\phi = \frac{d^2v}{dx^2} \tag{A.1}$$

subject to the boundary conditions, Eqs. (1.76), we obtain the displacement–deformation relations

$$u_b = u_a + u_{ab} \tag{A.2a}$$

$$v_b = v_a + \theta_a L + v_{ab} \tag{A.2b}$$

$$\theta_b = \theta_a + \theta_{ab} \tag{A.2c}$$

where the deformations are

$$u_{ab} = \int_0^L \varepsilon \, dx \tag{A.3a}$$

$$v_{ab} = \int_0^L (L - x)\phi \, dx \tag{A.3b}$$

$$\theta_{ab} = \int_0^L \phi \, dx \tag{A.3c}$$

In Eqs. (A.3), ε and ϕ are the extensional strain and the curvature of the centroidal axis (Figures 1.10a, 1.11).

PROOF. Equation (A.2a) follows from Eqs. (1.31) and (1.66). From Eqs. (A.1) and (1.76) we obtain the integral equation

$$\frac{dv(x)}{dx} = \theta_a + \int_0^x \phi \, dx \qquad (A.4)$$

which becomes Eq. (A.2c) for $x = L$. The integration of Eq. (A.4) yields

$$v(x) = v_a + \theta_a x + \int_0^x \int_0^x \phi \, dx \, dx \qquad (A.5)$$

Reducing the double integral in Eq. (A.5) to a single integral by the relation (Hildebrand, 1965)

$$\underbrace{\int_0^x \cdots \int_0^x \phi(x) \, dx \cdots dx}_{n \text{ times}} = \frac{1}{(n-1)!} \int_0^x (x - \xi)^{n-1} \phi(\xi) \, d\xi \qquad (A.6)$$

and evaluating the result at $x = L$, we obtain Eq. (A.2b).   □

Matrix Equation

By denoting the six element-end displacements d_1–d_6 and the three element deformations e_1–e_3 (Figure 1.17), we can express Eqs. (A.2) in the matrix form

$$\begin{bmatrix} d_4 \\ d_5 \\ d_6 \end{bmatrix} = \begin{bmatrix} 1 & 0 & 0 \\ 0 & 1 & L \\ 0 & 0 & 1 \end{bmatrix} \begin{bmatrix} d_1 \\ d_2 \\ d_3 \end{bmatrix} + \begin{bmatrix} e_1 \\ e_2 \\ e_3 \end{bmatrix} \qquad (A.7)$$

where

$$e_1 = \int_0^L \varepsilon \, dx$$

$$e_2 = \int_0^L (L - x)\phi \, dx \qquad (A.8)$$

$$e_3 = \int_0^L \phi \, dx$$

Equation (A.7) is identical to Eq. (1.131).

Displacement Computations

Equations (A.2), (A.3) can be used to compute displacements at specific points of elements. The sign conventions are consistent with those defined in Sections 1.5 and 1.7 relative to the local reference frame. Specifically, deflections are positive in the directions of the coordinate axes, rotations are positive in a counterclockwise sense, strains are positive in tension, and curvatures are positive concave up (Figure A.1a).

The displacement formulations are analogous to those of the *moment–area method* (Popov, 1968), with one important difference: They do not depend on a sketch of the deformed configuration. Moreover, flexural and axial displacements[1] can be computed.

Flexural Displacements

The following set of equations applies to continuous beam segments (Figure A.1a):

$$\theta_b = \theta_a + \theta_{ab} \tag{A.9a}$$

$$v_b = v_a + \theta_a L + v_{ab} \tag{A.9b}$$

$$v_a = v_b - \theta_b L + v_{ba} \tag{A.9c}$$

where according to Eqs. (A.3b, c) and Figure A.1b

$$\theta_{ab} = \int_0^L \phi \, dx = \mathscr{A} \tag{A.10a}$$

$$v_{ab} = \int_0^L (L - x)\phi(x) \, dx = \mathscr{A}\bar{x}_b \tag{A.10b}$$

$$v_{ba} = \int_0^L x\phi(x) \, dx = \mathscr{A}\bar{x}_a \tag{A.10c}$$

Equation (A.9c) can be derived, analogous to Eq. (A.9b), by integrating Eq. (A.1) from b to a.

By Eqs. (A.9) we can compute the displacements at a point on either side of a reference point with known displacements. Specifically, we can use Eqs. (A.9a, b) compute the displacements at b in terms of the displacements at a, and Eqs. (A.9a, c) to compute the displacements at a in terms of the displacements at b.

---

[1] The extension of this formulation to axial displacements was suggested by Dr. A. E. Somers, Jr.

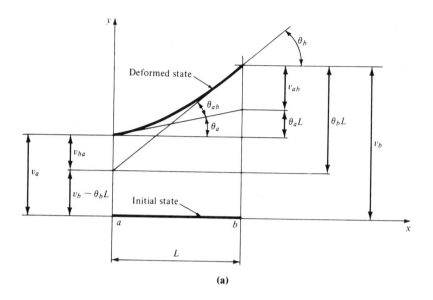

(a)

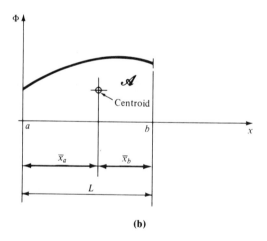

(b)

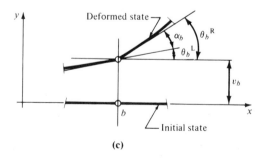

(c)

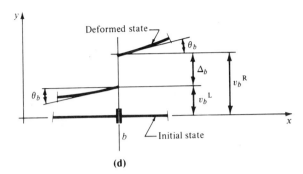

(d)

FIGURE A.1 Flexural displacements. **(a)** Continuous beam segment; **(b)** curvature diagram; **(c)** moment release; **(d)** shear release.

Discontinuities at moment and shear releases (Figures A.1c, d) can be computed by the relations

$$\alpha_b = \theta_b^R - \theta_b^L \tag{A.11a}$$

$$\Delta_b = v_b^R - v_b^L \tag{A.11b}$$

A geometric interpretation of Eqs. (A.9) is given in Figure A.1a. The element deformations θ_{ab}, v_{ab}, v_{ba} reflect the effect of the curvature ϕ, the measure of pointwise deformation.

Example 1. The deflection v_d and the hinge rotation α_b in Figure A.2a are computed. The curvature diagram in Figure A.2b determines the shapes of the continuous beam segments ab and bd. The boundary conditions are as follows. For beam segment ab,

$$v_a = 0, \quad \theta_a = 0 \tag{A.12a}$$

For beam segment bd,

$$\text{continuity of } v_b, \quad v_c = 0 \tag{A.12b}$$

Computation of v_d (Figure A.2a)
By Eqs. (A.9b), (A.12a)

$$v_b = v_a + \theta_a l + v_{ab} = v_{ab} \tag{A.13}$$

By Eqs. (A.9c), (A.12b), (A.13)

$$v_b = v_c - \theta_c l + v_{cb} = -\theta_c l + v_{cb} = v_{ab} \tag{A.14}$$

Thus,

$$\theta_c = \frac{1}{l}(v_{cb} - v_{ab}) = -\frac{4}{3}\mathcal{A} \tag{A.15}$$

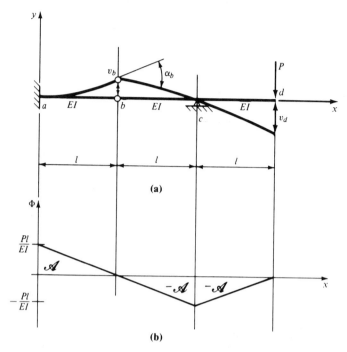

FIGURE A.2 Flexural displacements. **(a)** Beam with moment release; **(b)** curvature diagram.

By Eqs. (A.9b), (A.12b), (A.15)

$$v_d = v_c + \theta_c l + v_{cd} = \theta_c l + v_{cd} = -2l\mathscr{A}$$

or

$$v_d = -\frac{Pl^3}{EI} \tag{A.16}$$

The required deformations follow from Eqs. (A.10b, c) and Fig. A.2b:

$$v_{ab} = \frac{2}{3} l\mathscr{A} = -v_{cb} = -v_{cd}, \quad \mathscr{A} = \frac{Pl^2}{2EI} \tag{A.17}$$

Computation of α_b (Figure A.2a)
By Eqs. (A.9a), (A.10a), (A.11a), (A.12a), (A.15)

$$\theta_b^L = \theta_a + \theta_{ab} = \theta_{ab} = \mathscr{A} \tag{A.18}$$

$$\theta_b^R = \theta_c - \theta_{bc} = -\tfrac{4}{3}\mathscr{A} + \mathscr{A} = -\tfrac{1}{3}\mathscr{A} \tag{A.19}$$

$$\alpha_b = \theta_b^R - \theta_b^L = -\frac{4}{3}\mathscr{A} = -\frac{2}{3}\frac{Pl^2}{EI} \tag{A.20}$$

The deformed configuration in Figure A.2a provides a qualitative check on the results: Specifically, the hinge rotation is negative (clockwise), and the deflection at d is negative (in the negative direction of the y axis).

An alternative formulation for v_d is obtained by using b as reference point:

$$v_d = v_b + \theta_b^R 2l + v_{bd} \tag{A.21}$$

Example 2. The deflection v_d and the discontinuity Δ_b at the shear release in Figure A.3a are computed. The boundary conditions of the continuous beam segments ab and bd, whose shapes are defined by the curvature diagram in Figure A.3b, are as follows. For beam segment ab:

$$v_a = 0, \quad \theta_a = 0 \tag{A.22a}$$

For beam segment bd:

$$\text{continuity of } \theta_b, \quad v_c = 0 \tag{A.22b}$$

Computation of Δ_b (Figure A.3a)

$$v_b^L = v_a + \theta_a l + v_{ab} = v_{ab} = -\tfrac{1}{2}\mathscr{A}l \tag{A.23}$$

$$\theta_b = \theta_a + \theta_{ab} = -\mathscr{A} \tag{A.24}$$

$$v_c = v_b^R + \theta_b l + v_{bc} = 0 \tag{A.25}$$

FIGURE A.3 Flexural displacements. **(a)** Beam with shear release; **(b)** curvature diagram.

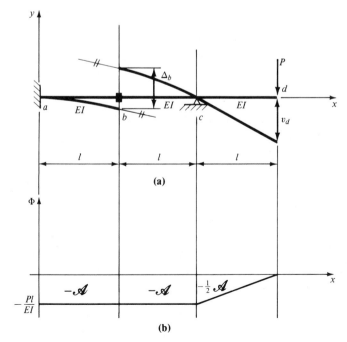

Thus,

$$v_b^R = -\theta_b l - v_{bc} = \mathcal{A}l + \tfrac{1}{2}\mathcal{A}l = \tfrac{3}{2}\mathcal{A}l \tag{A.26}$$

$$\Delta_b = v_b^R - v_b^L = 2\mathcal{A}l = \frac{2Pl^3}{EI} \tag{A.27}$$

Computation of v_d (Figure A.3a)

$$\theta_c = \theta_b + \theta_{bc} = -2\mathcal{A}$$

$$v_d = v_c + \theta_c l + v_{cd} = -2\mathcal{A}l - \tfrac{1}{3}\mathcal{A}l = -\tfrac{7}{3}\mathcal{A}l \tag{A.28}$$

or

$$v_d = -\frac{7}{3}\frac{Pl^3}{EI} \tag{A.29}$$

The deformed configuration in Figure A.3a reflects the signs of the deflections: Δ_b is up and v_d is down.

Axial Displacements

It follows from Eqs. (A.2a), (A.3a), and Figure A.4 that

$$u_b = u_a + u_{ab} \tag{A.30}$$

where

$$u_{ab} = \int_0^L \varepsilon\, dx = \mathcal{A} \tag{A.31}$$

FIGURE A.4 Axial displacements. **(a)** Element; **(b)** strain diagram.

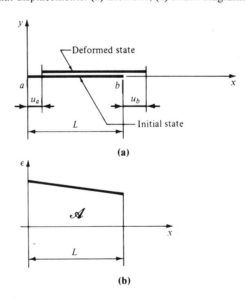

Example 3. The deflections u_b and u_c in Figure A.5a are computed. The boundary condition is

$$u_a = 0 \tag{A.32}$$

Thus,

$$u_b = u_a + u_{ab} = u_{ab} = \mathscr{A} = \frac{Pl}{EA} \tag{A.33}$$

and

$$u_c = u_a + u_{ac} = u_{ac} = 0 \tag{A.34}$$

Singularity Functions

We can represent discontinuous loads by functions from the family of singularity functions (Crandall et al., 1978; Budynas, 1977; Pilkey and Pilkey, 1974; Shames, 1975)

$$\langle x - a \rangle^n \tag{A.35}$$

where a is contained in some domain $\mathscr{D}$; we define $\mathscr{D}$ as the interval $0 \le x \le L$. For $n = -1$, Eq. (A.35) becomes the Dirac delta function (Figure A.6a)

$$\langle x - a \rangle^{-1} \tag{A.36}$$

which is also known as the unit impulse function or the unit concentrated load function. The Dirac delta function has the property

$$\langle x - a \rangle^{-1} = 0, \quad x \ne a \tag{A.37}$$

$$\int_0^x \langle x - a \rangle^{-1} \, dx = \langle x - a \rangle^0 = \begin{cases} 0, & x < a \\ 1, & x \ge a \end{cases} \tag{A.38}$$

FIGURE A.5 Axial displacements. **(a)** Continuous beams; **(b)** strain diagram.

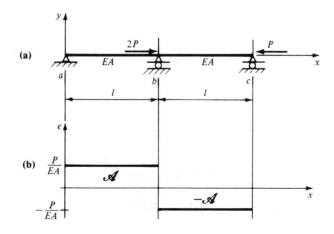

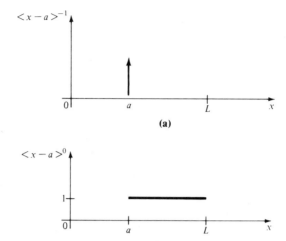

FIGURE A.6 Singularity functions. **(a)** Dirac delta function; **(b)** unit step function.

The expression $\langle x - a \rangle^0$ is called the unit step function (Figure A.6b). Thus,

$$\int_0^L \langle x - a \rangle^{-1} \, dx = 1 \tag{A.39}$$

and for some function $g(x)$

$$\int_0^L g(x)\langle x - a \rangle^{-1} \, dx = g(a) \tag{A.40}$$

because $g(x)\langle x - a \rangle^{-1} = g(a)\langle x - a \rangle^{-1}$.

For $n \geq 0$, the singularity functions have the property

$$\langle x - a \rangle^n = \begin{cases} 0, & x < a \\ (x - a)^n, & x \geq a \end{cases} \tag{A.41}$$

$$\int_0^x \langle x - a \rangle^n \, dx = \frac{1}{n + 1} \langle x - a \rangle^{n+1} \tag{A.42}$$

Fixed-End Forces

Equations (1.177) and (1.183) provide a simple way of computing fixed-end forces for a variety of element actions. These equations are exact solutions of the differential equations, Eqs. (1.65) and (1.75), respectively.

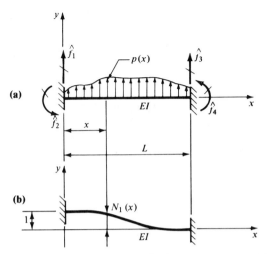

FIGURE A.7 Formulation of fixed-end force $\hat{f}_1$. (a) Force system 1; (b) force system 2.

Element Loads

The fixed-end forces caused by distributed loads (Figures A.7a, A.8a) can be computed by the formula

$$\hat{f}_i = -\int_0^L N_i(x)\, p(x)\, dx \qquad (A.43a)$$

or

$$\hat{f}_i = -L \int_0^1 N_i(\xi)\, p(\xi)\, d\xi, \quad \xi = \frac{x}{L} \qquad (A.43b)$$

FIGURE A.8 Formulation of fixed-end force $\hat{f}_1$. (a) Force system 1; (b) force system 2.

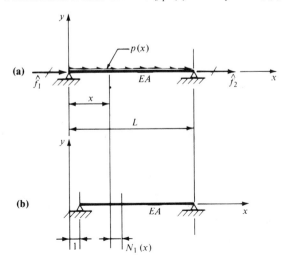

where the N_i's are defined in Eqs. (1.177) for axial loads and in Eqs. (1.183) for transverse loads.

PROOF. Equation (A.43a) is derived by Betti's law for $\hat{f}_1$ in Figures A.7a and A.8a. Betti's law can be expressed as

$$W^{i,j} = W^{j,i} \qquad (A.44)$$

where $W^{i,j}$ is the work done by force system i acting through the displacements produced by force system j. The application of Betti's law to the force systems in Figure A.7 or to those in Figure A.8 yields

$$W^{1,2} = W^{2,1} \qquad (A.45)$$

where

$$W^{1,2} = \hat{f}_1 \cdot 1 + \int_0^L N_1(x)\, p(x)\, dx$$
$$W^{2,1} = 0 \qquad (A.46)$$

Thus,

$$\hat{f}_1 = -\int_0^L N_1(x)\, p(x)\, dx \qquad (A.47)$$

$\square$

Example 4. By the Dirac delta function, Eq. (A.36), the concentrated load in Figure A.9 can be expressed as

$$p(x) = P\langle x - aL\rangle^{-1} \qquad (A.48)$$

Equations (A.40), (A.43a), (A.48), and (1.183) yield the fixed-end forces

$$\hat{f}_i = -\int_0^L N_i(x)P\langle x - aL\rangle^{-1}\, dx = -PN_i(\xi = a) \qquad (A.49)$$

FIGURE A.9 Fixed-end forces.

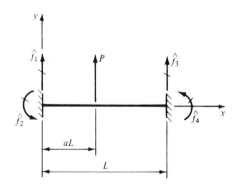

or

$$\hat{\mathbf{f}} = P \begin{bmatrix} -1 - a^2(2a - 3) \\ -La(1 - a)^2 \\ a^2(2a - 3) \\ La^2(1 - a) \end{bmatrix} \tag{A.50}$$

Example 5. By Eqs. (A.43b) and (1.183) we obtain the fixed-end forces of Figure A.10

$$\hat{f}_i = -pL \int_a^1 N_i(\xi) \, d\xi \tag{A.51}$$

Thus,

$$\hat{\mathbf{f}} = pL \begin{bmatrix} -\frac{1}{2}(1 - a^4 + 2a^3 - 2a) \\ -\dfrac{L}{12}(1 - 3a^4 + 8a^3 - 6a^2) \\ -\frac{1}{2}(1 + a^4 - 2a^3) \\ \dfrac{L}{12}(1 + 3a^4 - 4a^3) \end{bmatrix} \tag{A.52}$$

Example 6. The concentrated load in Figure A.11 and the resulting fixed-end forces can also be defined by Eqs. (A.48) and (A.49). Thus, Eqs. (A.49) and (1.177) yield

$$\hat{\mathbf{f}} = -P \begin{bmatrix} 1 - a \\ a \end{bmatrix} \tag{A.53}$$

FIGURE A.10 Fixed-end forces.

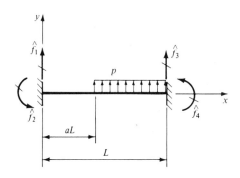

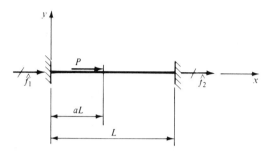

FIGURE A.11 Fixed-end forces.

Unit Displacements

The fixed-end forces caused by the unit displacements at the internal releases in Figures A.12a, b can be computed as follows: For the unit rotation in Figure A.12a at $x = aL$

$$\hat{f}_i = \frac{EI}{L^2} \frac{d^2 N_i(a)}{d\xi^2} \tag{A.54}$$

For the unit deflection in Figure A.12b at $x = aL$

$$\hat{f}_i = \frac{EI}{L^3} \frac{d^3 N_i(a)}{d\xi^3} \tag{A.55}$$

where the functions N_i are defined in Eqs. (1.183).

Equations (A.54) and (A.55) can be verified for $i = 1$ by applying Betti's law to Figures A.12a, c and A.12b, c, respectively. The expressions for the bending moment and the shear force at $x = aL$ in Figure A.12c follow from Eqs. (1.54a), (1.54b), and (1.81).

Example 7. By Eqs. (1.183) and (A.54) we obtain the fixed-end forces for Figure A.12a:

$$\hat{\mathbf{f}} = \alpha \begin{bmatrix} 6L(2a - 1) \\ 2L^2(3a - 2) \\ -6L(2a - 1) \\ 2L^2(3a - 1) \end{bmatrix}, \quad \alpha = \frac{EI}{L^3} \tag{A.56}$$

The corresponding configuration is

$$\hat{v} = L[(2a - 1)\xi^3 - (3a - 2)\xi^2], \quad 0 \le \xi \le a$$

$$\hat{v} = L[(2a - 1)\xi^3 - (3a - 2)\xi^2 - \xi + a], \quad a \le \xi \le 1 \tag{A.57}$$

$$\xi = \frac{x}{L}$$

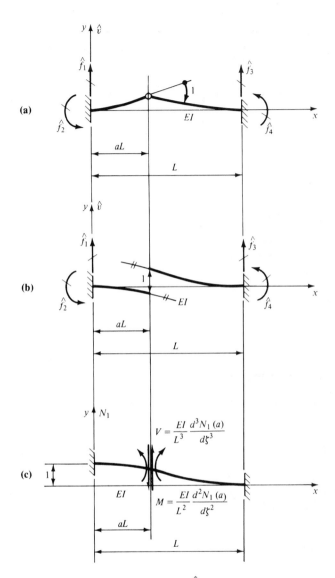

FIGURE A.12 Formulation of fixed-end force $\hat{f}_1$.

Equations (A.57) are of interest in the construction of influence lines (Section 4.10). They can be verified by solving Eq. (A.1) subject to the boundary conditions in Figure A.12a; ϕ can be obtained from Eqs. (A.56).

Example 8. Equations (1.183) and (A.55) yield the fixed-end forces for Figure A.12b:

$$\hat{f} = \alpha \begin{bmatrix} 12 \\ 6L \\ -12 \\ 6L \end{bmatrix}, \quad \alpha = \frac{EI}{L^3} \tag{A.58}$$

The corresponding configuration is

$$\hat{v} = 2\xi^3 - 3\xi^2, \quad 0 \le \xi \le a$$

$$\hat{v} = 2\xi^3 - 3\xi^2 + 1, \quad a \le \xi \le 1 \tag{A.59}$$

$$\xi = \frac{x}{L}$$

PROBLEMS

A.1 Consider the fixed-end element in Figure A.9:
  a. Integrate the pointwise conditions of equilibrium [Eqs. (1.48a)]

$$\frac{d\hat{V}}{dx} = p, \quad \frac{d\hat{M}}{dx} = \hat{V} \tag{A.60}$$

where p is defined in Eq. (A.48), subject to the boundary conditions (Figures 1.8b and A.9)

$$\hat{V}(0) = \hat{f}_1, \quad \hat{M}(0) = -\hat{f}_2 \tag{A.61}$$

to obtain the bending moment function

$$\hat{M}(x) = \hat{f}_1 x - \hat{f}_2 + P\langle x - aL \rangle^1 \tag{A.62}$$

By Eqs. (1.54c) and (A.62) we can compute the stress at any point in the beam of Figure A.9 as

$$\hat{\sigma}(x, y) = -\frac{\hat{M}(x)}{I} y \tag{A.63}$$

  b. Solve the differential equation [Eq. (1.54a)]

$$\frac{d^2\hat{v}}{dx^2} = \frac{\hat{M}}{EI} \tag{A.64}$$

subject to the boundary conditions (Figure A.9)

$$\hat{v}(0) = 0, \quad \frac{d\hat{v}(0)}{dx} = 0 \tag{A.65}$$

for the transverse deflection function

$$\hat{v}(x) = \frac{1}{EI}\left(\hat{f}_1\frac{x^3}{6} - \hat{f}_2\frac{x^2}{2} + \frac{P}{6}\langle x - aL\rangle^3\right) \tag{A.66}$$

A.2 Consider the fixed-end element in Figure A.11:
  a. Integrate the pointwise condition of equilibrium [Eq. (1.33)]

$$\frac{d\hat{N}}{dx} = -p \tag{A.67}$$

where p is defined in Eq. (A.48), subject to the boundary condition (Figures 1.7d and A.11)

$$\hat{N}(0) = -\hat{f}_1 \tag{A.68}$$

to obtain the normal force function

$$\hat{N}(x) = -\hat{f}_1 - P\langle x - aL\rangle^0 \tag{A.69}$$

By Eq. (1.35), the resulting stress is

$$\hat{\sigma}(x) = \frac{\hat{N}(x)}{A} \tag{A.70}$$

  b. Solve the differential equation [Eq. (1.36)]

$$\frac{d\hat{u}}{dx} = \frac{\hat{N}}{EA} \tag{A.71}$$

subject to the boundary condition (Figure A.11)

$$\hat{u}(0) = 0 \tag{A.72}$$

for the axial deflection function

$$\hat{u}(x) = \frac{1}{EA}(-\hat{f}_1 x - P\langle x - aL\rangle^1) \tag{A.73}$$

A.3 The transverse load of the fixed-end element can be defined by the function

$$p(x) = p_a\langle x - aL\rangle^0 + s\langle x - aL\rangle^1 - p_b\langle x - bL\rangle^0 - s\langle x - bL\rangle^1 \tag{A.74}$$

where

$$s = \frac{p_b - p_a}{L(b - a)} \tag{A.75}$$

Use the procedure outlined in Problem A.1 to obtain
  a. the bending moment function

$$\hat{M}(x) = \hat{f}_1 x - \hat{f}_2 + \frac{p_a}{2}\langle x - aL\rangle^2 + \frac{s}{6}\langle x - aL\rangle^3$$
$$- \frac{p_b}{2}\langle x - bL\rangle^2 - \frac{s}{6}\langle x - bL\rangle^3 \tag{A.76}$$

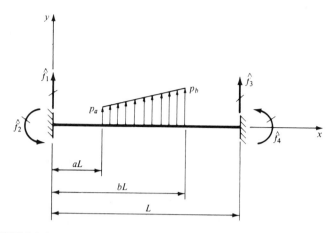

PROBLEM A.3

b. the transverse deflection function

$$\hat{v}(x) = \frac{1}{EI} \left(\hat{f}_1 \frac{x^3}{6} - \hat{f}_2 \frac{x^2}{2} + \frac{p_a}{24} \langle x - aL \rangle^4 + \frac{s}{120} \langle x - aL \rangle^5 \right.$$

$$\left. - \frac{p_b}{24} \langle x - bL \rangle^4 - \frac{s}{120} \langle x - bL \rangle^5 \right)$$

(A.77)

REFERENCES

Budynas, R. G. 1977. *Advanced Strength and Applied Stress Analysis.* McGraw-Hill, New York.

Crandall, S. H., N. C. Dahl, and T. J. Lardner. 1978. *An Introduction to the Mechanics of Solids,* 2nd ed. McGraw-Hill, New York.

Hildebrand, F. B. 1965. *Methods of Applied Mathematics.* Prentice-Hall, Englewood Cliffs, NJ.

Pilkey, W. D., and O. H. Pilkey. 1974. *Mechanics of Solids.* Quantum, New York.

Popov, E. P. 1968. *Introduction to Mechanics of Solids.* Prentice-Hall, Englewood Cliffs, NJ.

Shames, I. H. 1975. *Introduction to Solid Mechanics.* Prentice-Hall, Englewood Cliffs, NJ.

SLOPE–DEFLECTION METHOD

The slope–deflection method (see references cited in Section 1.6) is presented in a form, Figure B.1, that facilitates the transition to the matrix displacement method. In addition, the effects of element loads, which are represented in the slope–deflection equations by the fixed-end moments, are considered explicitly to illustrate how element actions in general can be incorporated in matrix methods of analysis.

The key steps of the slope–deflection method are (Figure B.1a) the following:

1. The *unknown* joint displacements are selected as the independent variables.
2. The *element models* are formulated. They are moment–displacement relations, called slope–deflection equations, which can be expressed as (Figure B.1b)

$$\overline{M}_{ab} = M_{ab} + \hat{M}_{ab}, \quad \overline{M}_{ba} = M_{ba} + \hat{M}_{ba} \qquad \text{(B.1a)}$$

where

$$M_{ab} = \eta(2\theta_a + \theta_b - 3\Psi_{ab}) \\ M_{ba} = \eta(2\theta_b + \theta_a - 3\Psi_{ab}) \qquad \text{(B.1b)}$$

$$\Psi_{ab} = \frac{1}{L}(v_b - v_a), \quad \eta = \frac{2EI}{L} \qquad \text{(B.1c)}$$

In Eqs. (B.1), M_{ab} and M_{ba} are the end moments caused by the end displacements [see Eqs. (1.86)]; $\hat{M}_{ab}$ and $\hat{M}_{ba}$ are the fixed-end moments caused by element loads; and $\overline{M}_{ab}$, $\overline{M}_{ba}$ are the total end moments. Consistent with the conventions in Section 1.7, the counterclockwise sense of the end rotations, the chord rotation, and the end moments is

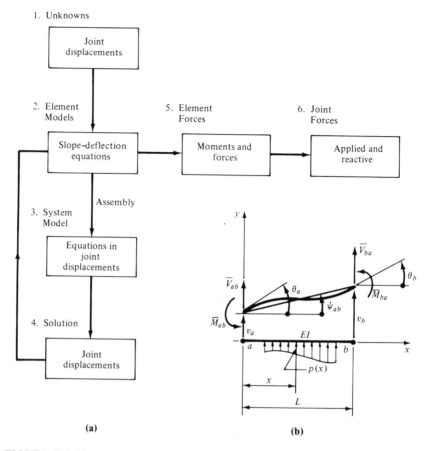

1. Unknowns

Joint displacements

2. Element Models

Slope–deflection equations

Assembly

3. System Model

Equations in joint displacements

4. Solution

Joint displacements

5. Element Forces

Moments and forces

6. Joint Forces

Applied and reactive

(a)

(b)

FIGURE B.1 Slope–deflection method. **(a)** Flow chart; **(b)** beam element.

positive (Figure B.1b). Note that the end forces $\overline{V}_{ab}$, $\overline{V}_{ba}$ can be expressed in terms of the end moments $\overline{M}_{ab}$, $\overline{M}_{ba}$ and the element load by conditions of equilibrium.

3. The element models are assembled into the *system model*, the model of the structure, by imposing conditions of compatibility and equilibrium. The conditions of compatibility consist of continuity conditions between element and joint displacements, including joint constraints. The conditions of equilibrium balance the applied joint forces and the element-end forces in the directions of the unknown joint displacements. This assures equilibrium of the assemblage (Section 2.7). The system model represents a set of simultaneous algebraic equations in the unknown joint displacements.

4. The system model is *solved* for the joint displacements.

5. The *element-end moments* are computed by substituting the values of the joint displacements into the slope–deflection equations. The *element-end* forces—the shear forces and the axial forces—are obtained from conditions of equilibrium.[1]
6. The *joint forces* are computed from conditions of joint equilibrium.

To focus on the effects of element loads in the slope–deflection analysis, the slope–deflection method is applied to structures with and without element loads.

Without Element Loads

The continuous beams in Figure B.2a are analyzed by the slope–deflection method defined in Fig. B.1. The analysis units are kilonewton (kN), meter (m), and radian (rad).

1. Unknowns

$$\theta_b, \theta_c \tag{B.2}$$

2. Element Models

$$M_{ab} = \eta(2\theta_a + \theta_b - 3\Psi_{ab}), \quad M_{ba} = \eta(2\theta_b + \theta_a - 3\Psi_{ab})$$
$$M_{bc} = \eta(2\theta_b + \theta_c - 3\Psi_{bc}), \quad M_{cb} = \eta(2\theta_c + \theta_b - 3\Psi_{bc}) \tag{B.3}$$

3. System Model

Compatibility (Figures B.1b and B.2a):

$$\theta_a = \Psi_{ab} = \Psi_{bc} = 0 \tag{B.4}$$

The remaining conditions of continuity, for example, the slope continuity at joint *b*, are satisfied implicitly because the element-end displacements and the corresponding joint displacements are denoted by the same symbols.

The substitutions of Eqs. (B.4) into Eqs. (B.3) yield the element-end moments expressed in terms of the unknown joint displacements

$$M_{ab} = \eta(\theta_b), \qquad M_{ba} = \eta(2\theta_b)$$
$$M_{bc} = \eta(2\theta_b + \theta_c), \quad M_{cb} = \eta(2\theta_c + \theta_b) \tag{B.5}$$

Equilibrium (Figure B.2b)

$$M_{ba} + M_{bc} = 14, \quad M_{cb} = 0 \tag{B.6}$$

---

[1] This presumes that the structure is statically determinate with respect to the axial element forces. Otherwise, the extensional deformations of the element centroidal axes, which are neglected in the slope-deflection analysis, must be taken into account.

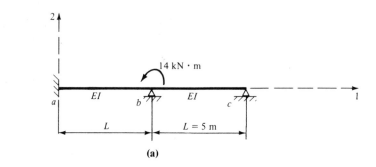

(a)

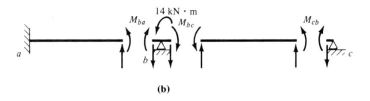

(b)

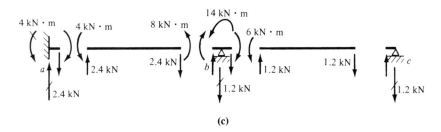

(c)

FIGURE B.2 Slope–deflection analysis. **(a)** Continuous beams; **(b)** joint equilibrium; **(c)** free-body diagrams.

The substitutions of Eqs. (B.5) into Eqs. (B.6) yield the system model

$$\eta(4\theta_b + \theta_c) = 14, \quad \eta(\theta_b + 2\theta_c) = 0 \tag{B.7a}$$

where η is defined in Eqs. (B.1c). To provide a reference for the matrix displacement analysis (Section 3.2), Eqs. (B.7a) are expressed in matrix form as

$$\alpha \begin{bmatrix} 8L^2 & 2L^2 \\ 2L^2 & 4L^2 \end{bmatrix} \begin{bmatrix} \theta_b \\ \theta_c \end{bmatrix} = \begin{bmatrix} 14 \\ 0 \end{bmatrix}, \quad \alpha = \frac{EI}{L^3} \tag{B.7b}$$

4. Solution

The solution to Eqs. (B.7) is

$$\theta_b = \frac{4}{\eta}, \quad \theta_c = -\frac{2}{\eta} \tag{B.8}$$

5. Element Forces

From Eqs. (B.5) and (B.8), we obtain the element-end moments (Figure B.2c)

$$M_{ab} = 4, \quad M_{ba} = 8, \quad M_{bc} = 6, \quad M_{cb} = 0 \tag{B.9}$$

The conditions of element equilibrium result in the shear forces shown in Figure B.2c.

6. Joint Forces

The conditions of joint equilibrium yield the applied and reactive joint forces shown in Figure B.2c.

With Element Loads

Consider the continuous beams in Figure B.3a. Instead of applying the slope–deflection method directly to Figure B.3a—as is customary—we decompose the actual load, $\bar{\mathscr{L}}$, into the component loads, $\hat{\mathscr{L}}$ and $\mathscr{L}$, shown in Figures B.3b and c. In general,

$$\bar{\mathscr{L}} = \mathscr{L} + \hat{\mathscr{L}} \tag{B.10}$$

where $\bar{\mathscr{L}}$ represents the actual loads (they may consist of element and joint loads); $\hat{\mathscr{L}}$ are the fixed-joint loads (they are composed of the *actual element loads* and *joint loads required to prevent joint displacements*); and $\mathscr{L}$ are equivalent joint loads. Since the joints corresponding to $\hat{\mathscr{L}}$ do not experience displacements, the element-end forces are fixed-end forces that can be computed from formulas developed for specific element loads. The joint forces required to prevent joint displacements can then be computed from conditions of joint equilibrium.

It follows from Eq. (B.10) that

$$\mathscr{L} = \bar{\mathscr{L}} - \hat{\mathscr{L}} \tag{B.11}$$

Thus, the equivalent joint loads can be obtained by subtracting the fixed joint loads from the actual loads. Since the structure subjected to $\mathscr{L}$ is without element loads, the element-end moments can be defined by Eqs. (B.1b) and computed by the slope–deflection method.

The superposition of the element-end moments of $\mathscr{L}$ and $\hat{\mathscr{L}}$, consistent with Eq. (B.10), yields the actual element-end moments defined in Eqs. (B.1a). Finally, the actual joint forces can be computed from conditions of joint equilibrium.

Note that by the principle of superposition, the joint displacements corresponding to $\bar{\mathscr{L}}$ and $\mathscr{L}$ are identical because, by definition, the joint displacements corresponding to $\hat{\mathscr{L}}$ are zero. This is illustrated in Figures B.3a–c, where

$$\bar{\theta}_b = \theta_b + \hat{\theta}_b = \theta_b$$

and

$$\bar{\theta}_c = \theta_c + \hat{\theta}_c = \theta_c \tag{B.12}$$

388

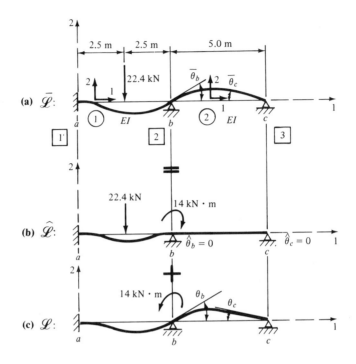

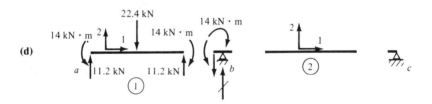

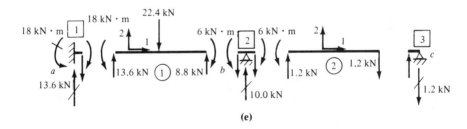

(e)

FIGURE B.3 Slope–deflection analysis. (a) $\overline{\mathscr{L}}$; (b) $\widehat{\mathscr{L}}$; (c) $\mathscr{L}$; (d) joint equilibrium for $\widehat{\mathscr{L}}$; (e) free-body diagrams.

Following this procedure, the element and joint forces of the continuous beams in Figure B.3a are now computed. The analysis units are kilonewton (kN), meter (m), and radian (rad).

Analysis of $\hat{\mathscr{L}}$ (Figure B.3b)

$$\hat{M}_{ab} = \frac{PL}{8} = \frac{22.4(5)}{8} = 14 = -\hat{M}_{ba}$$

$$\hat{M}_{bc} = \hat{M}_{cb} = 0$$

(B.13)

The fixed-end moments are shown in Figure B.3d. Note that the moment supplied by element ab to joint b must be balanced by a clockwise moment of value 14 kN·m applied to joint b. No applied moment is required at joint c because in the absence of element loads element bc does not develop any fixed-end moments. The required joint moment is shown in Figure B.3b.

Analysis of $\mathscr{L}$ (Figure B.3c)
The values of the element-end moments of the structure in Figure B.3c, which is identical to that in Figure B.2a, are given in Eqs. (B.9).

Analysis of $\overline{\mathscr{L}}$ (Figure B.3a)
Equations (B.1a), (B.9), and (B.13) yield the actual element-end moments

$$\overline{M}_{ab} = 18, \quad \overline{M}_{ba} = -6, \quad \overline{M}_{bc} = 6, \quad \overline{M}_{cb} = 0 \qquad (B.14)$$

which are shown, together with the element-end forces required to establish element equilibrium, in Figure B.3e. The actual joint forces (Figure B.3e) are obtained from conditions of joint equilibrium. Note that the actual joint forces could also be obtained from the joint forces of $\mathscr{L}$ and $\hat{\mathscr{L}}$, but this would require the formulation of joint equilibrium for two structures —for $\mathscr{L}$ and $\hat{\mathscr{L}}$—and the superposition of the component joint forces.

PROBLEMS

B.1 Use the slope–deflection method (Figure B.1) to analyze (a) the frame of Problem 2.6 and (b) the continuous beams in Figure 3.5a. Draw free-body diagrams of elements and joints.

B.2 Use a displacement method analogous to the slope–deflection method in Figure B.1a to analyze the continuous beams experiencing only axial deformations. The required element models are defined by Eqs. (1.71) (Figure 1.10a) as

$$N_{ab} = \gamma(u_a - u_b), \quad N_{ba} = \gamma(u_b - u_a), \quad \hat{\gamma} = \frac{EA}{L}$$

Note that the unknown joint displacement is u_b. Draw free-body diagrams of elements and joints.

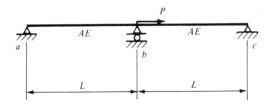

PROBLEM B.2

B.3 Use the slope–deflection method (Figure B.1) to analyze the continuous beams of Problem 2.9:
  a. Apply the slope–deflection method directly.
  b. Resolve the actual loads as defined by Eq. (B.10), analyze the continuous beams for the component loads, and superpose the results. Draw free-body diagrams of elements and joints.

B.4 Analyze the continuous beams by the slope–deflection method (Figure B.1):
  a. Apply the method directly.
  b. Resolve the actual loads as defined by Eq. (B.10), analyze the continuous beams for the component loads, and superpose the results. Draw free-body diagrams of elements and joints.

PROBLEM B.4

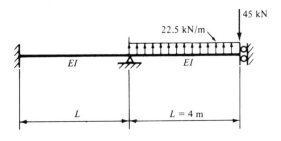

COORDINATE TRANSFORMATIONS

Rotation Matrix: General

Figure C.1a depicts two right-handed orthogonal coordinate systems (Thomas, 1956) with a common origin. The solid lines represent local coordinate axes and the dashed lines represent global coordinate axes (see conventions in Section 2.3). The vectors along the coordinate axes are defined as follows:

$$\mathbf{I}_j = \begin{cases} \text{unit vector on global } j \text{ axis defined relative} \\ \text{to the global coordinate system} \end{cases} \quad \text{(C.1a)}$$

Thus

$$\mathbf{I}_1 = [1 \quad 0 \quad 0]$$
$$\mathbf{I}_2 = [0 \quad 1 \quad 0] \quad \text{(C.1b)}$$
$$\mathbf{I}_3 = [0 \quad 0 \quad 1]$$

and

$$\mathbf{i}_j = \begin{cases} \text{unit vector on local } j \text{ axis defined relative} \\ \text{to the global coordinate system} \end{cases} \quad \text{(C.2a)}$$

It follows from Figure C.1b that

$$\mathbf{i}_j = [l_{j1} \quad l_{j2} \quad l_{j3}], \quad j = 1, 2, 3 \quad \text{(C.2b)}$$

where

$$l_{jk} = \cos \alpha_{jk}, \quad k = 1, 2, 3 \quad \text{(C.3)}$$

The component l_{jk} is the direction cosine and α_{jk} is the direction angle (Hohn, 1972) between the local j axis and the global k axis.

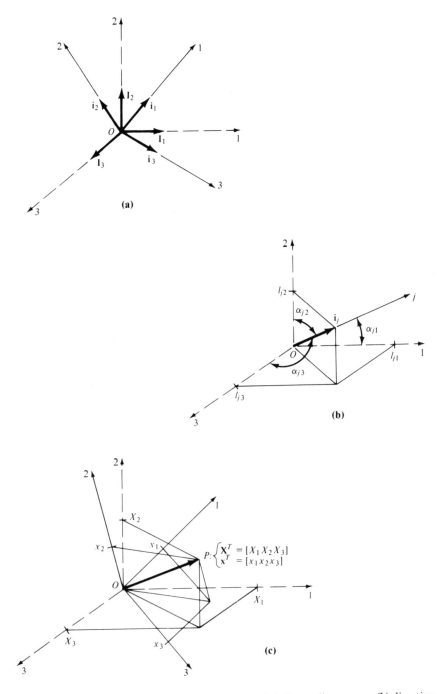

FIGURE C.1 Coordinate systems. **(a)** Local and global coordinate axes; **(b)** direction cosines; **(c)** local and global vectors.

Since the coordinate systems are orthogonal and right handed, the scalar products and cross products of the unit vectors have the following properties (Thomas, 1956)

$$\mathbf{i}_j \mathbf{i}_k^T = \delta_{jk} = \begin{cases} 1 & \text{if } j = k \\ 0 & \text{if } j \neq k \end{cases} \tag{C.4}$$

where δ_{jk} is the Kronecker delta, and

$$\mathbf{i}_1 \times \mathbf{i}_2 = \mathbf{i}_3$$

$$\mathbf{i}_2 \times \mathbf{i}_3 = \mathbf{i}_1 \tag{C.5}$$

$$\mathbf{i}_3 \times \mathbf{i}_1 = \mathbf{i}_2$$

Moreover, Eqs. (C.1b) and (C.2b) yield

$$\mathbf{i}_j \mathbf{I}_k^T = l_{jk} \tag{C.6}$$

Consider the point P in Figure C.1c, which can be located by the local coordinates x_1, x_2, x_3 or the global coordinates X_1, X_2, X_3. The corresponding local and global vectors can be defined as (Thomas, 1956)

$$\mathbf{x} = x_1 \mathbf{i}_1^T + x_2 \mathbf{i}_2^T + x_3 \mathbf{i}_3^T \tag{C.7}$$

$$\mathbf{X} = X_1 \mathbf{I}_1^T + X_2 \mathbf{I}_2^T + X_3 \mathbf{I}_3^T \tag{C.8}$$

Since the vectors $\mathbf{x}$ and $\mathbf{X}$ represent the same directed line segment $\overrightarrow{OP}$ and are expressed through the unit vectors [Eqs. (C.1a) and (C.2a)] relative to the same reference frame, they satisfy the equality

$$\mathbf{x} = \mathbf{X} \tag{C.9}$$

If we premultiply Eq. (C.9) by $\mathbf{i}_j$, that is, if we write

$$\mathbf{i}_j \mathbf{x} = \mathbf{i}_j \mathbf{X} \tag{C.10}$$

and use Eqs. (C.4) and (C.6), we obtain the coordinate transformation

$$x_j = l_{j1} X_1 + l_{j2} X_2 + l_{j3} X_3, \quad j = 1, 2, 3 \tag{C.11}$$

which becomes in matrix form

$$\begin{bmatrix} x_1 \\ x_2 \\ x_3 \end{bmatrix} = \begin{bmatrix} l_{11} & l_{12} & l_{13} \\ l_{21} & l_{22} & l_{23} \\ l_{31} & l_{32} & l_{33} \end{bmatrix} \begin{bmatrix} X_1 \\ X_2 \\ X_3 \end{bmatrix} \tag{C.12}$$

or

$$\mathbf{x} = \lambda \mathbf{X} \tag{C.13}$$

The *rotation matrix* λ defines the transformation from global to local coordinates. According to Eqs. (C.2b), λ can be partitioned as

$$\lambda = \begin{bmatrix} \mathbf{i}_1 \\ \mathbf{i}_2 \\ \mathbf{i}_3 \end{bmatrix} \tag{C.14}$$

The rotation matrix is said to be *orthogonal* (Hohn, 1972) because it has the property

$$\lambda\lambda^T = \lambda^T\lambda = \mathbf{I} \tag{C.15}$$

where $\mathbf{I}$ is the identity matrix; Eq. (C.15) follows from Eqs. (C.4) and (C.14). Premultiplication of Eq. (C.13) by λ^T and utilization of Eq. (C.15) yields

$$\mathbf{X} = \lambda^T\mathbf{x} \tag{C.16}$$

Accordingly, λ^T defines the transformation from local to global coordinates.

Rotation Matrix: Plane Element

The rotation matrix can be specialized for plane elements (see conventions in Section 2.3) by computing the unit vectors in Eq. (C.14). This is illustrated by two approaches, each suitable for a specific situation.

Hand Computation

The local coordinate axes are uniquely located by the angle θ in Figure C.2a because the counterclockwise rotation of the global 3 axis through the angle θ transforms the global coordinate axes into the local coordinate axes. By Eq. (C.2a), the projections of a unit length from the local j axis onto the global axes are the components of $\mathbf{i}_j$. Accordingly, Eq. (C.14) and Figure C.2a yield

$$\lambda = \begin{bmatrix} \cos\theta & \sin\theta & 0 \\ -\sin\theta & \cos\theta & 0 \\ 0 & 0 & 1 \end{bmatrix} \tag{C.17}$$

If desirable, Eq. (C.17) can be expressed in terms of the direction cosines of the local 1 axis, which are denoted by

$$c_i = \cos\alpha_{1i}, \quad i = 1, 2, 3 \tag{C.18}$$

According to Figure C.2a, the corresponding direction angles are $\alpha_{11} = \theta$, $\alpha_{12} = \pi/2 - \theta$, and $\alpha_{13} = \pi/2$. Thus,

$$c_1 = \cos\theta, \quad c_2 = \sin\theta, \quad c_3 = 0 \tag{C.19}$$

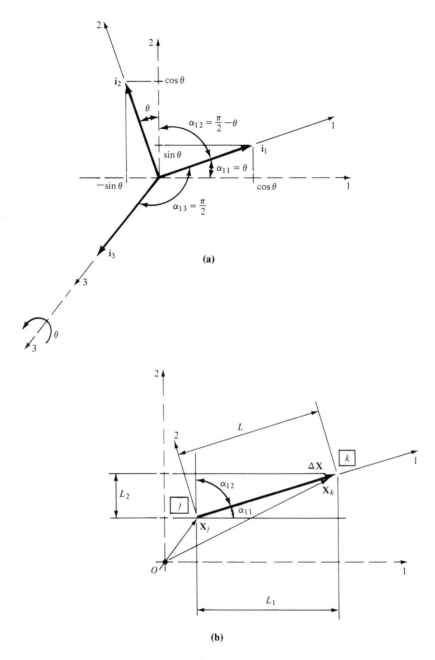

FIGURE C.2 Plane element. **(a)** Local and global coordinate axes; **(b)** element.

and Eq. (C.17) becomes

$$\lambda = \begin{bmatrix} c_1 & c_2 & 0 \\ -c_2 & c_1 & 0 \\ 0 & 0 & 1 \end{bmatrix} \tag{C.20}$$

Computer Computation

In computer analysis (Chapter 7), the direction cosines in Eq. (C.20) are expressed in terms of the coordinates of the joints to which the element is incident. Consider the element in Figure C.2b, which is directed from joint j to joint k. Accordingly, the direction cosines of the local 1 axis with respect to the global 1 and 2 axes are

$$c_i = \frac{L_i}{L}, \quad i = 1, 2 \tag{C.21a}$$

where

$$\Delta \mathbf{X} = \mathbf{X}_k - \mathbf{X}_j = [L_1 \quad L_2] \tag{C.21b}$$

and

$$L = |\Delta \mathbf{X}| = (L_1^2 + L_2^2)^{1/2} \tag{C.21c}$$

In Eqs. (C.21), $\mathbf{X}_k$ and $\mathbf{X}_j$ are the joint coordinate vectors at the b end and a end, respectively, of the element; the difference vector $\Delta \mathbf{X}$ represents the element of length L; and L_i is the projection of L onto the global i axis.

The comparison of Eqs. (C.14) and (C.20) suggests a relation among the unit vectors, which is illuminated in preparation for the derivation of the rotation matrix for space elements:

1. The orientation of the local 1 axis determines

$$\mathbf{i}_1 = [c_1 \quad c_2 \quad 0] \tag{C.22a}$$

2. Since the local and global 3 axes coincide, that is, $\alpha_{31} = \alpha_{32} = \pi/2$ and $\alpha_{33} = 0$,

$$\mathbf{i}_3 = [0 \quad 0 \quad 1] \tag{C.22b}$$

3. And since the unit vectors are mutually orthogonal, the right-hand rule [Eqs. (C.5)] yields

$$\mathbf{i}_2 = \mathbf{i}_3 \times \mathbf{i}_1 = \begin{vmatrix} \mathbf{I}_1 & \mathbf{I}_2 & \mathbf{I}_3 \\ 0 & 0 & 1 \\ c_1 & c_2 & 0 \end{vmatrix} = [-c_2 \quad c_1 \quad 0] \tag{C.22c}$$

Rotation Matrix: Space Element

The rotation matrix of a space element assumes special forms depending on the orientations of the principal plane of the element defined by the local 1, 2 axes and the centroidal axis of the element, which coincides with the local 1 axis (see conventions in Sections 2.3 and 5.3).

Principal Plane is Vertical

Consider Figure C.3a, which shows the local 1, 2 axes and the global 2 axis contained in a vertical plane. The unit vectors of the rotation matrix in Eq. (C.14) can be obtained as follows:

1. According to Figure C.3b in which the element is directed from joint j to joint k,

$$\mathbf{i}_1 = \frac{\Delta \mathbf{X}}{|\Delta \mathbf{X}|} \tag{C.23}$$

where

$$\Delta \mathbf{X} = \mathbf{X}_k - \mathbf{X}_j = [L_1 \quad L_2 \quad L_3]$$
$$L = |\Delta \mathbf{X}| = (L_1^2 + L_2^2 + L_3^2)^{1/2} \tag{C.24}$$

Thus,

$$\mathbf{i}_1 = [c_1 \quad c_2 \quad c_3]$$

and $\hspace{8cm}$ (C.25)

$$c_i = \frac{L_i}{L}, \quad i = 1, 2, 3$$

In Eqs. (C.23)–(C.25), $\mathbf{X}_k$ and $\mathbf{X}_j$ are the joint coordinate vectors at the b end and a end, respectively, of the element; the difference vector $\Delta \mathbf{X}$ represents the element of length L; L_i is the projection of L onto the global i axis; and c_i is the direction cosine of the local 1 axis relative to the global i axis.

2. The local 3 axis is perpendicular to the vertical plane that contains the vectors $\mathbf{i}_1$ and $\mathbf{I}_2$. Hence,

$$\mathbf{i}_3 = \frac{\mathbf{v}_3}{|\mathbf{v}_3|} \tag{C.26}$$

where

$$\mathbf{v}_3 = \mathbf{i}_1 \times \mathbf{I}_2 = \begin{vmatrix} \mathbf{I}_1 & \mathbf{I}_2 & \mathbf{I}_3 \\ c_1 & c_2 & c_3 \\ 0 & 1 & 0 \end{vmatrix} = [-c_3 \quad 0 \quad c_1] \tag{C.27a}$$

398

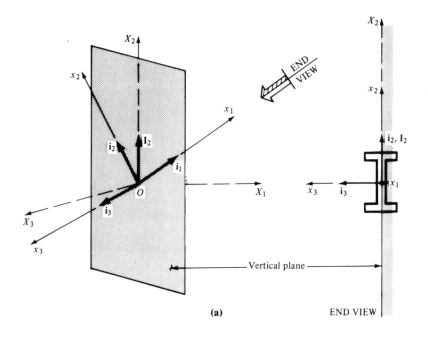

(a)

END VIEW

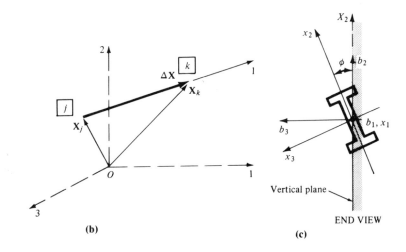

(b)

(c)

END VIEW

FIGURE C.3 Space element. **(a)** Principal plane is vertical; **(b)** element; **(c)** principal plane is not vertical.

and

$$l = |\mathbf{v}_3| = (c_1^2 + c_3^2)^{1/2} \tag{C.27b}$$

Thus,

$$\mathbf{i}_3 = \left[-\frac{c_3}{l} \quad 0 \quad \frac{c_1}{l} \right] \tag{C.28}$$

3. According to Eqs. (C.5),

$$\mathbf{i}_2 = \mathbf{i}_3 \times \mathbf{i}_1 = \begin{vmatrix} \mathbf{I}_1 & \mathbf{I}_2 & \mathbf{I}_3 \\ -\dfrac{c_3}{l} & 0 & \dfrac{c_1}{l} \\ c_1 & c_2 & c_3 \end{vmatrix}$$

which yields

$$\mathbf{i}_2 = \left[-\frac{c_1 c_2}{l} \quad l \quad -\frac{c_2 c_3}{l} \right] \tag{C.29}$$

Combining Eqs. (C.14), (C.25), (C.28), and (C.29), we obtain the rotation matrix

$$\lambda = \begin{vmatrix} c_1 & c_2 & c_3 \\ -\dfrac{c_1 c_2}{l} & l & -\dfrac{c_2 c_3}{l} \\ -\dfrac{c_3}{l} & 0 & \dfrac{c_1}{l} \end{vmatrix}, \quad l = (c_1^2 + c_3^2)^{1/2} \tag{C.30}$$

which defines the transformation from global to local coordinates [Eq. (C.13)] for a space element whose principal plane corresponding to the local 1, 2 axes is vertical. Thus,

$$\mathbf{x} = \lambda \mathbf{X} \tag{C.31}$$

Note that if the local and global 3 axes coincide, $c_3 = 0$, $c_1 = l$, and Eq. (C.30) reduces to Eq. (C.20).

Principal Plane Is Not Vertical

If the principal plane of the element defined by the local x_1, x_2 axes is not vertical (Figure C.3c), we can formulate the rotation matrix in two steps: First, we regard the transformation defined by the rotation matrix in Eq. (C.30) as an intermediate state and denote it by

$$\mathbf{b} = \lambda_{bX} \mathbf{X} \tag{C.32}$$

Accordingly, the b_1 axis coincides with the local x_1 axis of the element and the b_1, b_2 axes lie in a vertical plane as shown in Figure C.3c. Next, we rotate the b_1 axis in a counterclockwise (right-handed) sense through the angle ϕ until the b_2 axis coincides with the x_2 axis. This transformation can be expressed as

$$\mathbf{x} = \lambda_{xb} \mathbf{b} \tag{C.33}$$

where

$$\lambda_{xb} = \begin{bmatrix} 1 & 0 & 0 \\ 0 & \cos\phi & \sin\phi \\ 0 & -\sin\phi & \cos\phi \end{bmatrix} \tag{C.34}$$

Equation (C.34) can be obtained, analogous to Eq. (C.17), by projecting unit lengths from the local axes onto the b_1, b_2, b_3 axes. Specifically, the components of the ith row of λ_{xb} are the projections of a unit length from the x_i axis onto the b_1, b_2, b_3 axes. Combining Eqs. (C.32) and (C.33), we obtain

$$\mathbf{x} = \lambda_{xb} \lambda_{bX} \mathbf{X} = \lambda \mathbf{X} \tag{C.35}$$

where

$$\lambda = \lambda_{xb} \lambda_{bX} \tag{C.36}$$

The substitutions of Eq. (C.34) for λ_{xb} and Eq. (C.30) for λ_{bX} yield

$$\lambda = \begin{bmatrix} c_1 & c_2 & c_3 \\ -\dfrac{c_1 c_2}{l}\cos\phi - \dfrac{c_3}{l}\sin\phi & l\cos\phi & -\dfrac{c_2 c_3}{l}\cos\phi + \dfrac{c_1}{l}\sin\phi \\ \dfrac{c_1 c_2}{l}\sin\phi - \dfrac{c_3}{l}\cos\phi & -l\sin\phi & \dfrac{c_2 c_3}{l}\sin\phi + \dfrac{c_1}{l}\cos\phi \end{bmatrix} \tag{C.37}$$

Note that if $\phi = 0$, $\cos\phi = 1$, $\sin\phi = 0$, and Eq. (C.37) reduces to Eq. (C.30).

Centroidal Axis is Vertical

If the centroidal axis of the element is vertical, the local 1 axis and the global 2 axis are parallel. In this case, $\mathbf{i}_3$ cannot be determined by Eqs. (C.26) and (C.27) since

$$\mathbf{i}_1 = [0 \quad c_2 \quad 0] \tag{C.38}$$

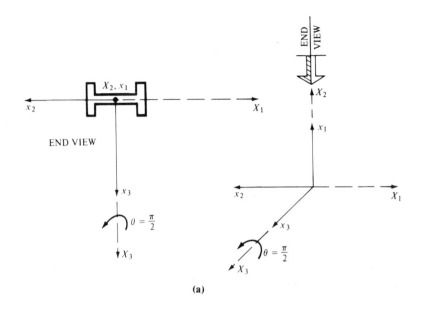

(a)

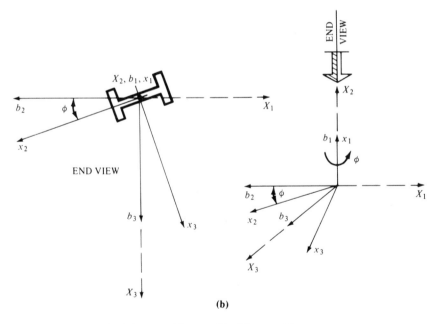

(b)

FIGURE C.4 Vertical element. **(a)** $x_3 = X_3$; **(b)** $x_3 \neq X_3$.

and $\mathbf{v}_3 = \mathbf{0}$; $c_2 = \pm 1$, depending on the relative directions of the global 2 axis and the local 1 axis. This is reflected by Eq. (C.30), which becomes indeterminate for $c_1 = c_3 = l = 0$.

The rotation matrix can be determined with the aid of Figure C.4, which depicts two cases:

1. If the local and global 3 axes coincide as shown in Figure C.4a, the rotation matrix can be obtained from Eqs. (C.19) and (C.20) by setting $\theta = \pm \pi/2$. Thus,

$$\lambda = \begin{bmatrix} 0 & c_2 & 0 \\ -c_2 & 0 & 0 \\ 0 & 0 & 1 \end{bmatrix}, \quad c_2 = \pm 1 \tag{C.39}$$

Equation (C.39) can be verified by projecting unit lengths from the local axes onto the global axes.

2. If the local and global 3 axes do not coincide as shown in Figure C.4b, we can use the two-step approach characterized by Eqs. (C.32)–(C.36) to obtain the rotation matrix. Specifically, λ_{xb} is defined by Eq. (C.34) and λ_{bX} is defined by Eq. (C.39). Thus, Eq. (C.36) yields

$$\lambda = \begin{bmatrix} 0 & c_2 & 0 \\ -c_2 \cos \phi & 0 & \sin \phi \\ c_2 \sin \phi & 0 & \cos \phi \end{bmatrix} \tag{C.40}$$

Again Eq. (C.40) can be verified directly by projecting unit lengths from the local axes onto the global axes.

The special equations for λ matrices of space elements are summarized in Table C.1.

TABLE C.1   Equations for λ Matrices

x_1 Axis Not Vertical		x_1 Axis Vertical	
x_1-x_2 Plane Vertical	x_1-x_2 Plane Not Vertical	$x_3 = X_3$	$x_3 \neq X_3$
(C.30)	(C.37)	(C.39)	(C.40)

PROBLEMS

In the following problems determine the λ matrix, sketch the local and global coordinate axes, and show the unit local vectors $\mathbf{i}_1, \mathbf{i}_2, \mathbf{i}_3$.

C.1 $\mathbf{i}_2 = \begin{bmatrix} -\dfrac{1}{\sqrt{2}} & 0 & \dfrac{1}{\sqrt{2}} \end{bmatrix}$, $\mathbf{i}_3 = \begin{bmatrix} 0 & -1 & 0 \end{bmatrix}$, $\phi = ?$

C.2 $\mathbf{X}_k = [0 \quad 1 \quad 0]$, $\mathbf{X}_j = [0 \quad 0 \quad 1]$, $\phi = 0°$

C.3 $\mathbf{X}_k = [1 \quad 0 \quad 1]$, $\mathbf{X}_j = \mathbf{0}$;
 a. $\phi = 0°$;
 b. $\phi = 45°$.

C.4 $\mathbf{i}_1 = \mathbf{I}_1$, $\phi = 30°$

C.5 $\mathbf{X}_k = [1 \quad 1 \quad 1]$, $\mathbf{X}_j = \mathbf{0}$, $\phi = 0°$

C.6 $\mathbf{i}_1 = \mathbf{I}_2$, $\phi = 45°$

C.7 $\mathbf{i}_1 = -\mathbf{I}_2$, $\phi = 45°$

REFERENCES

Hohn, F. E. 1972. *Elementary Matrix Algebra*, 3d ed. Macmillan, New York.

Thomas, G. B., Jr. 1956. *Calculus and Analytic Geometry*, 2nd ed. Addison-Wesley, Reading, MA.

PRINCIPLES AND CONCEPTS OF ANALYTICAL MECHANICS

The science of mechanics developed along two main lines (Lanczos, 1970; Langhaar, 1962). One branch, called *vectorial mechanics* [or Newtonian mechanics (Synge and Griffith, 1959)], is based on Newton's laws. The other branch, called *analytical mechanics*, is based on the principle of virtual work. The principles, concepts, and mathematics required in the application of the principle of virtual work to discrete conservative systems are presented.

Basic Laws (Langhaar, 1962)

For an adiabatic process, the *law of conservation of energy* can be expressed as

$$W_e = \Delta T + U \tag{D.1}$$

where W_e is the work done by external forces, ΔT is the change in kinetic energy, and U is the strain energy, the change in the internal energy of an elastic system. The *law of kinetic energy* states that

$$W_e + W_i = \Delta T \tag{D.2}$$

where W_i is the work done by internal forces. Equations (D.1) and (D.2) yield

$$W_i = -U \tag{D.3}$$

If the body is rigid, the internal forces perform no work, hence, $U = 0$. If the external forces are applied gradually, at infinitesimal speed, $\Delta T = 0$ and Eq. (D.1) reduces to

$$W_e = U \tag{D.4}$$

Equation (D.4) is a statement of the law of conservation of energy for an adiabatic process with static loading.

Strain Energy

If the material behavior is represented by Hooke's law

$$\sigma = E\varepsilon \tag{D.5}$$

The strain energy can be expressed as

$$U = \iiint_V \overline{U} \, dV \tag{D.6}$$

where the strain energy density is (Figure D.1a)

$$\overline{U} = \tfrac{1}{2}\sigma\varepsilon = \tfrac{1}{2}E\varepsilon^2 \tag{D.7}$$

The strain energy is a function of the configuration of the model (Langhaar, 1962), which is determined by the generalized displacements (Section 1.4). The strain energy can be formulated indirectly by Eq. (D.4) or directly by Eqs. (D.6) and (D.7). In fact, the derivation of Eq. (D.7) is based on Eq. (D.4) (Popov, 1968).

Example 1. For the axial deformation element in Figure 1.10b, the strain energy is formulated (1) by Eq. (D.4) and (2) by Eqs. (D.6), (D.7).
  1. The axial force–deformation relation is defined as [Eq. (1.72)]

$$P = \gamma e, \quad \gamma = \frac{EA}{L} \tag{D.8}$$

The external work done is (Figure D.1b)

$$W_e = \int_0^e P \, de = \int_0^e \gamma e \, de = \tfrac{1}{2}\gamma e^2 = \tfrac{1}{2}Pe \tag{D.9}$$

FIGURE D.1 Axial deformation. **(a)** Stress–strain diagram; **(b)** equilibrium path.

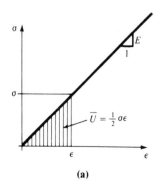
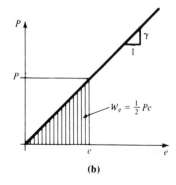

By Eqs. (D.4) and (D.9)

$$U = \tfrac{1}{2}\gamma e^2 \tag{D.10}$$

2. The strain of the uniformly deformed element is

$$\varepsilon = \frac{e}{L} \tag{D.11}$$

Equations (D.7) and (D.11) yield

$$\overline{U} = \frac{1}{2} E\left(\frac{e}{L}\right)^2 = \text{constant} \tag{D.12}$$

Thus,

$$U = \overline{U} \iiint\limits_{V} dV = \overline{U}AL = \tfrac{1}{2}\gamma e^2 \tag{D.13}$$

Expressed in matrix form, Eq. (D.10) can be linked to the axial deformation model [Eq. (1.96b)]

$$\mathbf{f} = \mathbf{kd} \tag{D.14}$$

as follows: By Eqs. (1.74) and (1.95)

$$e = -d_1 + d_2 = \begin{bmatrix} -1 & 1 \end{bmatrix}\begin{bmatrix} d_1 \\ d_2 \end{bmatrix} = \overline{\mathbf{B}}\mathbf{d} \tag{D.15}$$

Thus, in matrix form

$$U = \tfrac{1}{2}\gamma e^T e = \tfrac{1}{2}\gamma \mathbf{d}^T \overline{\mathbf{B}}^T \overline{\mathbf{B}}\mathbf{d} \tag{D.16}$$

or

$$U = \tfrac{1}{2}\mathbf{d}^T \mathbf{kd} \tag{D.17}$$

where

$$\mathbf{k} = \gamma \begin{bmatrix} 1 & -1 \\ -1 & 1 \end{bmatrix} \tag{D.18}$$

Virtual Displacements (Lanczos, 1970)

Virtual displacements are *arbitrary variations* in displacements consistent with constraints. A variation is an infinitesimal change. However, unlike the actual variation d in elementary calculus, virtual variations can be imposed at will. Following Lagrange, we use the symbol δ to denote virtual variations.

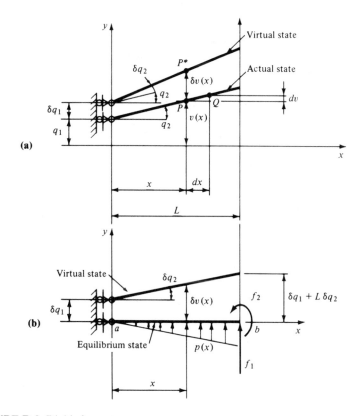

FIGURE D.2 Rigid element.

Example 2. Consider the rigid element in Figure D.2a with the generalized displacements q_1, q_2. If the rotation q_2 is infinitesimal, the actual configuration can be expressed as (Section 1.4)

$$v(x) = q_1 + xq_2 \tag{D.19}$$

If we impose the variations δq_1 and δq_2, the element assumes the virtual configuration in the neighborhood of the actual configuration. The variation of the configuration δv can be computed by the rules of elementary calculus:

$$\delta v = \frac{\partial v}{\partial q_1} \delta q_1 + \frac{\partial v}{\partial q_2} \delta q_2 \tag{D.20}$$

Thus, by Eqs. (D.19) and (D.20)

$$\delta v = \delta q_1 + x \, \delta q_2 \tag{D.21}$$

Figure D.2a illustrates the fundamental difference between the virtual variation δv and the actual variation dv; dv is caused by the infinitesimal change dx, while δv is an infinitesimal change of v at constant x that produces the new function $v + \delta v$. Accordingly, $\delta x = 0$ in virtual variations.

Note that for configurations expressed in terms of generalized displacements, the virtual configuration obtained by the rules of elementary calculus automatically satisfies the geometric constraints. Specifically, Eq. (D.21) is consistent with the boundary condition and the rigidity constraint of the element in Figure D.2a.

Virtual Work

Consider a system with n degrees of freedom and the generalized displacements $q_1, q_2, \ldots, q_n$. If we impose the variations $\delta q_1, \delta q_2, \ldots, \delta q_n$, we obtain the virtual configuration defined by $q_1 + \delta q_1$, $q_2 + \delta q_2$, $\ldots$, $q_n + \delta q_n$. The work done by all the forces acting on the system during the virtual displacement can be expressed in a Taylor series as (Langhaar, 1962; Rubinstein, 1970)

$$\Delta W = \delta W + \frac{1}{2!} \delta^2 W + O(\delta^3) \tag{D.22}$$

where

$$\Delta W = W(q + \delta q) - W(q) \tag{D.23a}$$

$$\delta W = \sum_{i=1}^{n} \frac{\partial W(q)}{\partial q_i} \delta q_i = \sum_{i=1}^{n} R_i \, \delta q_i \tag{D.23b}$$

$$\delta^2 W = \sum_{i=1}^{n} \sum_{j=1}^{n} \frac{\partial^2 W(q)}{\partial q_i \, \partial q_j} \delta q_i \, \delta q_j \tag{D.23c}$$

and $O(\delta^3)$ is the remainder. The variations δW and $\delta^2 W$ are linear and quadratic functions of the variations δq_i, respectively; they are called the first and second variations of W. The first variation δW is the *virtual work* and the coefficients R_i are the *generalized forces* (Section 1.4).

The virtual work can be separated into the virtual work of external and internal forces:

$$\delta W = \delta W_e + \delta W_i \tag{D.24}$$

where δW_e is a linear function in the δq_i's of the form

$$\delta W_e = \sum_{i=1}^{n} Q_i \, \delta q_i \tag{D.25}$$

the coefficients Q_i are called *generalized external forces*. Moreover, by Eq. (D.3)

$$\delta W_i = -\delta U \tag{D.26}$$

Since U is a function of the generalized displacements,

$$\delta U = \sum_{i=1}^{n} \frac{\partial U}{\partial q_i} \delta q_i \tag{D.27}$$

By Eqs. (D.24)–(D.27), the virtual work can be expressed as

$$\delta W = \delta W_e - \delta U = \sum_{i=1}^{n} R_i \, \delta q_i \tag{D.28}$$

where

$$R_i = Q_i - \frac{\partial U}{\partial q_i} \tag{D.29}$$

Example 3. For an illustration of Eqs. (D.22), (D.28), and (D.29), consider the element in Figure D.3a. Suppose that the load Q is increased gradually until point 1 on the equilibrium path in Figure D.3b is reached. The equilibrium path is the set of all points in the load–deflection plane satisfying the equation of equilibrium

$$Q = \gamma q \tag{D.30}$$

The corresponding point 1 on the stress–strain curve of the uniformly strained element is shown in Figure D.3c.

Let us impose the variation δq. If equilibrium is maintained, the corresponding variation in the load is, by Eq. (D.30),

$$\delta Q = \gamma \, \delta q \tag{D.31}$$

and point 2 in Figure D.3b is the new equilibrium point. Similarly, δq causes the variation in the strain $\delta \varepsilon$ and the variation in the stress $\delta \sigma = E \, \delta \varepsilon$ (Figure D.3c). Note that unless the virtual displacement is a rigid-body displacement, as in Example 2, it causes a virtual strain. Specifically, for the uniformly strained element

$$\varepsilon = \frac{q}{L} \tag{D.32}$$

and, hence,

$$\delta \varepsilon = \frac{\delta q}{L} \tag{D.33}$$

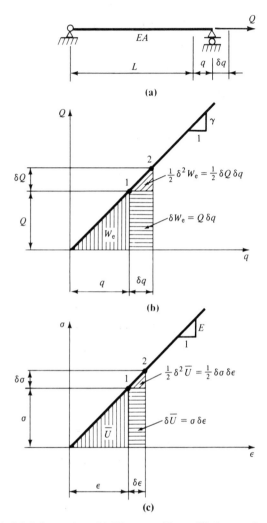

FIGURE D.3 Axial deformation. (a) Element; (b) equilibrium path; (c) stress–strain diagram.

The resulting increment in the external work is equal to the area under the equilibrium path between q and $q + \delta q$ in Figure D.3b. Thus,

$$\Delta W_e = \delta W_e + \tfrac{1}{2}\,\delta^2 W_e \tag{D.34}$$

where the external virtual work is

$$\delta W_e = Q\,\delta q \tag{D.35}$$

and

$$\tfrac{1}{2}\,\delta^2 W_e = \tfrac{1}{2}\,\delta Q\,\delta q = \tfrac{1}{2}\gamma\,\delta q^2 \tag{D.36}$$

Note that δW_e and $\delta^2 W_e$ are linear and quadratic functions in δq; $O(\delta^3) = 0$ because the equilibrium path is linear.

From Eq. (D.35) we can infer that the external virtual work is the product of the applied forces and the corresponding virtual displacements.

The change in the strain energy density is equal to the area under the stress–strain curve between ε and $\varepsilon + \delta\varepsilon$ in Figure D.3c. Thus,

$$\Delta \overline{U} = \delta \overline{U} + \tfrac{1}{2} \delta^2 \overline{U} \tag{D.37}$$

where the first variation of $\overline{U}$ is

$$\delta \overline{U} = \sigma \, \delta\varepsilon = E\varepsilon \, \delta\varepsilon \tag{D.38}$$

and

$$\tfrac{1}{2} \delta^2 \overline{U} = \tfrac{1}{2} \delta\sigma \, \delta\varepsilon = \tfrac{1}{2} E \, \delta\varepsilon^2 \tag{D.39}$$

From Eq. (D.6) we obtain the first variation of the strain energy

$$\delta U = \delta \iiint_V \overline{U} \, dV = \iiint_V \delta \overline{U} \, dV \tag{D.40}$$

Equation (D.40) indicates that variation and integration are permutable operations (see mathematical operations at the end of this appendix). By Eqs. (D.38) and (D.40), the first variation of the strain energy can be expressed in the form

$$\delta U = \iiint_V \delta\varepsilon^T \sigma \, dV \tag{D.41}$$

which is used in the formulation of finite element models in Section 1.9. Equations (D.32), (D.33), (D.38), and (D.40) yield

$$\delta U = \gamma q \, \delta q \tag{D.42}$$

where

$$\gamma = \frac{EA}{L} \tag{D.43}$$

By Eqs. (D.28), (D.35), and (D.42) the virtual work can be expressed as

$$\delta W = \delta W_e - \delta U = R \, \delta q \tag{D.44}$$

where the generalized force

$$R = Q - \gamma q \tag{D.45}$$

Principal of Virtual Work (Lanczos, 1970; Langhaar, 1962; Synge and Griffith, 1959)

A system[1] is in a configuration of equilibrium if and only if the virtual work of all forces is zero for any virtual displacement. Accordingly,

$$\delta W = 0 \leftrightarrow \text{equilibrium} \tag{D.46}$$

Equation (D.46) is a necessary and sufficient condition for equilibrium. By Eq. (D.28) the principle of virtual work can be expressed as

$$\delta W = \sum_{i=1}^{n} R_i \, \delta q_i = 0 \tag{D.47}$$

which yields the conditions of equilibrium

$$R_i = Q_i - \frac{\partial U}{\partial q_i} = 0, \quad i = 1, 2, \ldots, n \tag{D.48}$$

Thus, a configuration for which all generalized forces vanish is an equilibrium configuration (Langhaar, 1962).

For example, if we set the generalized force in Eq. (D.45) equal to zero, we obtain the equation of equilibrium in Eq. (D.30).

Example 4. The rigid element in Figure D.2b is subjected to the concentrated forces f_1, f_2 and the linearly distributed load

$$p(x) = \frac{p_0}{L} x \tag{D.49}$$

The conditions of equilibrium are formulated by the principle of virtual work.

The variations δq_1, δq_2 are imposed on the equilibrium state, which corresponds to the initial state $q_1 = 0$, $q_2 = 0$ (Figure D.2b). The resulting virtual state is defined by Eq. (D.21). The virtual work done during the virtual displacement is

$$\delta W = \delta W_e - \delta U = \delta W_e \tag{D.50}$$

because for the rigid-body displacement $\delta U = 0$. According to Figure D.2b,

$$\delta W_e = f_1(\delta q_1 + L \, \delta q_2) + f_2 \, \delta q_2 + \int_0^L p(x) \, \delta v(x) \, dx \tag{D.51}$$

where by Eqs. (D.21) and (D.49)

$$\int_0^L p(x) \, \delta v(x) \, dx = \int_0^L \frac{p_0}{L} x(\delta q_1 + x \, \delta q_2) \, dx = \frac{1}{2} p_0 L \left(\delta q_1 + \frac{2}{3} L \, \delta q_2 \right) \tag{D.52}$$

---

[1] A system with reversible displacements—that is, δq can be replaced by $-\delta q$—is meant.

Thus,

$$\delta W_e = (f_1 + \tfrac{1}{2}p_0 L)\,\delta q_1 + (f_1 L + f_2 + \tfrac{1}{3}p_0 L^2)\,\delta q_2 = \sum_{i=1}^{2} Q_i\,\delta q_i \quad \text{(D.53)}$$

where the generalized external forces are

$$Q_1 = f_1 + \tfrac{1}{2}p_0 L$$
$$Q_2 = f_1 L + f_2 + \tfrac{1}{3}p_0 L^2 \quad \text{(D.54)}$$

By setting the generalized external forces equal to zero, we obtain the equations of equilibrium

$$f_1 + \tfrac{1}{2}p_0 L = 0 = \sum F_y$$
$$f_1 L + f_2 + \tfrac{1}{3}p_0 L^2 = 0 = \sum M_a \quad \text{(D.55)}$$

which are equivalent to Newton's first law.

Mathematical Operations

We have seen that variational operations follow the rules of elementary calculus. We need to supplement these operations with rules for matrix differentiation and integration and some properties of the δ process before we can apply them in the formulation of finite elements (Section 1.9).

Differentiation and Integration of Matrices

Consider the matrices $\mathbf{A}$ and $\mathbf{B}$ whose elements are functions of x, that is,

$$\mathbf{A} = [A_{ij}(x)] \quad \text{and} \quad \mathbf{B} = [B_{ij}(x)] \quad \text{(D.56)}$$

Then

$$\frac{d\mathbf{A}}{dx} = \left[\frac{dA_{ij}}{dx}\right] \quad \text{(D.57a)}$$

$$\frac{d}{dx}(\mathbf{A} + \mathbf{B}) = \frac{d\mathbf{A}}{dx} + \frac{d\mathbf{B}}{dx} \quad \text{(D.57b)}$$

$$\frac{d}{dx}(\mathbf{AB}) = \frac{d\mathbf{A}}{dx}\mathbf{B} + \mathbf{A}\frac{d\mathbf{B}}{dx} \quad \text{(D.57c)}$$

$$\int_a^b \mathbf{A}\,dx = \left[\int_a^b A_{ij}\,dx\right] \quad \text{(D.57d)}$$

Properties of δ Process (Lanczos, 1970)

Variation is permutable with respect to differentiation and integration. Thus,

$$\delta \frac{dv}{dx} = \frac{d}{dx} \delta v \tag{D.58a}$$

and

$$\delta \int_a^b \overline{U} \, dx = \int_a^b \delta \overline{U} \, dx \tag{D.58b}$$

Example 5. The strain energy of the axial deformation element is [Eq. (D.17)]

$$U = \tfrac{1}{2} \mathbf{d}^T \mathbf{k} \mathbf{d} \tag{D.59}$$

Let us determine the first variation of the strain energy.

By Eqs. (D.57c) and (D.59)

$$\delta U = \tfrac{1}{2} \delta \mathbf{d}^T \mathbf{k} \mathbf{d} + \tfrac{1}{2} \mathbf{d}^T \delta \mathbf{k} \mathbf{d} + \tfrac{1}{2} \mathbf{d}^T \mathbf{k} \, \delta \mathbf{d} \tag{D.60}$$

However,

$$\delta \mathbf{k} = [\delta k_{ij}] = \mathbf{0}$$

because **k** is constant. Moreover,

$$\tfrac{1}{2} \mathbf{d}^T \mathbf{k} \, \delta \mathbf{d} = \tfrac{1}{2} \delta \mathbf{d}^T \mathbf{k}^T \mathbf{d} = \tfrac{1}{2} \delta \mathbf{d}^T \mathbf{k} \mathbf{d} \tag{D.61}$$

Thus,

$$\delta U = \delta \mathbf{d}^T \mathbf{k} \mathbf{d} \tag{D.62}$$

In Eqs. (D.61), the first equality follows from the symmetry property of inner products. Specifically, let $\mathbf{r} = \mathbf{k} \, \delta \mathbf{d}$, then

$$\mathbf{d}^T \mathbf{k} \, \delta \mathbf{d} = \mathbf{d}^T \mathbf{r} = \mathbf{r}^T \mathbf{d} = \delta \mathbf{d}^T \mathbf{k}^T \mathbf{d} \tag{D.63}$$

The second equality follows from the symmetry of **k**.

Example 6. The axial deformation model, Eq. (1.96b), is formulated by the principle of virtual work.

The configuration of the constant strain element (Figure D.4) can be expressed by Eqs. (1.7), (1.8), (1.95) as

$$u = [N_1 \quad N_2] \begin{bmatrix} d_1 \\ d_2 \end{bmatrix} = \mathbf{N} \mathbf{d} \tag{D.64}$$

where

$$N_1 = 1 - \xi, \quad N_2 = \xi, \quad \xi = \frac{x}{L} \tag{D.65}$$

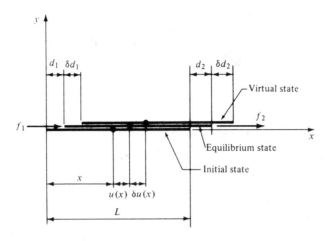

FIGURE D.4 Axial deformation element

The condition of compatibility can be expressed by Eqs. (1.31), (D.64) in the form

$$\varepsilon = \frac{1}{L}\frac{du}{d\xi} = \frac{1}{L}\frac{d\mathbf{N}}{d\xi}\mathbf{d} = \mathbf{B}\mathbf{d} \qquad (D.66)$$

where

$$\mathbf{B} = \frac{1}{L}[-1 \quad 1] \qquad (D.67)$$

Thus, the virtual strain is

$$\delta\varepsilon = \mathbf{B}\,\delta\mathbf{d} \qquad (D.68)$$

By Eqs. (D.5), (D.41), (D.66), (D.68) we obtain the first variation of the strain energy

$$\delta U = \delta\mathbf{d}^T\left(\iiint_V \mathbf{B}^T E\mathbf{B}\,dV\right)\mathbf{d} = \delta\mathbf{d}^T\mathbf{k}\mathbf{d} \qquad (D.69)$$

where

$$\mathbf{k} = \iiint_V \mathbf{B}^T E\mathbf{B}\,dV = \gamma\begin{bmatrix} 1 & -1 \\ -1 & 1 \end{bmatrix}, \quad \gamma = \frac{EA}{L} \qquad (D.70)$$

According to Figure D.4, the external virtual work is

$$\delta W_e = [\delta d_1 \quad \delta d_2]\begin{bmatrix} f_1 \\ f_2 \end{bmatrix} = \delta\mathbf{d}^T\mathbf{f} \qquad (D.71)$$

Thus, the principle of virtual work yields

$$\delta W = \delta W_e - \delta U = \delta \mathbf{d}^T(\mathbf{f} - \mathbf{kd}) = 0 \qquad (D.72)$$

which results in Eq. (1.96b).

REFERENCES

Lanczos, C. 1970. *The Variational Principles of Mechanics*, 4th ed. University of Toronto Press, Toronto.

Langhaar, H. L. 1962. *Energy Methods in Applied Mechanics*. Wiley, New York.

Popov, E. P. 1968. *Introduction to Mechanics of Solids*. Prentice-Hall, Englewood Cliffs, NJ.

Rubinstein, M. F. 1970. *Structural Systems—Statics, Dynamics and Stability*. Prentice-Hall, Englewood Cliffs, NJ.

Synge, J. L., and B. A. Griffith. 1959. *Principles of Mechanics*, 3d ed. McGraw-Hill, New York.

IMPOSITION OF CONSTRAINTS ON SYSTEM MODEL

We consider the imposition of joint constraints on the system model, an alternative to imposing them on element models before they are assembled (Section 3.1).

Example. To facilitate a comparison of the two procedures of imposing joint constraints, the support settlement problem in Figure 4.23a of Section 4.7 is reanalyzed: This time all prescribed joint displacements are imposed on the system model. Accordingly, all joint displacements, constrained and unconstrained, are initially treated as degrees of freedom (Figure E.1a). The analysis could be based on Figure 3.2. However, we select the notation of Figure 3.9, since in Section 4.7 prescribed joint displacements are regarded as element actions.

1. *Unknowns*

$$q_k, \quad k = 1, \ldots, 6 \tag{E.1}$$

2. *Element Models*

$$\bar{\mathbf{f}}^i = \mathbf{k}\bar{\mathbf{d}}^i \tag{E.2}$$

where

$$\mathbf{k} = \alpha \begin{bmatrix} 12 & 6L & -12 & 6L \\ 6L & 4L^2 & -6L & 2L^2 \\ -12 & -6L & 12 & -6L \\ 6L & 2L^2 & -6L & 4L^2 \end{bmatrix}, \quad \alpha = \frac{EI}{L^3} \tag{E.3}$$

According to Figures E.1a, b the member code matrix is

$$\mathbf{M} = \begin{bmatrix} 1 & 3 \\ 2 & 4 \\ 3 & 5 \\ 4 & 6 \end{bmatrix} \tag{E.4}$$

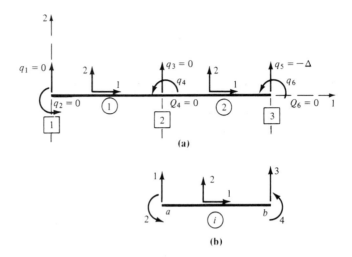

FIGURE E.1 Continuous beams. **(a)** Assembly; **(b)** element.

3. System Model

By Eqs. (3.23), (3.24), (E.3), and (E.4) we obtain the unconstrained system model

$$\alpha \begin{bmatrix} 12 & 6L & -12 & 6L & 0 & 0 \\ 6L & 4L^2 & -6L & 2L^2 & 0 & 0 \\ -12 & -6L & 24 & 0 & -12 & 6L \\ 6L & 2L^2 & 0 & 8L^2 & -6L & 2L^2 \\ 0 & 0 & -12 & -6L & 12 & -6L \\ 0 & 0 & 6L & 2L^2 & -6L & 4L^2 \end{bmatrix} \begin{bmatrix} q_1 \\ q_2 \\ q_3 \\ q_4 \\ q_5 \\ q_6 \end{bmatrix} = \begin{bmatrix} Q_1 \\ Q_2 \\ Q_3 \\ Q_4 \\ Q_5 \\ Q_6 \end{bmatrix} \quad \text{(E.5)}$$

which is subject to the joint constraints

$$q_1 = q_2 = q_3 = 0, \quad q_5 = -\Delta \quad \text{(E.6)}$$

and the joint loads

$$Q_4 = Q_6 = 0 \quad \text{(E.7)}$$

From Eqs. (E.5)–(E.7) we obtain the constrained system model

$$\alpha \begin{bmatrix} 8L^2 & 2L^2 \\ 2L^2 & 4L^2 \end{bmatrix} \begin{bmatrix} q_4 \\ q_6 \end{bmatrix} = \begin{bmatrix} -6L\alpha\Delta \\ -6L\alpha\Delta \end{bmatrix} \quad \text{(E.8)}$$

which, aside from different displacement subscripts, is identical to Eq. (4.199). It is not practical in computer analysis to reduce Eq. (E.5) to Eq. (E.8) directly. Instead Eq. (E.8) is modified as follows to reflect the boundary conditions.

$$
\alpha
\begin{bmatrix}
1 & 0 & 0 & 0 & 0 & 0 \\
0 & 1 & 0 & 0 & 0 & 0 \\
0 & 0 & 1 & 0 & 0 & 0 \\
0 & 0 & 0 & 8L^2 & 0 & 2L^2 \\
0 & 0 & 0 & 0 & 1 & 0 \\
0 & 0 & 0 & 2L^2 & 0 & 4L^2
\end{bmatrix}
\begin{bmatrix}
q_1 \\ q_2 \\ q_3 \\ q_4 \\ q_5 \\ q_6
\end{bmatrix}
=
\begin{bmatrix}
0 \\ 0 \\ 0 \\ -6\alpha L\Delta \\ -\Delta \\ -6\alpha L\Delta
\end{bmatrix}
\tag{E.9}
$$

4. *Solution*

The solution to Eq. (E.9) yields all joint displacements

$$
\mathbf{q} =
\begin{bmatrix}
0 \\
0 \\
0 \\
-\dfrac{3\Delta}{7L} \\
-\Delta \\
-\dfrac{9\Delta}{7L}
\end{bmatrix}
\tag{E.10}
$$

5. *Element Forces*

We can compute the element-end forces by Eqs. (E.2), (E.3), (E.4), and (E.10). For example,

$$
\mathbf{d}^2 =
\begin{bmatrix}
0 & 3 \\
-\dfrac{3\Delta}{7L} & 4 \\
-\Delta & 5 \\
-\dfrac{9\Delta}{7L} & 6
\end{bmatrix},
\quad
\bar{\mathbf{f}}^2 = \frac{M}{7}
\begin{bmatrix}
\dfrac{2}{L} \\
2 \\
-\dfrac{2}{L} \\
0
\end{bmatrix},
\quad M = 6\alpha L\Delta
\tag{E.11}
$$

Note that the computations of $\bar{\mathbf{f}}^i$ by Eqs. (4.193) and (E.2) are equivalent. Specifically,

$$
\bar{\mathbf{f}}^i = \mathbf{k}\mathbf{d}^i + \mathbf{k}\hat{\mathbf{d}}^i = \mathbf{k}(\mathbf{d}^i + \hat{\mathbf{d}}^i) = \mathbf{k}\bar{\mathbf{d}}^i
\tag{E.12}
$$

6. Joint Forces

All joint forces, applied and reactive, can be computed by substituting Eq. (E.10) into the unconstrained system model, Eq. (E.5). Accordingly,

$$\mathbf{Q} = \frac{M}{7} \begin{bmatrix} -\dfrac{3}{L} \\[2mm] -1 \\[2mm] \dfrac{5}{L} \\[2mm] 0 \\[2mm] \mathbf{1}_5 \\[2mm] -\dfrac{2}{L} \\[2mm] 0 \end{bmatrix} \tag{E.13}$$

which agrees with the joint forces in Eqs. (4.202).

Algorithm. In general, the joint constraints can be imposed on the system model $\mathbf{Kq} = \mathbf{Q}$ in the following manner: For each constrained joint displacement

$$q_j = \Delta \tag{E.14}$$

replace

$$Q_i \quad \text{by} \quad Q_i - K_{ij}\, \Delta \tag{E.15}$$

and set

$$K_{ij} = K_{ji} = 0$$
$$K_{jj} = 1, \quad Q_j = \Delta \tag{E.16}$$

where $i = 1, 2, \ldots, n; i \neq j$. If $\Delta = 0$, the modification of Q_i in Eq. (E.15) is not required. Moreover, Eq. (E.15) need only be applied to Q_i's that correspond to degrees of freedom [for example, Q_4 and Q_6 in Eq. (E.5)].

Merits of Procedures

An advantage of imposing the joint constraints before the elements are assembled is that the order of the system model is equal only to the number of unknown joint displacements. This has a positive effect on storage requirements and solution effort (Chapter 6). An advantage of imposing the joint constraints after the elements are assembled is that, provided we keep a copy of the unconstrained system stiffness matrix, we can use it to compute all joint forces and to analyze identical assemblies with distinct boundary conditions. However, the first procedure is more convenient in production programs, where **K** and **Q** are not stored entirely in core (Felippa and Clough, 1970).

REFERENCE

Felippa, C. A., and R. W. Clough. 1970. "The Finite Element Method in Solid Mechanics," *Numerical Solution of Field Problems in Continuum Physics*, SIAM–AMS Proceedings, Vol. II, pp. 210–252. American Mathematical Society, Providence, RI.

INDEX

A

Accuracy of solution, 316–322
Actions, 2, 75, 134, 161
  decomposition of, 162
Actions and Responses
  antisymmetric, 162
  asymmetric, 172
  symmetric, 161
Assemblages of distinct elements, 190

B

Band matrix
  fixed, 280
  variable, 282
Band storage, 284
Bandwidth, 281
Betti's law, 38, 48, 147, 239, 255, 376

C

Center of twist, 255
Cholesky method, 300
  modified, 301
Compatibility and equilibrium, 74, 80, 99
Complex elements, substructures, or
      superelements, 196, 199, 200, 209
Condensation, 196, 294
Condition number, 317
Configuration (state), 1,3
Connectivity array, 108, 285
Constraints on system model, 417
Contragredient transformations, 38, 41, 99,
      101, 114
Control structures, 329
  alternative (if-then-else), 330
  case, 331
  loop (while-do), 330
  sequence, 330

Convergence criteria (finite element), 63
  compatibility (conformity), 65
  completeness, 65
Coordinate axes (reference frames)
  global, 75, 79, 90
  joint, 232
  local, 28, 79, 90
Coordinate transformations, 90, 391
  frame element, 90, 262, 394–402
  grid element, 272
  truss element, 95, 258
Cramer's rule, 288

D

Deformation
  axial, 5, 12, 254
  flexural, 7, 15, 254
  parameters, 38, 40
  shearing, 18, 253
  torsional, 19, 254
  vector, 40, 217
Degrees of freedom, 3, 40, 75, 254, 255, 256
  external, 196
  internal, 196
Destination vectors, 108
Determinants, 288, 300, 317
Dirac delta function, 141, 147, 148, 373, 376
Discrete elements
  beam element, 29
  beam element with internal hinge, 201
  complex element (substructure or
        superelement), 197, 199, 200, 204, 209
  frame element, 29, 254, 263
  frame element with internal hinge, 44
  grid element, 271
  rotational spring element, 190
  tapered beam element, 56
  truss element, 29, 254, 257

E

Element actions, 134, 143, 365
Element response, 140, 365
Element stiffness matrix
  generalized, 111, 114, 115
  global, 111, 114, 126, 131, 259, 266, 272
  in joint coordinates, 234, 235
  local, 30, 31, 57, 58, 109, 119, 125, 130, 199, 265, 272
Equilibrium conditions, 13, 15, 17, 19, 35, 83, 88
Error analysis, 320
Error detection, 318
  diagonal decay test, 318
  equilibrium check, 319
  roundoff errors, 318
  truncation errors, 318
Error source, 321

F

Finite element discretization, 49
Finite element formulations, 51, 143
Finite element method, 48
Finite element strains and stresses, 152
Finite elements, 51–58
Fixed-end forces, 135, 137, 138, 218, 224, 374
Flexibility matrix, 45
Flexible joint, 190
Frame program, 337
  design, 337
  function, 337
  input data, 340
  structure, 337
  symbolic names, 338
  test problem, 357
Frames
  braced, 195
  orthogonal, 176
  plane, 76, 125
  space, 262
Frontal solution method, 312
Function interpolation, 49, 51, 59

G

Gaussian elimination, 290
  algorithm, 292
  band solvers, 302
  compact, 296
  condensation, 294
  operation count, 306
  order of decomposition (fractorization), 299
  pivoting, 293
  standard, 291
  symmetry, 293

symmetry and bandedness, 294
  variable band solvers, 307
Generalized displacements, 1, 3, 8, 75, 82, 110, 134, 253
Generalized forces, 3, 8, 82, 83 110
  principle of, 10, 82
Geometric imperfections, 214
  crooked beam, 218
  of elements, 217, 219
  fabrication errors, 214
  support settlements, 214, 417
Grids, 270

H

Half bandwidth, 281, 339, 344
Hermite interpolation, 61–63

I

Improvement of solution accuracy
  double precision arithmetic, 320
  iterative improvement, 319, 322
  modification of model, 320
Independent displacements, 41
Independent forces, 38, 256
Inextensible centroidal axes, 130, 176, 385
Infinitesimal
  deformation, 17
  rotation, 4, 5, 40
  work, 8
Influence lines, 238–246
  of frames and continous beams, 240
  of trusses, 245
Internal constraints, 176
Internal releases, 184, 256
Interpolation functions, 49, 51, 59, 146

J

Joint constraints, 75, 232, 256
Joint coordinate axes, 233, 236
Joint loads
  actual, 135
  equivalent, 135
  fixed, 135

K

Kronecker delta, 32, 48, 59, 62, 116

L

Lagrange interpolation, 59–61, 66
Law of conservation of energy, 404
Law of kinetic energy, 404

M

Mathematical operations, 413
Matrix
  band, 280
  ill conditioned, 317
  lower triangular, 289
  member code, 81, 108
  nonsingular, 38, 43, 45, 118, 288
  partitioned, 30, 31, 34, 125
  positive definite, 118, 280, 288
  rank, 36, 118
  rotation, 91, 391–402
  singular, 36, 38
  sparse, 280
  symmetric, 36, 37, 47, 116, 280, 288
  triangular, 289
  unit lower triangular, 290, 297
  unit upper triangular, 290, 300
  upper triangular, 290
  variable band (skyline, profile, envelope),
Matrix displacement analysis of
  continuous beams, 119
  frames, 125, 267
  grids, 272
  trusses, 130, 259
Matrix displacement method, 74, 108, 113, 136
Member code matrix, 81, 108
Model(s)
  actions, 2, 134
  axial deformation, 12, 22
  components of, 1, 2, 12
  components of compatibility, 1, 2
  components of constitutive law, 1, 2, 3
  components of equilibrium, 1, 2, 3
  configuration, 1, 3, 76, 77
  continuum, 1, 11
  degrees of freedom of, 1, 3, 11
  discrete, 1, 11, 22, 29
  flexural deformation, 15, 24
  generalized displacements of, 1, 3, 8, 75,
    134, 253
  linear, 5, 12, 21, 28, 133
  mathematical, 1, 2
  physical, 2
  response of, 2
  of a structure, 1, 2, 3
  torsional deformation, 19, 27
Modeling process, 11
Moment-area method, 367
  axial displacements, 372
  flexural displacements, 367

N

Nassi–Schneiderman diagrams, 329
Notation and conventions, 13, 16, 28, 29, 79,
  256
Numerical integration, 141

P

Principle of
  analytical mechanics, 404
  generalized forces, 10, 82
  Muller-Breslau, 239
  superposition, 21, 22
  virtual work, 51, 52, 99, 100, 143, 144,
    151, 412
Program correctness, 332, 334
  testing, 335
  verification, 335
Program development, 328

R

Rank of a matrix, 36
Rigid-body displacements, 3, 37, 38, 40

S

Shear center, 255
Sign conventions
  axial deformation model, 13
  discrete models, 29, 79, 80, 89
  flexural deformation model, 16
  torsional deformation model, 20
Singularity functions, 141, 373
Skeletal structures
  continuous beams, 77
  frames, 76, 262
  grids, 270
  trusses, 75, 256
Skyline storage, 285
Slope-deflection
  equations, 26, 32, 383
  method, 383
Solution errors, 316
  roundoff, 317, 318, 322
  truncation, 317, 318, 321
Solution of linear algebraic equations, 280,
  288
  band solvers, 302
  Cholesky method, 289, 300, 324
  computer programs, 307, 312, 316
  direct methods, 288
  frontal solutions, 312
  Gaussian elimination, 290
  iterative methods, 288
  modified Cholesky method, 289,
    301
  variable band solvers, 307
Stable equilibrium, 118
Stable structure, 256
Static determinacy, 256
Stiffness coefficients, 31, 116
  element, 32
  system, 116

Stiffness matrix
  axial deformation element, 30, 264
  axial and flexural deformation element, 31
  beam element, 119
  condensed, 200, 212, 296
  element with hinge, 45, 201
  flexural deformation element, 30, 265
  grid element, 272
  plane frame element, 125
  rotational spring element, 192
  space frame element, 265
  tapered beam element, 57, 58
  torsional deformation element, 264
  truss element, 130, 259
Strain-displacement relations
  axial, 12
  flexural, 17
  torsional, 19
Strain energy, 405
Structural analysis, 2
Structural engineering, 2
Structured flow charts (Nassi–Schneiderman
     diagrams), 329
Structured programming, 328
  bottom-up approach, 332
  control structures, 329
  data structure, 333
  efficiency, 336
  incremental testing, 335
  key principle, 329
  languages, 334
  methods of modularization, 331
  reduction of complexity, 329
  top-down approach, 332
Stub (dummy program), 335
Substructuring, 196, 203
Symmetric
  actions and responses, 161

frame, 166
matrix, 36, 37, 47, 116, 280, 288
structures, 161, 168
truss, 164
Symmetry, 160, 253
Symmetry axiom, 163
System model, 110–112, 150
System stiffness matrix, 112, 114

T
Temperature changes, 143, 149, 222
Thermal strain, 149, 222
Transformation
  displacement, 82
  force, 83
  stiffness matrix, 114
Trusses
  plane, 75
  space, 256

U
Unique solution, 118, 288
Unit displacements, 239, 378

V
Variable band (skyline, profile, envelope),
     282
Virtual displacements, 50, 52, 145, 406
Virtual work, 8, 52, 145, 408

W
Warping, 253, 255
Work, 8, 42, 405
Work of external forces, 404
Work of internal forces, 404